FUNDAMENTALS OF RADIOCHEMISTRY

FUNDAMENTALS OF RADIOCHEMISTRY

Jean-Pierre Adloff, D.Sc.
Professor
Université Louis Pasteur
Centre de Recherches Nucléaires
Chimie Nucléaire
Strasbourg, France

Robert Guillaumont, D.Sc.
Professor
Université Paris XI
Institut de Physique Nucléaire
Groupe de Radiochimie
Orsay, France

CRC Press
Boca Raton Ann Arbor London Tokyo

Library of Congress Cataloging-in-Publication Data

Adloff, J.-P. (Jean-Pierre)
Fundamentals of radiochemistry / Jean-Pierre Adloff, Robert Guillaumont.
p. cm.
Includes bibliographical references and index.
ISBN 0-8493-4244-9
1. Radiochemistry. I. Guillaumont, R. (Robert) II. Title.
QD601.2.A35 1993
541.3'8--dc20 92-29544
CIP

Direct all inquiries to CRC Press, Inc., 2000 Corporate Blvd., N.W., Boca Raton, Florida 33431.

International Standard Book Number 0-8493-4244-9

Library of Congress Card Number 92-29544

Printed in the United States of America 1 2 3 4 5 6 7 8 9 0

Printed on acid-free paper

FOREWORD

Very few books devoted to the basis of radiochemistry have been published in the last two decades. During this period the orientation of the subject has changed very substantially. A new work, with a different approach is to be welcomed wholeheartedly.

The classical treatises on this subject devoted much of their text to the properties of radiation and the measurement of radioactivity. The behavior of species present at tracer concentrations was usually treated in a rather descriptive manner aimed mainly at the separation and concentration of the trace species.

In this book it is assumed that the methods of measurement of radioactivity are known. Its principal objective is to demonstrate the usefulness of radiochemistry in most areas of chemistry and to describe the physicochemical behavior of systems containing one or more compounds at tracer or sub-tracer concentration.

The authors have carried out a critical and thorough analysis of the principles and applications of radiochemical methods and have emphasized many aspects of radioactivity and chemistry that are often overlooked.

This may be the first book to treat the question of what chemical information can be obtained with sub-tracer amounts, i.e., systems containing a few, perhaps only one, radioactive atom.

The topic of speciation, identification of chemical form, of radionuclides present at tracer concentrations, has achieved considerable importance in recent years, because of its relevance to the control of nuclear wastes. In this book, speciation of the radioactive component is treated as the main objective of radiochemical operations.

The pedigree of this book is impeccable: Professor Adloff was a pupil of Marguerite Perey at the University of Strasbourg and Professor Guillaumont studied at the Institut du Radium and at the Arcueil Laboratories in Paris under the guidance of Moïse Haissinsky and Georges Bouissières.

The authors have succeeded in their ambition to present a modern and original view of radiochemistry and their extensive practical experience ensures that their approach to the subject is successful.

Alfred G. Maddock
University Chemical Laboratory
Cambridge, U.K.

PREFACE

Many books have been written on the subjects of the nucleus and radioactivity, delineating their importance in various fields of fundamental or applied science. Some of these treatises deal specifically with relationships between radioactivity and chemistry, introducing with various degrees of pertinence and precision the term "radiochemistry" in an effort to establish useful interdisciplinary connections. Unfortunately, the result frequently leads to confusion; as yet, only a few authors have effectively succeeded in outlining the true scope of radiochemistry and defining its position within the larger context of chemistry in general. Aware of this need, we have attempted to overcome at least part of the difficulty in the present book. This original treatise is the result of a critical consideration of the fields of chemistry covered by radiochemistry, which is examined in its own right on the basis of its fundamental principles, objectives, and methods.

Modern radiochemistry is defined as the study of chemical properties of matter under conditions for which the classical methods of physical chemistry no longer apply and recourse to techniques of radioactivity is required. This is particularly true for elements that are present at extremely high dilution and which can only be detected and investigated with the aid of characteristic radiations emitted by representative radionuclides. In other terms, radiochemistry embodies the use of radioactivity in the elucidation of chemical phenomena.

The fields and concepts described are essentially those that are less familiar to chemists in general, and examples are provided that include the chemistry of ephemeral radionuclides, actinides of high atomic number, positronium, and thermodynamic and kinetic considerations at the tracer level and at the scale of only one or several atoms.

Radiochemical methods are defined in terms of analysis of chemical information conveyed by radiation; they are considered and discussed in a unified manner adapted primarily to situations encountered in radiochemistry, but which can also be extended to other fields. It is shown how this information has already contributed to various aspects of fundamental physical chemistry and how it can be applied to problems of speciation in systems relative to the environment.

This book is addressed to a wide range of chemists and environmental scientists not having specialized knowledge in the field. Characteristics of radioactivity and radioactive matter are presented that are useful to the chemist. The treatise is also recommended for practitioners in radiochemistry requiring to update their knowledge. It is of particular interest to research workers dealing with fundamental and applied aspects of matter at extreme dilution as encountered in the dissemination of stable and radioactive wastes, in living organisms and in space. The material has been presented in a form suitable for use as a textbook or aid for special courses in physical and analytical chemistry.

Our conception of radiochemistry is the result of numerous discussions and exchanges with our colleagues from the international community of radiochemistry and nuclear chemistry. One of us acknowledges the continuing intellectual support of close collaborators of the Group of Radiochemistry at the Institut de Physique Nucléaire in Orsay.

We wish to express our deep gratitude to Professor Alfred G. Maddock for having accepted, with enthusiasm, revision of the manuscript. His critical insight and numerous constructive comments have largely contributed to the achievement of our endeavor.

Throughout the two years during the preparation of the manuscript we were assisted by Dr. John MacCordick. Tirelessly, with a rare degree of patience and conscientiousness, he corrected and amended the successive versions and helped us to avoid many errors. It is our pleasant duty to thank him for his linguistic expertise and for many fruitful discussions. Whatever errors remain are, of course, the responsibility of the authors.

The sections on positron and positronium have been prepared with competence by Dr. Gilles Duplâtre, a recognized expert in the field. We gratefully acknowledge his valuable contribution.

Special thanks are due to Dr. Xavier Talmont for providing recent data and to Mrs. Simone Bouby for careful preparation of the illustrations.

Jean-Pierre Adloff
Strasbourg, France
Robert Guillaumont
Orsay, France

THE AUTHORS

Jean-Pierre Adloff, D.Sc., is titular Professor of Nuclear Chemistry at the University Louis Pasteur in Strasbourg, France and former Vice-Director of the Nuclear Research Center and Director of the Laboratory of Nuclear Chemistry in Strasbourg.

He received a diploma in Chemical Engineering from the Higher School of Chemistry in Strasbourg in 1951 and a D.Sc. from the University of Strasbourg in 1958. He prepared his doctorate with Marguerite Perey, discoverer of the element francium, and former assistant of Marie Curie.

J.-P. Adloff is author or co-author of 130 papers on radiochemistry, hot atom chemistry, radiation chemistry, and Mössbauer spectroscopy. His interest encompasses all aspects of fundamental and applied radiochemistry and education in these fields. He is co-author with M. Haissinsky of *Radiochemical Survey of the Elements* and with I. and Z. Draganic of *Radiation and Radioactivity on Earth and Beyond*. He is co-editor of the *Handbook of Hot Atom Chemistry* and of the journal *Radiochimica Acta*.

As a participant in international educative programs of UNESCO and IAEA, he has given extensive courses on radiochemistry in Mexico, Brazil, Paraguay, Ecuador, Kenya, and Thailand. He has acted as an expert of IAEA for development of research in radiochemistry at the Universities of Chiang Mai in Thailand and of Asuncion in Paraguay. He was a UNESCO expert at the University of Viscosa in Brazil.

J.-P. Adloff is a member of the division of Nuclear Chemistry of the American Chemical Society, of the French Chemical Society, and of the French Society for Nuclear Energy. He is a member of several international committees for conferences on Radiochemistry and Hot Atom Chemistry, and of the committee of the biennial Mexican Symposium on Radiochemistry.

J.-P. Adloff has received an award from the French Academy of Science and has been nominated Dr. Honoris Causa of the University of Asuncion. He is presently Chairman of the IUPAC commission on Radiochemistry and Nuclear Techniques.

Robert Guillaumont, D.Sc., graduated from the Faculty of Science of the University of Paris in 1957 and in 1966 earned his D.Sc. prepared at the Laboratoire Curie. Since 1970 he has been titular Professor of Chemistry at the University Paris-Sud in Orsay. He is in charge of the French national postgraduate course, "Radiation, Radioelements, Radiochemistry".

In 1988, R. Guillaumont was elected Corresponding Member in the section of Chemistry of the French Academy of Sciences and since 1991 he has been an honorary member of the Academy of Natural Science of Russia. He has been appointed chairman or active member of various French national scientific committees on behalf of the Ministers for Education, Research and Industry, of the Atomic Energy Commission, and of the Academy of Science. He is member of several international committees and received several awards and distinctions.

R. Guillaumont is the former Director of the Radiochemistry group of the Institute of Nuclear Physics at Orsay. He has presented many lectures at international meetings on Radiochemistry and Nuclear Chemistry and as a guest at universities and institutes over the world and has published 150 original papers. His major interests include chemistry at the tracer level, chemistry of actinides, in particular thermodynamics and spectroscopy, and the scientific basis and safety aspects of radwaste management.

CONTENTS

CHAPTER **1**

THE ROLE OF CHEMISTRY IN THE EMERGENCE AND DEVELOPMENT OF NUCLEAR SCIENCE

I. **The Beginning** ... 1

II. **From Polonium and Radium to the Dawn of Artificial Radioactivity** ... 3
- A. The Discovery of Polonium ... 3
- B. The Discovery of Radium ... 6
- C. The Discovery of Actinium ... 7
- D. The Radioactive Emanations ... 8
- E. Recognition of the Nature of Radioactivity ... 8
- F. The Radioelements, the Periodic Chart, and Isotopes ... 9
- G. The Discovery of the Last Natural Elements ... 14

III. **Chemical Proof of Artificial Radioactivity** ... 15

IV. **Chemists Discover Fission** ... 16

V. **Extension of the Periodic Table** ... 20

VI. **The Completion of the Periodic Table** ... 22

VII. **The Chemistry of Nucleogenic Atoms** ... 24

VIII. **The Intrusion of Radioactivity in Chemistry** ... 26

IX. **About Semantics** ... 27

References ... 29
Further Reading ... 31
Appendix 1.1. Nobel Prizes Related to Radiochemistry ... 32

CHAPTER **2**

RADIOCHEMISTRY TODAY

I. **From Radioactivity to the Topics of Nuclear Science** 33

II. **Radiation Chemistry** 33

III. **Radiochemistry** 34

IV. **Nuclear Chemistry** 36

V. **Frontiers of Radiochemistry** 37
- A. Radioactive Tracers 37
- B. Radioanalytical Chemistry 38
- C. Nuclear Industry 38

VI. **Radiochemistry and Periodic Table of the Elements** 39

VII. **Radiochemistry and Chemistry** 40

CHAPTER **3**

ESSENTIALS OF RADIOACTIVITY

I. **Introduction** 43

II. **Radioactivity and Radioactive Decay Modes** 44
- A. Radioactive Decay Modes 45
  - 1. Emissions from the Nucleus 45
  - 2. Emission Involving the Nucleus and the Orbital Electrons 46
    - a. Electron Capture 46
    - b. Internal Conversion 46
  - 3. Competing Modes 46
- B. Secondary Emissions 47
  - 1. Shake-Off 47
  - 2. Inner-Shell Ionization 47
- C. The Immediate Products of Radioactive Decay 49

III. **The Origin of Primary Emissions** 50
- A. Nuclear Properties 50
- B. Nuclear Decay 52
  - 1. Nuclear and Atomic States 53
  - 2. Energy Balance of the Transition 54
- C. Stable Atoms 56

IV. **Probability of Radioactive Decay** ... 57
A. The Parameter λ ... 57
B. The Radioactive Decay Law ... 58
C. Activity ... 63

V. **Decay Schemes** ... 65
A. γ Emission and Internal Conversion ... 65
1. Energy of γ Photons ... 65
2. Character of γ Radiation ... 66
3. De-Excitation Rate ... 67
4. Internal Conversion ... 69
B. β Emission and Electron Capture ... 70
1. Decay Energy ... 70
2. Decay Rate ... 72
C. α Decay ... 72
1. Decay Energy ... 72
2. Decay Rate ... 74
D. Spontaneous Fission ... 74

VI. **Nuclear Reactions** ... 76
A. Introduction ... 76
B. Energetics of a Nuclear Reaction ... 78
C. Cross Section ... 79
D. Excitation Function ... 80

VII. **The Kinetics of Nuclear Transformations** ... 81
A. Kinetics in a Filiation ... 82
1. The Half-Life of the Parent Exceeds that of the Daughter: $T_1 > T_2$ ... 83
2. The Half-Life of the Daughter Exceeds that of the Parent: $T_2 > T_1$ ... 84
3. The Parent and the Daughter have Similar Half-Lives ... 86
B. Build-Up Kinetics ... 87

General References ... 90

CHAPTER **4**

RADIOACTIVE MATTER

I. **Activity of Radioactive Matter** ... 91
A. Total and Partial Activities ... 92
B. Specific Activity ... 92

II. Non-Genuine Radiation Emitted by Radioactive Matter 93
A. Nuclear Reactions 93
B. Emission of Electromagnetic Radiation 95
C. Emission Involving Positrons 95

III. Interactions of Radiation in Radioactive Matter 96
A. Interaction of Electromagnetic Radiation 97
B. Interaction of Charged Particles 99
C. Interaction of Electrons 101
D. Absorption of Neutrons 101
E. Spatial Distribution of Electrons 103
F. Quantification of the Absorbed Energy 103
G. Heat from Radioactive Matter 106

IV. Chemical Changes in Radioactive Matter 106
A. Radiolysis and Transmutation Effects 106
1. Autoradiolysis 107
a. Gases 107
b. Aqueous Solutions 108
c. Labeled Molecules 111
d. Solid Compounds 111
2. Transmutation Effects 112
a. α Disintegration 112
b. β^- Disintegration 113
c. Electron Capture 114
B. Isotope Exchange Reactions 116

General References 117

CHAPTER **5**

BRICKS AND TOOLS FOR RADIOCHEMISTRY

I. The Periodic Table Viewed by the Radiochemist 119

II. Natural Radioelements 124
A. The Natural Radioactive Families 125
1. ^{238}U or 4n + 2 Family 125
2. ^{235}U or 4n + 3 Family 126
3. ^{232}Th or 4n Family 129
B. Atmospheric Radioactivity 130

III. Artificial Radioelements 130

IV. Radionuclides Found in Nature ... 139
A. Natural Origin ... 139
1. Primary Radionuclides ... 139
2. Isotopic Composition and Atomic Weights of Primary Radionuclides and Their Daughters ... 140
3. Radionuclides from Reactions of Cosmic Radiation ... 141
4. Neutron-Induced Nuclear Reactions ... 143
B. Anthropogenic Sources of Radionuclides in Nature ... 143

V. How to Obtain Radionuclides ... 145
A. Natural Sources of Radionuclides ... 146
B. Nuclear Reactions in a Neutron Flux ... 147
C. Nuclear Reactions with Charged Particles ... 149
D. Radionuclide Generators ... 153
E. Purity of Radionuclides ... 155

VI. Preparation of Radioactive Matter for Measurement ... 158
A. Electrodeposition ... 158
B. Nuclear Recoil ... 160
C. Absorption and Self-Absorption of the Emitted Particles ... 161

VII. Activity Measurements ... 162
A. Approximations of the Binomial Distribution Law ... 163
B. Activity Measurements and Precision ... 164
1. Poisson Distribution Law ... 165
2. Laplace-Gauss Distribution Law ... 166
C. Errors in Combined Measurements ... 167
D. Detection Limits ... 167

References ... 169
Detectors and Radiochemical Techniques ... 171
Appendix 5.1. Derivation of the Poisson and Laplace-Gauss Distributions ... 172

CHAPTER 6
THE RADIOCHEMIST'S VIEW OF CHEMICAL REACTIONS

I. From Macrochemistry to Tracer Chemistry ... 175

II. Distribution of Elements Among Various Species ... 177
A. Aqueous Solutions ... 178
B. Non-Aqueous Solutions ... 180

III. Macrocomponents and Microcomponents 180
A. Definitions 180
B. Graphical Representation of a Microcomponent Solution 181

IV. The Nature of Reactions 182
A. Reaction Pathways 182
B. Observable and Observed Reactions 184

V. The Limit of Knowledge of a Chemical System with Conventional Methods 185

References 188
Appendix 6.1. Expressions for the Graphical Representation of a Microcomponent Solution 190

CHAPTER 7
CHEMISTRY AT THE TRACER LEVEL

I. Experimental Systems 195

II. Solutions Containing Radiocolloids 196

III. Kinetics of Reactions at Very Low Concentration 199
A. Reactions 3A and 3B between Microcomponents 202
1. Isotopic Exchange at the Tracer Level 203
2. Disproportionation 204
3. Other Cases 205
B. Reactions 2A and 2B between Micro- and Macrocomponents 205

IV. Thermodynamics of Tracer Level Reactions 206
A. Acid-Base Equilibrium 206
B. Redox Equilibrium 207

V. The Experimental Approach to Tracer Level Chemistry 208

General References 210
References 210
Appendix 7.1. Equilibrium Concentrations 212
Appendix 7.2. Reaction Half-Times 214
Appendix 7.3. Acid-Base Equilibrium at the Tracer Level 217
Appendix 7.4. Redox Equilibrium 218

CHAPTER **8**

RADIOCHEMICAL METHODS. ADVANTAGES AND LIMITATIONS

Part One — Chemical Information Carried by Radiation

I. **Introduction** ... 221

II. **Information from Energy Change of Radiation with the Electronic Environment** ... 222
- A. Internal Conversion Electron Spectroscopy ... 223
- B. Mössbauer Emission Spectroscopy ... 224
- C. Positron Annihilation ... 228

III. **Information from Change in Radiation Characteristics with the Electronic Environment** ... 229
- A. Change in Half-Lives ... 229
- B. Change in the Relative Intensity of X-Rays ... 232
- C. Change in the Relative Intensity of Conversion Electrons ... 232
- D. Change in the Angular Correlation of γ Photons ... 233

Part Two — Partition and Transport Methods

I. **General Considerations** ... 236

II. **Static Partition** ... 238
- A. Distribution Coefficients ... 238
- B. Kinetics of Transfer ... 239
- C. Thermodynamics of Transfer ... 242

III. **Dynamic Partition** ... 244

IV. **Formulation of the Species Involved in Partition Methods** ... 247
- A. Aqueous Phase ... 247
- B. Non-Aqueous Phases ... 250

V. **Determination of the Equilibrium Distribution Coefficient** ... 251
- A. Solvent Extraction ... 251
  - 1. Extraction of Complexes ... 252

2. Extraction of Solvates 257
3. Extraction of Ion Pairs 259
B. Ion Exchange 261
C. Coprecipitation 262
1. Syncrystallization 263
2. Adsorption 269

VI. Determination of Non-Equilibrium Distribution Coefficients 272
A. Spontaneous Deposition 272
B. Electrolysis at Controlled Potential 274
C. Radiopolarography 276
D. Radiocoulometry 277

VII. Radiochromatography 278
A. Chromatographic Extraction 279
1. Paper Chromatography 279
2. Extraction Chromatography 279
3. Chromatography on Resin 280
B. Thermochromatography 280

VIII. Transport Methods 282
A. Diffusion and Migration 282
B. Electromigration and Electrophoresis 284

IX. Perturbation of a System by Radiochemical Methods 286

X. How to Formulate the Entities 286

XI. Separation of Radionuclides Based on Radiochemical Methods 287

General References 289
References 290
Appendix 8.1. Hyperfine Interactions 295
Reference 296
Appendix 8.2. Transfer Kinetics at the Tracer Level 297
Appendix 8.3. Thermodynamic Activity Coefficients 300
References 303
Appendix 8.4. Concentration-Dependent Partition of a Microcomponent 304
References 309
Appendix 8.5. Transfer of a Microcomponent into a Metal 310

CHAPTER **9**

CHEMISTRY WITH A FEW ATOMS

Part One — Radioactive Properties

I. Disintegration of a Single Radioactive Atom ... 314
- A. Δt and t Are Random ... 315
- B. Δt Is Replaced by dt ... 315

II. Disintegration of Several Radioactive Atoms ... 317
- A. The Value of the Half-Life Is Known ... 318
- B. The Half-Life Is Unknown ... 321

III. Activity Measurements ... 326

Part Two — Chemical Behavior of a Hundred Atoms or Less

I. Problems Encountered ... 327

II. Strategy for a Solution ... 328

III. Working Out the Calculation ... 329

IV. Application to Specific Cases ... 331
- A. The Number of Atoms Increases ... 332
- B. The Number of Atoms Decreases ... 336
- C. Fluctuations ... 338

V. Experimental Approach ... 338

VI. Beyond Radiochemistry ... 339

References ... 340
Appendix 9.1. Limiting Value of $(1-\lambda\ t/n)^n$ ... 341
Appendix 9.2. Integration of P (t_s = t) When N Is Not Constant During the Observation Time ... 342
Reference ... 343
Appendix 9.3. Mean Concentration for a Very Small Number of Atoms ... 344
Appendix 9.4. Canonic Probability for Realizing the States of a System ... 345
Appendix 9.5. Calculation of ρ Ratios ... 347
Appendix 9.6. Example of a Forbidden Reaction ... 353

CHAPTER **10**

RADIOCHEMISTRY IN ACTION

I. **Introduction** 355

II. **Radiochemistry and Chemistry of Actinides and Transactinides** 356
- A. Hydrolysis and Complexation 356
- B. Speciation of Actinides in the Environment 360
- C. Chemical Identification of Elements 363
- D. Structure of Aqua Ions 366
- E. Determination of Oxidation States 368
- F. Oxidation Number in the Metallic State 370

III. **Radiochemistry and Nuclear Waste Management** 371
- A. Radionuclides in the Packages 372
- B. The Distribution Coefficient K_d and the Migration of Radionuclides 373

IV. **Radon and Its Progeny** 375
- A. The Behavior of Radon 375
- B. The Descendants of Radon 377

V. **Recoil-Induced Synthesis of Radiolabeled Molecules** 379
- A. ^{11}C Compounds 381
- B. ^{13}N Compounds 382
- C. ^{15}O Compounds 382
- D. ^{18}F Compounds 383

VI. **Positrons in Chemistry** 383
- A. Determination of Association Constants 384
- B. Physicochemical Properties of Species in Solution 386
- C. Physicochemical Properties of Liquids and Molecular Solids 387

References 388

Conclusion 395

Index 399

Dedicated
to
the Memory of
Moïse Haissinsky

CHAPTER 1

THE ROLE OF CHEMISTRY

IN THE EMERGENCE AND DEVELOPMENT OF NUCLEAR SCIENCE

I. THE BEGINNING

Although radioactivity is primordial in the universe and is associated with the origin of all things such as the chemical elements, stars, the solar system, and the earth, less than 100 years have elapsed since the discovery of *uranic rays* by Henri Becquerel (1852–1908) on Saturday, February 29, 1896, a date which marks the beginning of a new era in the history of mankind. Historians contending for *scripta manent* may prefer the following Monday, March 2, when Becquerel reported at the weekly session of the Academy of Sciences in Paris on the ''invisible radiation emitted by phosphorescent substances''.[1] This paper announced that uranium salts spontaneously blacken a photographic plate, even without the stimulus of solar light, which was initially thought to be at the origin of the phenomenon.

Less than 4 months before, on Friday, November 9, 1895, Wilhelm Konrad Roentgen (1845–1923) had discovered X-rays (X for ''unknown'') generated by an electric discharge in a glass tube containing a gas at low pressure.[2] These strange radiations became visible from fluorescence of the surrounding glass or from luminescence excited by their impact on various materials. The suggestion made by the mathematician Henri Poincaré (1854–1912) that any luminescence, whatever its origin, may be associated with X-rays had prompted the experiments of Becquerel, who was acquainted with the phosphorescence of uranium salts.

The correlation between the two discoveries is obvious. A long time before Becquerel, in 1867, Niepce de Saint-Victor (1805–1870), the cousin of Nicéphore Niepce (1765–1833), inventor of photography, had published in the

Comptes Rendus of the Academy a series of reports "on a new action of light" in which he notes that " . . . porous or rough-surfaced compounds exposed to light maintain the property of reducing silver salts in the dark, in the same way as if the salts themselves had been acted upon by light." In his sixth contribution, Niepce noticed that the phenomenon was particularly pronounced when the photographic plate was in contact with tartaric acid or with *urane azotate*.[3] The reduction of silver salts by the chemical action of the acid is easily explained, but the effect of the uranium salt was obviously associated with the uranic rays first discovered by Becquerel 29 years later. Niepce's observations, at a time prior to that of Roentgen, were obviously premature.

The discovery of radioactivity by a chemical reaction initiated by ionizing radiations, namely, the reduction of silver ions in microcrystallites of sodium bromide, foreshadowed the role of chemistry in the history of nuclear science not only in its beginning, but also throughout its evolution up to the present time. Curiously, however, the first pioneers were not chemists; Becquerel was professor of physics at the Museum of Natural History in the Jardin des Plantes in Paris, Pierre Curie (1859–1906) was a famous physicist, and Marie Curie (née Sklodowska) (1867–1934) held a B.Sc. in mathematics and physics.

The first observed chemical effect of ionizing radiation that led to the discovery of radioactivity is still widely used in detectors and for dosimetry. The mechanism of the blackening of the photographic plate exposed to uranium is now well established. The incident radiation ionizes the silver bromide, forming electron-hole pairs. During the ensuing diffusion, electrons are trapped at crystal defects and impurity atoms, while the positive holes are trapped at interstitial silver atoms. A latent image is formed, that can be developed into a visible image by a chemical process for which a mechanism has only recently been elucidated.[4]

Since their first successful use in 1896, photographic emulsions have been extensively applied in nuclear physics, in particular for recording new types of particles and cosmic radiations.

For a long time, X-rays aroused enormous popular interest, mainly by their use in numerous gadgetries, such as in removal of superfluous hair or in shoe-fitting fluoroscopes, a practice in use until the 1940s. On the other hand, Becquerel's discovery was on the verge of being forgotten. A reminder came only 2 years later, however, when Gerhardt C. Schmidt (1865–1949) announced on February 2, 1898, that thorium salts shared the peculiar properties of uranium compounds in acting on a photographic plate and ionizing air.[4] Marie Curie reported the same observation on April 12 of that year.[5]

The time had come to seek the origin of the mysterious radiation. It is certain that little progress would have been made at this point without Pierre Curie's genius.[7] Curie decided to measure the intensity of the uranic radiation by its effect on the conductivity of air and invented an electrical device that permitted the quantitative and accurate measurement of the radiation. The detector consisted of an *ionization chamber* with two parallel metallic electrodes and a charge compensation system based on the piezoelectric effect which Pierre Curie had discovered together with his brother Jacques in 1882. Curie's invention replaced the crude detection method based on the blackening of a photographic plate or the discharge rate of an electroscope, and his technique remained in use until the early 1960s because it was better than more recent devices for measuring intense radioactive sources.

With the new detector, Marie Curie was able to measure the intensity of the radiations emitted by uranium compounds and to search systematically for other emitters. She found that the activity was proportional to the uranium content of the samples, a result which represents the first example of "radioanalytical chemistry". Of greater importance was her observation that two uranium minerals, pitchblende (uranium oxide) and chalcolite, a uranyl and copper phosphate now called *tobernite*, showed higher activity than uranium itself and suggested that " . . . these minerals may contain an element much more active than uranium."[6]

This hypothesis marks the real beginning of chemistry dealing with unstable elements. In the following section, the role of chemistry will be emphasized first in the early history of radiochemistry, with the discovery of the natural radioelements and the unravelling of the 44 radionuclides which are the descendants of uranium and thorium. The chemical proof of the first man-made transmutation of a stable element into a radioactive one opened the era of artificial radioactivity. The two major events of the discovery of nuclear fission and of transuranium elements belong preeminently to the realm of chemistry. That nuclear transformations could induce atomic and molecular, i.e., chemical, effects was recognized as early as 1934. The role of chemistry in the evolution of nuclear science has been reciprocated by an intrusion of radioactivity into the whole field of chemistry.

II. FROM POLONIUM AND RADIUM TO THE DAWN OF ARTIFICIAL RADIOACTIVITY

A. The Discovery of Polonium

To examine pitchblende for the hypothetical element which manifested more activity than pure uranium itself, the Curies hired Georges Bémont (1867–1932), Head of Laboratory at the Paris Municipal School of Industrial Chemistry and Physics.

The chemical procedure used was classical, but Marie Curie in a biography of Pierre Curie published in 1924 explained, " . . . the method we have used is a new one for chemical research based on radioactivity. It consists in separations performed with the ordinary procedures of analytical chemistry, and in the measurement of the radioactivity of all compounds separated. In

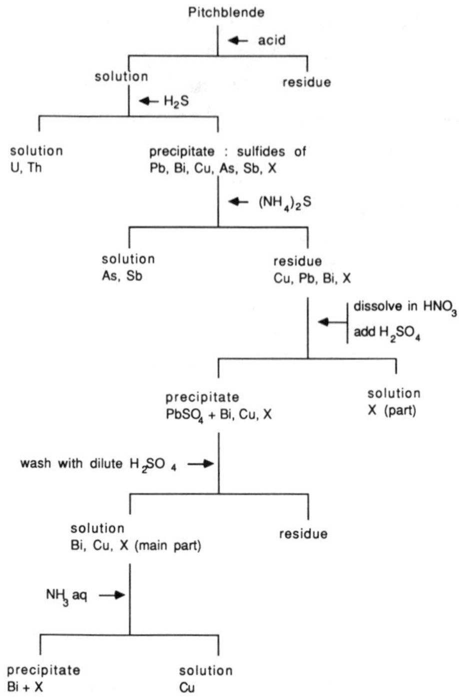

FIGURE 1.1. Flow chart of the chemical separations which led to the discovery of the element X, polonium. The Curies were unable to separate Po from Bi in a wet procedure, but accomplished a partial separation of the sulfides by fractional sublimation.

this way, one can recognize the chemical character of the radioactive element sought; the latter concentrates in fractions which become increasingly radioactive in the course of the separation."[7] This method was to be adopted by generations of chemists and is still employed in the most recent searches of new elements.

The treatment began with acidic dissolution of the mineral; the procedure as described in the original paper is shown in Figure 1.1.[8] The Curies found that the bismuth fraction was enriched by a radioactive substance that precipitated as the sulfide and could be volatilized, probably as the metal.

Marie Curie's second paper (in which the signature initial is still S., for Sklodowska) in July 1898 was the first joint publication with Pierre Curie. The paper ends with the statement: "We believe . . . that the substance we recovered from pitchblende contains a heretofore unknown metal, similar to bismuth in its analytical properties. If the existence of this new metal is confirmed, we propose that it be named *polonium* (Po), in honor of the native land of one of us."

The name polonium with the symbol Po appears in the laboratory notebook for the first time on July 15, in the handwriting of Pierre Curie. Before polonium, four elements were given names referring to countries: germanium, ruthenium, gallium, and scandium. However, the denomination of polonium had a provocative significance since at that time Poland did not exist, being parcelled out between the Austrian Empire, Germany, and Russia.

The prudent announcement of the discovery is understandable; it had become customary that no claim for the existence of a new element was considered valid until a pure substance had been isolated, the atomic weight of the element determined, and its spectral lines measured.[9] Let us suppose that the Curies treated 1 kg of pure UO_2; the amount of polonium extracted in a complete recovery would not exceed 100 ng, a quantity naturally invisible to the eye and below the sensitivity of spectroscopical detection available at that time. In their paper, the Curies indicated, probably with considerable disappointment, that Eugène Anatole Demarçay (1852–1904), the leading spectroscopist in France at the time, was unable to find a characteristic spectral line, " . . . a fact which does not favor the idea of the existence of a new metal."

The title of the publication is, "On a New Radio-Active Substance Contained in Pitchblende". This is the very first mention of the word *radioactive*, from which it can be stated that Marie Curie is the first *radiochemist* and performed the first *radiochemical* separation.

The isolation of polonium from the uranium ore was successful, although the Curies were unaware of the genetic relationship between the two elements (they considered the whole material as a mixture) and of the radioactive decay law. In that sense, it was a matter of luck since the separation was accomplished in a sufficiently short time with respect to the 138 d half-life of polonium. It was only a few years later that the Curies noticed with astonishment and great perplexity that polonium was progressively disappearing. Moreover, the Curies could not know that they were dealing with *tracer* amounts of polonium. By simple reasoning, the substance in the ore, more active than uranium itself, must be present in a very small amount because it had not yet been found. However, it was not immediately evident that this amount was below the limit of any chemical means of detection.

Without knowing it, the Curies had invented the use of *carriers*; in the final step of the separation the polonium was effectively *carried* by bismuth. The first macroscopic sample of polonium weighing 100 μg was prepared in 1910.[10]

The assignment of polonium in the Periodic Table of Mendeleev does not seem to have preoccupied the discoverers. The final part of the table was unoccupied beyond $_{83}Bi$, with the exception of $_{90}Th$ and $_{92}U$. The element next to bismuth was predicted to be analogous to tellurium, with an atomic weight of 212. The Curies failed in their attempt to separate polonium from bismuth and thus had no reason to place it in the *dvi-tellurium* position. After it was found that the activity decreased with time, they even doubted that polonium was a real element.

This question was settled only 4 years later when W. Marckwald claimed the discovery of *radiotellurium* in pitchblende.[11] The latter species showed chemical similarity with tellurium rather than with bismuth as observed by the Curies. After several years of controversy Marie Curie definitively proved the identity of polonium with radiotellurium, and the new element was correctly placed in the vacant position with atomic number 84 in the Table of Mendeleev.[12] Today, more than 25 isotopes of polonium, all radioactive, have been recorded, the longest lived being ^{209}Po with a half-life of about 100 years. The polonium identified by the Curies, ^{210}Po, is now readily prepared by irradiation of bismuth in a nuclear reactor.

B. The Discovery of Radium

After having concentrated their effort on the sulfide fraction precipitated from the soluble part of the pitchblende, the Curies searched the origin of another active substance in the solid ore residue. They announced together with Bémont on December 26, 1898 the discovery of a second element which showed all the chemical characteristics of barium.[13] It was found, however, that the new substance could be partially separated from barium by fractional recrystallization of the chlorides. The authors wrote: " . . . the new radioactive substance contains a new element for which we propose the name *radium* (Ra)." This time, Demarçay found in the spark spectrum of the new substance a line at 3818.8 Å which he could not assign to any known element.

This observation was not sufficient in itself to prove the existence of a new element. The available sample was too small for a precise determination of the atomic weight.

The sample of pitchblende from which the two radioelements were extracted came from a mine in Sankt Joachimsthal in Bohemia (now Jáchymov in Czechoslovakia). This mine had been in continuous operation since 1410, initially for the production of silver and in the 19th century for that of uranium for use in the glass industry. Pitchblende was a costly mineral, but the residue following separation of uranium had little commercial value and one ton was offered to the Curies by the Austrian government. The tedious processing of 20-kg batches of the minerals under primitive conditions had been popularized in numerous biographies, novels, and movies. In March 1902, Marie Curie isolated 120 mg of radium chloride. In 1907, a sample of higher purity, free from traces of barium, served for a first determination of the atomic weight of the element. The value obtained, 225.9, was remarkably close to the mass number 226 of the nuclide.[14]

Polonium and radium later understandably became the subject of Marie Curie's thesis entitled, "Researches on Radioactive Substances" defended at the Faculty of Science in Paris on June 23, 1903.[15]

With the foregoing discovery of polonium, the Curies had, oddly enough, begun with the most difficult part of the work. In its own right, radium had outstanding advantages; its concentration in the ore was about 5000 times greater than that of polonium and was thus present in weighable amounts; the half-life of ^{226}Ra is 1600 years; it is a true chemical analogue of barium from which it could be separated, and it could be readily assigned to its correct place in the periodic table.

The preparation of a macroscopic quantity of radium rendered a radioelement visible for the first time. Exactly 40 years later, in 1942, one was to see the first macroamount of an artificial element, namely, 1 μg of plutonium.

For decades, radium was the most important radioelement, and it was intimately related to the name of the Curies. For a time, its half-life was the basis of the unit of activity, defined as the number of decays occurring per second in one gram of radium, a unit which later served in the more precise definition *curie* (Ci). Radium was the first radioactive element to be used for medical uses denoted as *curietherapy*.

C. The Discovery of Actinium

The identification of an additional radioelement turned out to be much more difficult. After uranium, polonium, and radium had been removed from the pitchblende residues, the material still contained radioactive substances. In 1899, André Louis Debierne (1874–1949), a former student of Pierre Curie, announced that he had discovered a new fraction that had properties similar to those of titanium, but with an activity 100,000 times greater than that of uranium.[16] Later he found that the new substance more closely resembled thorium, but it was also much more active than this element. Since it behaved in a manner different from that of the known radioelements, Debierne claimed the discovery of *actinium* (Ac). However, he was puzzled by the erratic changes in the activity of some samples with time; in contact with barium, actinium seemingly produced a substance that behaved like barium.

In fact, Debierne's assertion was false. The substance separated by him was a mixture of two thorium isotopes, *ionium* (^{230}Th) and *radioactinium* (^{227}Th) with little, if any, of the element now known as actinium. The error was quite understandable, however, because it arose from the extreme complexity of the mixture of natural radioisotopes and at a time when the existence of isotopes, genetic relations between the daughter products of uranium and thorium, the concept of decrease in activity were still unknown.

A critical survey of the papers published during the first years of the present century leads to the conclusion that the real discoverer of actinium was probably Friedrich Oskar Giesel (1852–1927), who detected a substance that could be concentrated in the rare earth fraction from pitchblende and in which thorium was absent.[9,17] Nevertheless, most historians continue to credit Debierne with the discovery of actinium. The name *emanium* given by Giesel to the element was rather unfortunate since actinium emits no radiation which could have been detected at that time. Neither Debierne or Giesel ever measured actinium itself, but rather the radiations of the daughter *radiothorium*.

D. The Radioactive Emanations

Certain observations were particularly puzzling, especially the disappearance of activity in the course of the experiments and the fact that some samples appeared to regain their "lost" activity. In other cases the activity seemed to be transferred from one fraction to the next.

Apart from the French team, Ernest Rutherford (1871–1937), first at Cambridge and later at McGill University in Montreal, became interested in the new phenomenon of radioactivity. He noticed that the radiation of thorium was particularly erratic.[18] It seemed to be affected simply by opening of the door and windows of the laboratory. Reproducible measurements could only be made by enclosing the sample in an airtight box.

In 1899, Rutherford came to the conclusion that thorium compounds continuously release " . . . radioactive particles of some kind, which retain their radioactive power for several minutes. The *emanation*, as it will be termed for shortness . . . acts like an ordinary gas."[19] Moreover, any solid in the vicinity became temporarily radioactive, but here the activity decayed to half value in about 11 h instead of within minutes. Indeed, the Curies also observed an *induced activity* on objects close to radium, and Giesel noticed the same phenomenon with his rare earth fraction, which for this reason he had named *Emanationskörper* and which was later identified as the "true" actinium.

Rutherford was not a chemist but he was fortunate in benefiting from the collaboration of Frederick Soddy (1877–1956) during a term of two years. This team first found that the emanation did not come from thorium itself, but from another substance that they called thorium X (now ^{224}Ra). Soddy attempted both to decompose the emanation and to combine it chemically, but without success. Such chemical inertness was not altogether unfamiliar since the whole group of rare gases ranging from helium to xenon had been discovered in the course of previous years. Thus the behavior of the emanation was typical for that of a rare gas, and the authors named it *thoron*.

In this most unexpected and unprecedented event, everything pointed to the transformation of solid thorium into a gas.[20,21] This was probably the first recognition of the fact that radioactive decay is a process within the atom and involves its spontaneous transformation: the atom changes its chemical identity, and is *transmuted* into the atom of another element.

E. Recognition of the Nature of Radioactivity

At the beginning of the present century, observations of new examples of radioactivities were flourishing. An elusive "activity" similar to that of the emanations was discovered early in 1900 by Sir William Crookes (1832–1919) in uranium itself. In the course of purifying uranyl nitrate by its dissolution in ether and recrystallization, Crookes noticed a fugitive activity in the aqueous phase which he concentrated in a small, very radioactive precipitate. He concluded that " . . . the supposed radio-activity of uranium and its salts is not an inherent property of the element, but is due to the presence of a foreign body." He named the substance *uranium X*.[22]

The discovery of so many substances exhibiting the property of radioactivity was rather confusing and posed the problem of their identification. Rutherford proposed to systematize the nomenclature by letters, although this recommendation was not strictly followed; names such as *mesothorium* were also used. The first product formed from emanation was called "A", the second "B", and so on.

By the year 1910, most of the 44 natural radioactive substances were known and could be arranged in three natural radioactive families (Figure 1.2) ending with nonradioactive lead. Two of these had uranium as the parent element and one was headed by thorium. Although it appeared that two decay series descended from uranium, it was not until the mid-1930s that the connection was really understood. The actinium series was something of a mystery until the discovery of ^{235}U by Dempster in 1935.

An authoritative historical account of the unravelling of the mixture of natural radioelements has been written by Kirby.[9]

By now the basic features of radioactivity had been elucidated: (1) the distinction of α, β, and γ radiations, (2) the identification of α particles as helium, (3) the radioactive growth and decay laws, and (4) the fact that one radioactive entity can be transformed into another. Rutherford, who had played a leading role in the chemical investigations which progressively led to these concepts, also showed experimentally in 1911 that atoms possess a positively charged center concentrated in a very small region or *nucleus*, surrounded by electrons in rapid motion. Semantically, this event marks the birth of *nuclear* science.

Rutherford also observed the appearance of hydrogen atoms when nitrogen was bombarded with α particles. He concluded that the hydrogen atoms existed as positive ions inside the nucleus and called them προτον, the first, or proton. The new image of the atom was accordingly that of an assembly comprising a nucleus made up of protons and surrounded by electrons.

F. The Radioelements, the Periodic Chart, and Isotopes

Too much was too much! Three dozen new radioactive species had been definitely identified and characterized by their radiation, but only a limited number of places remained free in the periodic table of the elements set up in 1869 by Dmitri Ivanovich Mendeleev (1834–1907). Because its atomic weight was known and its properties very similar to those of barium, radium could be easily accommodated in the table at the vacant position corresponding to Z = 88 in the alkaline earth series. The three emanations that behaved

Uranium family

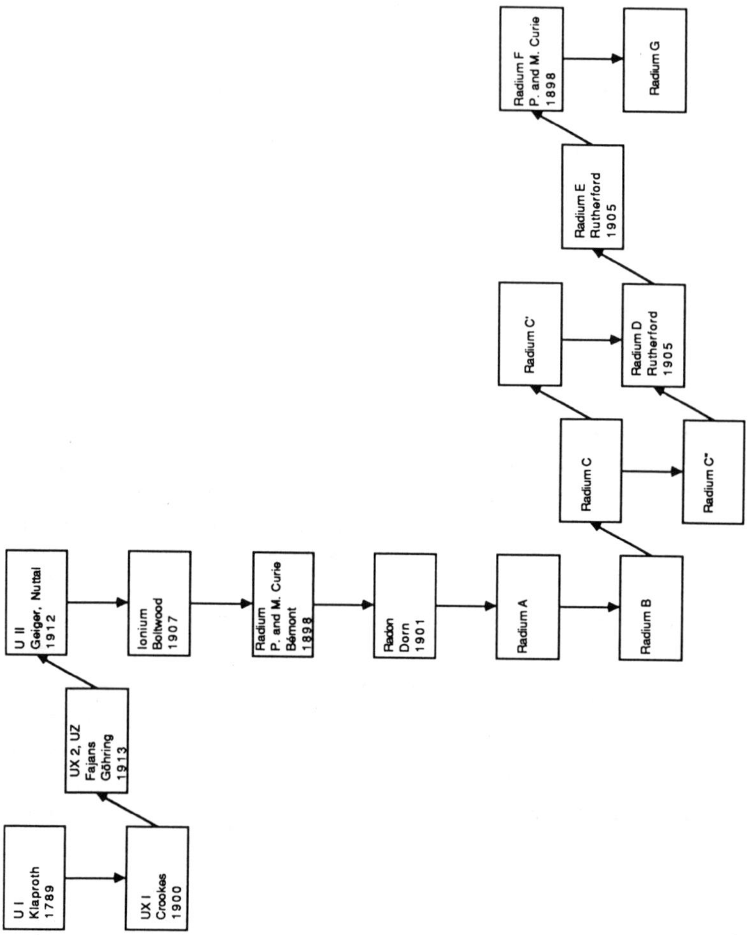

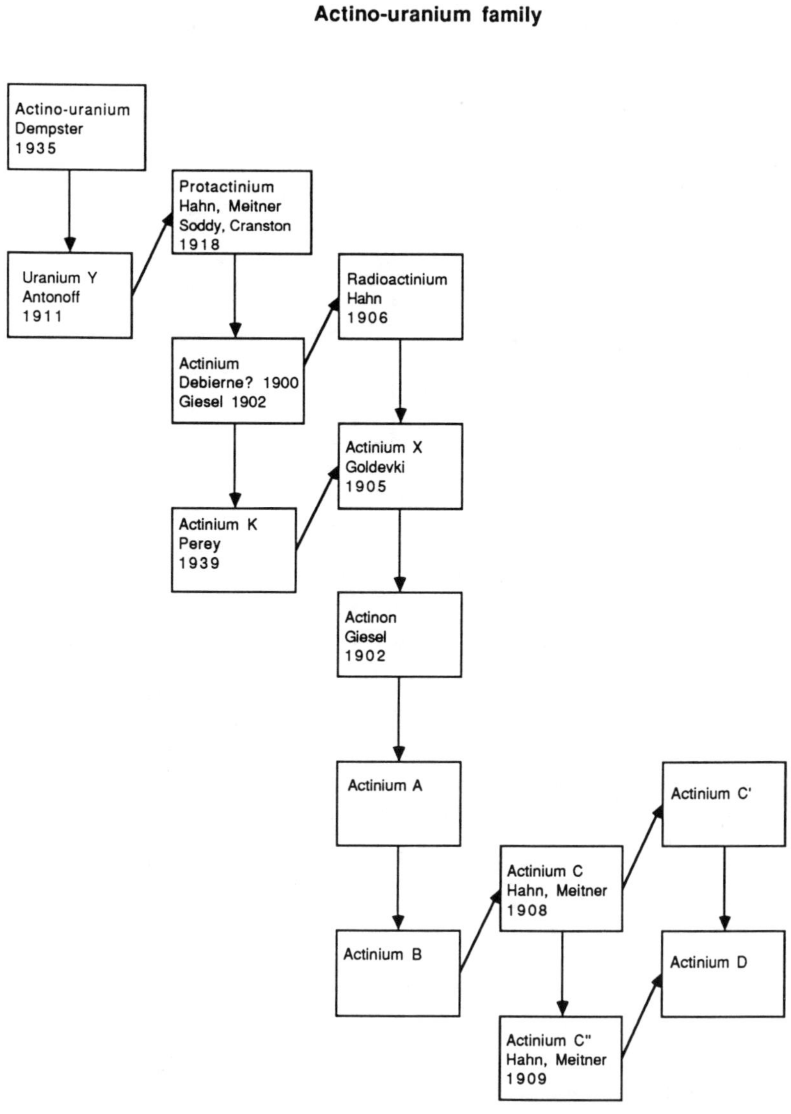

FIGURE 1.2. Historical names and year of identification of the members of the three natural radioactive families. Some dates are approximative. Several members were identified simultaneously by different authors and the assignment of original discovery is uncertain. The terms of the active deposits were found between 1899 and 1911. Until 1935, the actino-uranium family was considered as a branching of the uranium-radium family.

Thorium family

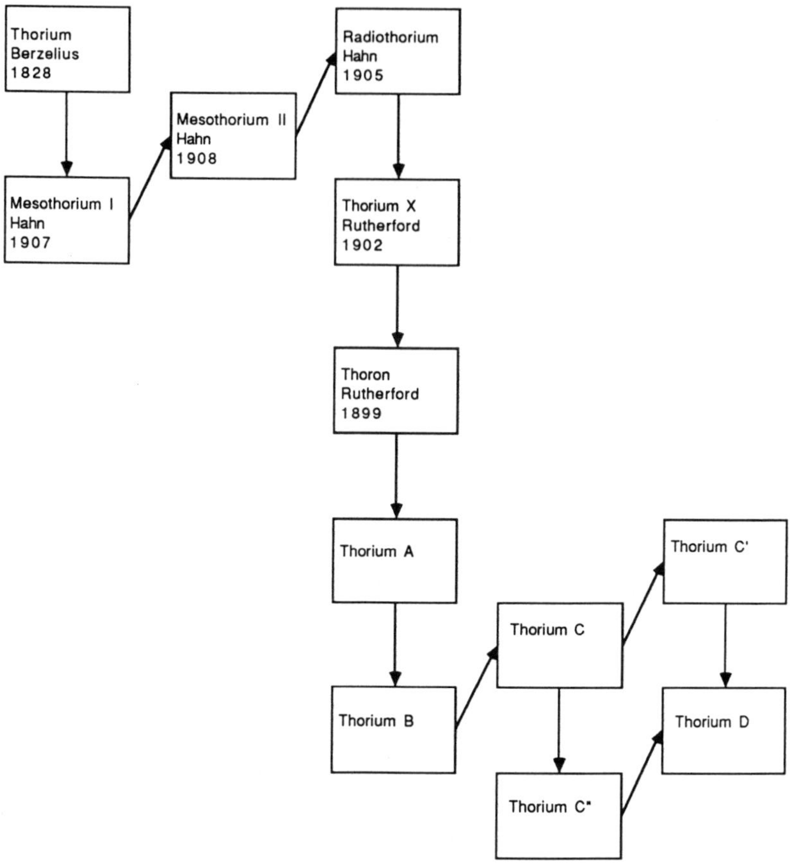

FIGURE 1.2 (continued).

like the five known rare gases undoubtedly belonged in the zero valency group in the eka-xenon position.

For the five remaining spaces between bismuth (Z = 83) and uranium, (Z = 92), there were still too many candidates. Polonium was not much of a problem because of its analogy with tellurium. While awaiting the determination of its atomic weight, it could be tentatively assigned to the eka-tellurium position with Z = 84.

For the remaining substances, it took several years and a great deal of work to settle the problem without violating Mendeleev's classification. At this point, the efficacy of the periodic table for recognizing the nature of new radioactive elements should be emphasized. In 1899, Mendeleev had already predicted the existence of dvi-tellurium, the future polonium, and described the properties of this element with remarkable accuracy. The reasoning on which Mendeleev had based his classification was found to be even more useful when the time came to put the transuranium elements into sequential order.

The idea emerged that some of the radioactive substances might represent varieties of the same element. In 1910, Soddy noted that thorium X, actinium X, mesothorium, and radium " . . . show a complete identity of chemical behavior and are members of the alkaline-earth group of elements." Other groups of radioactive substances also shared identical chemical properties. For instance, as shown by Alexander Fleck (1889–1968), uranium X could not be separated from thorium; the "B" products from the radioactive families were unseparable from lead, and the "C" products followed the chemistry of bismuth.[23] The characteristic radioactive behavior of these substances indicated, however, that they were not strictly the same.

For the modern chemist, the puzzle seems easy to solve but at that time the idea of chemically identical, but physically different, species was not straightforward. A great step ahead was made when Kasimir Fajans (1887–1975) found ("by intuition", according to the author) the *displacement laws* in 1912.[24,25] The emission of α and β particles causes a characteristic shift of the radioactive element in the periodic chart: two units to the left after α decay and one unit to the right following β decay. On the basis of Fajans' laws all known radioactive elements could be fitted into the periodic table and the atomic weight could be estimated. The validity of Mendeleev's table for predicting chemical properties was preserved.

It was now firmly established that several radioactive elements of different origin, with different atomic weights, could share the same place in the periodic table. Soddy[15] in 1914 reviewed the consequences of this very important conclusion and introduced the new word *isotope*, from the Greek ισοσ, same and τοποσ, place (actually, the name was coined by the physician and novelist Margaret Todd during a dinner party with Soddy).[9,24]

The neologism was used for the first time in a short note published in *Nature* in 1913, "The same algebraic sum of the positive and negative charges in the nucleus, when the arithmetical sum is different, gives what I call *isotopes* or *isotopic elements*, because they occupy the same place in the periodic table. They are chemically identical, and save only as regards the relatively few physical properties which depend upon atomic mass directly, physically identical also."[26]

The emergence of the concept of isotope is probably the most important aftereffect of the discovery of radioactivity. As emphasized by Kirby, " . . . the term was uniquely chemical in conception, for the place referred to was the position in the periodic table occupied by chemically indistinguishable elements."[9] The existence of isotopes was corroborated practically at the same time from the physical side when J. J. Thomson (1856–1940) with his first mass spectrograph found that " . . . what has been called neon is not a simple gas, but a mixture of two gases, one of which has an atomic weight of about 20 and the other of about 22.''[27] Thus, nonseparable elements were not necessarily heavy or radioactive.

The discovery of isotopes resurrected the hypothesis of William Prout (1785–1850) that the atoms of chemical elements are multiaggregates of hydrogen atoms, but now in a different image: the atomic weights of isotopes, and not of atoms should be a whole number multiple of the atomic weight of hydrogen.

G. The Discovery of the Last Natural Elements

According to Fajans' displacement law, the β decay of thorium should lead to an isotope of the element *eka*-tantalum, with atomic number 91. The two uranium families possess a β-emitting thorium isotope. One of these is the uranium X (^{234}Th) discovered by Crookes, whose direct known descendant was a uranium isotope, U II. However, the conversion of thorium to U II had to proceed via two β decays.

The dilemma was solved when Fajans showed in 1913, that uranium X was actually a mixture of two isotopes (UX1 and UX2) and isolated uranium X2, a short-lived (1.15 min) β emitter.[28] The genetic relationship between UX1 and UII was thus established and the new radioactive substance placed correctly in the table between thorium and uranium.

Owing to the short lifetime of UX2 (which, for that reason, was named *brevium*), no further progress was made on eka-tantalum, until the discovery in 1918 independently by two groups of authors, (Hahn and Meitner; Soddy and Cranston) of a second, long-lived (34,000 years) eka-tantalum in the residues of pitchblende. This radioelement is the β daughter of a thorium isotope, uranium Y (^{231}Th) in the actinium series. It links UY with actinium and for this reason, the element received the name *protoactinium* (meaning *engenders* actinium), later condensed to *protactinium*.[29]

In 1921, Hahn noted the existence of a nuclear isomer of UX2, which he named *uranium Z*. The couple UX2-UZ was the first known example of isomers and the only one existing among natural radioelements.

The last radioactive element to be found in nature was discovered in 1939 by Marguerite Perey (1909–1975).[30] It occupies the eka-cesium position in the periodic table with atomic number 87 and is the heaviest alkali element. As in the case of protactinium, it was possible to predict its genesis according to the Fajans' law, as the α daughter of actinium ($Z = 89$). It was eventually discovered as a branching of ^{227}Ac, which decays to the extent of 1.2% into eka-cesium which has a half-life of 22 min. The isotope was named *actinium K* and the element for which $Z = 87$ was called *francium* (Fr), in honor of the country where it was discovered. The nuclide ^{223}Fr is the longest-lived isotope, which makes francium the most unstable of the first 101 elements of the present-day table.

III. CHEMICAL PROOF OF ARTIFICIAL RADIOACTIVITY

The exciting first decade of the present century was followed by a quieter period during which the genealogy of the three radioactive families was definitely established and the properties of the radioelements and their radiations investigated in detail. At this point, Rutherford realized two major achievements: the recognition of the atomic nucleus, which at this stage was thought to consist of protons and electrons, and in 1919 the first man-made *nuclear reaction*, whereby nitrogen atoms bombarded with α particles were transmuted into oxygen:

$$^{14}_{7}\text{N} + ^{4}_{2}\text{He} \rightarrow ^{17}_{8}\text{O} + ^{1}_{1}\text{H}$$

The reaction was evidenced by the detection of the emitted protons in a cloud chamber. Unfortunately, the product of the reaction was a stable nuclide. The major event of the synthesis of a radioactive nuclide in a nuclear reaction came only 15 years later, in 1934.

In the meantime, in 1932, James Chadwick (1891–1974) had discovered the *neutron* (in fact its existence had been hypothesized by Rutherford in 1920). From then on, the basic structure of the atom was established: it consisted of a nucleus composed of neutrons and protons, surrounded by electrons. Soddy's concept of isotopes fitted perfectly with this nuclear model since, for instance, all "B" products in the radioactive families had the same number of protons, but differed by the number of neutrons, thus corresponding to the definition of isotopes.

Although the discovery of the neutron is attributed to Chadwick, several groups of experimenters had observed the emission of a radiation much more penetrating than γ photons when various targets of light elements were bombarded by α particles. These experiments required very intense sources of polonium, the most common α emitter without interference of γ rays.

The most appropriate place to prepare large amounts of this element was undoubtedly the *Institut du Radium* in Paris, founded by the discoverer of the element. In 1925, Marie Curie assigned this task to her assistant Frédéric Joliot (1900–1958). The following year Joliot married Irène Curie (1897–1956), Marie's daughter and close collaborator.

The research undertaken by the Joliots was guided by the discovery of a new particle, the *positive electron* or *positron*, first observed in cosmic rays in 1932, but also produced, together with neutrons, in the bombardment of light elements by α rays. Accordingly, the team noticed that both types of particles were emitted when aluminum was exposed to an intense (80 millicuries) polonium source. The mass number of Al is 27; since the mass of the positron is negligible, the emission signified the formation of a *phosphorus* atom according to the nucleon balance in the nuclear reaction

$$^{27}_{13}Al + ^{4}_{2}He \rightarrow ^{1}_{0}n + ^{30}_{15}P$$

The authors noticed, however, that the emission of positrons was delayed and that it continued even after withdrawal of the polonium source. Moreover, the positron emission decreased with a half-life of 3 min. This fact indicated that the phosphorus was not stable, but radioactive, decaying by the emission of positrons:

$$^{30}_{15}P \rightarrow ^{30}_{14}Si + e^{+} + \nu$$

This was confirmed chemically. The irradiated aluminum foil was dissolved in hydrochloric acid. The trapped phosphorus atoms are hereby reduced to P(-III) by nascent hydrogen and swept out from the solution as gaseous phosphine PH_3. The chemical separation of the radioactive phosphorus was the undeniable proof of the discovery of artificial radioactivity.[31-33] Until that day, all nuclear reactions had led to stable products. Similar results were reported for boron, transmuted into radioactive nitrogen, and magnesium transmuted into radioactive aluminum.

One may wonder to what extent the success of the Joliots' experiment was fortuitous. Had the irradiated aluminum been dissolved in nitric instead of hydrochloric acid, the radioactive phosphorous atoms trapped in the metal might have been oxidized to the phosphate anion, which would have remained in solution, as Joliot later remarked. The chemical identification of the short-lived nuclide would have been much more difficult, if indeed possible!

This discovery of artificial radioactivity on Friday, January 12, 1934 made an enormous impact on chemists and was followed by a fantastic increase in the number of synthetic radionuclides: 200 nuclides were produced in the course of 3 years. The number of new radioactive nuclides reached 1000 in 1945 and exceeds at present 2500 species. This proliferation was a consequence of the use of the neutron as an efficient projectile and the availability of high neutron fluxes in nuclear reactors. In addition, the development of accelerators also contributed towards the production of a great variety of energetic ions that were suitable for nuclear reactions.

IV. CHEMISTS DISCOVER FISSION

The conjunction of the nearly simultaneous events of the discoveries of the neutron and of artificial radioactivity was very fortunate. It was soon recognized that the neutron should be a very efficaceous transmutation agent because it could penetrate more easily into the nucleus than the positively charged α particle. Furthermore, it appeared from the Joliot-Curie experiments that most, if not all, elements might have radioactive isotopes.

These considerations prompted Enrico Fermi (1901–1954) to undertake a systematic investigation of the neutron-induced transmutation of all stable elements. It follows from the simple nucleon balance that a stable target nuclide of mass number A is transmuted after neutron capture into the A + 1 nuclide. If the latter were radioactive, it would be expected to be a β^- emitter and produce upon decay an isotope of the neighboring element in the periodic chart, as for example in the case of iodine,

$$^{127}_{53}\text{I} + {}^{1}_{0}\text{n} \rightarrow {}^{128}_{53}\text{I}\ \beta^-,\ 25\ \text{min} \rightarrow {}^{128}_{54}\text{Xe}$$

The application of this scheme to uranium was particularly challenging because the same process should lead to the synthesis of a *trans-uranium* element beyond the limit fixed by Mendeleev in his periodic table. Fermi was optimistic in this respect since the experiment had proved successful with most elements and had led to the discovery of about 40 new radionuclides.[34] Indeed, Fermi observed in neutron-irradiated uranium several β^- emitters with half-lives of 10 s, 40 s, 13 min, 23 min, 40 min, and 1 d.[35] On the basis of previous experiments, these isotopes should correspond to elements close to uranium, but with an atomic number $Z > 92$, and thus easily distinguishable from the natural radioactive daughters of uranium. Hence, the observed activities were evidence for the formation not only of an isotope of element 93 but also of successive β^- daughters of the latter:

$$^{238}_{92}\text{U} + {}^{1}_{0}\text{n} \rightarrow {}^{239}_{92}\text{U}\ \beta^- \rightarrow {}^{239}_{93}\text{X}\ \beta^- \rightarrow {}^{239}_{94}\text{Y}\ \beta^- \rightarrow {}^{239}_{95}\text{Z} \ldots$$

The chemical identification of these elements was based on the expected position of transuranium elements in the periodic table. In this classification, they should occupy positions below Re and the Group VIII of platinum metals according to:

Group	**IV**	**V**	**VI**	**VII**		**VIII**	
	Hf	Ta	W	Re	Os	Ir	Pt
	Th	Pa	U	93	94	95	96

The first transuranium element should be an eka-rhenium, and share the properties of Group VII whose lightest member is Mn. However, it was also known at this time that somewhere beyond uranium, if such elements existed, another group of rare earth-like (at least electronically) elements should exist.

Fermi found that the 13-min activity was carried by a precipitate of MnO_2, and, as an even more convincing proof, by rhenium sufide. On the other hand, the activity of 23 min was obviously that of an isotope of uranium, which should necessarily be ^{239}U.

The conclusion was straightforward: the chemical behavior of the induced activity was different from that of all known elements close to uranium.

Not unexpectedly, Fermi's claim aroused criticism and doubt and prompted numerous experiments of verification which have been reviewed in detail.[36] A very intensive study was undertaken by a group of famous radiochemists who had already played a major role in the characterization of the natural radioactive isotopes. This team, working in Berlin, was comprised of Otto Hahn (1879–1968), Liese Meitner (1878–1968), and Fritz Strassmann (1902–1980).

These scientists confirmed Fermi's results although they were aware that the chemical identification of the new substances was complex as a result of interference from the natural radioactivity of uranium and because of the steadily increasing number of newly discovered activities. It is significant that the Berlin group refrained from proposing names for the new elements, whereas Fermi and his co-workers in Rome were sufficiently confident to announce the discovery of *ausonium* (element 93) and *hesperium* (element 94).

Similar studies were undertaken on neutron-irradiated thorium. Here the goal was not to exceed Mendeleev's limit, but to search for the missing radioactive family. In the three known families, headed by uranium-238, uranium-235 and thorium-232, the mass numbers of the successive members can be divided by 4n, 4n + 2 and 4n + 3. Curiously, the 4n + 1 series was unknown and it was suspected that it might be possible to produce it artificially.

The activities induced in neutron-irradiated thorium also behaved in a very complicated manner and the resolution of the resulting mixture was not obvious. Today, this family is well known. It is, in fact, artificial and begins with neptunium-237, whose half-life is of the order of 2 million years.

A third team entered the scene and it was not surprising that it consisted of Parisian radiochemists who had initially discovered artificial radioactivity. Irène Curie and Pavle Savic (Paul Savitch) recorded an observation in October 1938 which happened to be a turning point in the research stimulated by Fermi's experiments. They discovered a radioactive species with a half-life of 3.5 h, which, after careful examination, was found to behave like lanthanum, since it could be precipitated with fluoride and oxalate carriers. This was a very unexpected result, since lanthanum is *far removed* from uranium in the periodic table. As it turned out later, it really *was* lanthanum and Irène Curie and Savic were within a hair's breadth of making a truly major discovery. Unfortunately, owing to difficulty in the interpretation of a residual activity in the irradiated sample (activity they considered as an indication of some subtle difference between the chemical behavior of the 3.5 h substance and that of lanthanum) they were unconvinced of the actual presence of the rare earth element.[37] The substance was accordingly thought to be either actinium or a transuranium element with properties very different from those known of elements 93 to 96.

The result of the Paris group was so unusual that it immediately prompted Hahn to verify the experiment, "despite his great respect for Irène". The only plausible explanation was that Curie and Savic had confused lanthanum with actinium formed in neutron-irradiated uranium, or that their production resulted from β^- decay of yet unknown isotopes of radium.

Within a few weeks the German trio solved the puzzle in a series of elegant and convincing radiochemical tracer experiments.[36] The *Cu-Sa* (for Curie-Savic) 3.5 h activity could be representative of radium or actinium possibly formed in the sequence

$$U + n \rightarrow \alpha + Th \qquad Th \rightarrow Ra + \alpha \qquad Ra \rightarrow Ac + \beta^-$$

The barium fraction precipitated from the solution of irradiated uranium carried several ''Ra'' isotopes, with activities of 25 min, 110 min, and several days. But the attempt to separate these ''Ra'' components from Ba by fractional recrystallization of the barium chlorides or bromides according to the well-established method of Marie Curie gave very strange results: it seemed that all of the fractions contained the same amount of ''Ra''.

At this point, Strassmann conceived what was probably the first ''indicator'' experiment. He added to the ''Ra'' a well-known radium isotope, *meso-thorium I* or ^{228}Ra, which had the function of what is today called a tracer. This radioelement behaved as expected in the fractional separation of Ra from Ba, while the ''Ra'' did not separate from Ba. The authors concluded that '' . . . our radium isotopes do not behave like Ra but like Ba.'' A similar trend was found for the ''Ac'' in neutron-irradiated uranium, which behaved like lanthanum rather than like actinium. Together, all facts favored the formation of Ba and La in neutron-irradiated uranium.

The report of this conclusion was submitted for publication December 22, 1938 and appeared on January 6, 1939.[38] This date is considered to mark the discovery of nuclear fission. The authors, however, were still very reluctant to claim *urbi et orbi* that uranium when bombarded with neutrons splits into two parts. They write '' . . . as chemists . . . we ought to replace here Ra, Ac, Th by the symbols Ba, La, Ce . . . but as *nuclear* chemists, closer to the realm of physics, we cannot decide to take this jump, which would be in contradiction with all present experience in nuclear physics.'' Later, in the same month, the authors submitted a second historical note in which they furnished new evidence for the formation of active barium isotopes from neutron-irradiated uranium and thorium.[39] The work described has been quoted as '' . . . the most careful and unambiguous ever carried out in radiochemistry.'' It provided the definitive proof that the uranium nucleus can split into light atoms.

The word *fission* appeared in a note by Liese Meitner and her nephew Otto Robert Frisch (1904–1979) at about the same time.[40] These authors proposed a preliminary explanation of the mechanism of the new phenomenon and indicated that uranium '' . . . divides itself into two nuclei . . . they will repel each other and gain a total kinetic energy of ca. 200 MeV.'' The term fission was suggested by a biochemist who saw the similarity between the division of uranium and the fission of cells.

Physical evidence of the phenomenon was given by Frisch, who recorded giant pulses produced by the fast fission fragments in an ionization chamber connected to a linear amplifier.[41]

The thrilling story of the discovery of fission has been elegantly related by Günter Herrmann.[36] Today, most of the radionuclides formed in the fission of uranium have been identified. The herculean task of resolving the mixture of more than 500 radionuclides was the most difficult problem ever encountered in radiochemistry.

V. EXTENSION OF THE PERIODIC TABLE

In the history of modern chemistry, two short periods were particularly important in witnessing fundamental discoveries: the first was 1896 to 1898, which marked the beginning of the era of radioactivity, and the second was during the years 1939 to 1940, which announced the ''nuclear'' era of humanity. Like the discovery of radium which echoed that of radioactivity, the discovery of fission was ineluctably followed by that of the first transuranium elements.

In fact, neither Fermi nor Hahn was wrong in announcing the formation of elements with $Z > 92$ in uranium bombarded by neutrons. Today the sequence

$$^{238}_{92}\text{U} + ^{1}_{0}\text{n} \rightarrow ^{239}_{92}\text{U}\ \beta^-\ 23.5\ \text{min} \rightarrow ^{239}_{93}\text{Np}\ \beta^-\ 2.3\ \text{d} \rightarrow$$

$$^{239}_{94}\text{Pu}\ \alpha\ 24{,}110\ \text{years} \rightarrow \ldots$$

is well established. It is also known that only the rare ^{235}U isotope (0.72% natural abundance) undergoes fission with thermal neutrons and that the fission cross section is 550 b. The neutron capture cross section of ^{238}U is 2.7 b. From these figures it can be calculated that in natural uranium the probability of fission is 15-fold higher than that of formation of element 93. The activity of ^{239}U was mixed with that of the various fission products with similar half-lives formed in the fission of ^{235}U. In fact, this isotope had been recognized by both Fermi and Hahn, and three of the new activities were attributed to ^{239}U, thus posing the enigma of a case of triple nuclear isomerism. Moreover, the chemists were looking for an eka-rhenium element and had no particular reason for seeking an isotope of uranium.

The discovery of fission was a ''by-product'' of the chemist's attempt to penetrate into the *terra incognita* beyond uranium. Despite the fascinating perspective opened by fission, the search for transuranium elements was not abandoned. On the contrary, the discovery of *neptunium* (Np), the element with $Z = 93$, was a consequence of the discovery of fission!

The two fragments produced in the scission of the uranium nucleus have very high initial recoil energies, i.e., up to tens of megaelectronvolts (MeV), owing to the coulombic repulsion between the two species bearing a high positive charge. They are readily expelled from a thin uranium layer. However, in the spring of 1939, E. M. McMillan and P. Abelson at the University of California at Berkeley, observed that one of the radioactive products formed in neutron-irradiated uranium did not follow this trend, and accordingly could not be formed in the fission of uranium. They suspected that it resulted from radiative neutron capture in the much more abundant ^{238}U, which confers a recoil energy that would not exceed 100 eV.[42]

A major step forward was made when the authors abandoned the idea that the next higher element beyond uranium should be similar to rhenium. Experiments did show in fact that it resembled uranium itself. This was the first experimental evidence for the existence of a horizontal group of elements sharing similar properties. This concept was a determinant factor in the history of the discovery and of the chemical identification of all transuranium elements up to Z = 103.

The actinide series is now familiar to all chemists. Initially, it was also known as the *5f-element* series since it was supposed that the inner 5f shell was progressively but irregularly filled in the elements from actinium (Z = 89) to *lawrencium* (Z = 103). This horizontal series was considered as analogous to the lanthanide series, ranging from lanthanum (Z = 57), to lutetium (Z = 71) and which corresponds to the filling of the 4f shell. The similarities in chemical properties of the elements in a series (much less marked in the 5f than in the 4f series) together with some vertical analogy between each actinide and its corresponding element in the lanthanide series (reflected *inter alia* in names, e.g., americium-europium) were vital concepts in the investigation of the transuranium elements. Today, it is established that protactinium is the first element with a 5f electron.

^{239}Np was probably the first transuranium isotope identified.[42] The discovery of the next element, *plutonium* (Pu), followed rapidly. A new isotope of neptunium, prepared by bombarding uranium with charged particles, decayed to an isotope of element 94:

$$^{238}_{92}U + ^{2}_{1}H \rightarrow ^{238}_{93}Np \ \beta^- \ 2.12\ d \rightarrow ^{238}_{94}Pu \ \alpha,\ 87.7 \text{ years}$$

The foremost pioneer in this field was Glenn T. Seaborg (born 1912), whose name is associated with the discovery of all transuranium elements up to *nobelium* (Z = 102).

Because of its ability to undergo thermal neutron fission, the isotope ^{239}Pu formed in neutron irradiated uranium is of major importance and it is now produced on a large scale.

The history of successive discoveries of the transuranium elements has been reviewed.[43] Most of the first preparations were based on nuclear reactions induced by charged particles; however, macroscopic amounts of these elements can only be obtained by intense neutron bombardment in nuclear reactors. Heavy projectiles such as oxygen ions are necessary for the preparation of isotopes of elements beyond Z = 101.

With increasing Z, the half-life of the most stable isotope decreases and the amount of the element produced becomes smaller at each step. The heaviest element that can yet be produced on a microscopic scale but in weighable amount is einsteinium (Z = 99).

With lawrencium, the last actinide, the filling of the 5f shell is complete. Additional electrons occupy the 6d shell. The last element so far produced has atomic number 109, and its atom is expected to have seven 6d electrons. At present, the chemical properties of the elements with $Z \geq 106$ (as well as those of future elements) can only be predicted since the very short lifetimes, in the millisecond range, preclude experimental investigation.

Seventeen transuranium elements have now been synthesized, representing an extension of nearly 20% to Mendeleev's periodic table. Even if only less than a dozen of these elements can be conveniently handled, radiochemistry has revolutionized the whole field of chemistry. The ultimate limit of the periodic table has not yet been set, since the possible existence of *superheavy elements*, although actually never proved to date, cannot be definitely ruled out.

VI. THE COMPLETION OF THE PERIODIC TABLE

Curiously, chemists had succeeded in synthesizing transuranium elements well before the classical periodic table had been completed. At the time artificial radioactivity was discovered, four elements below uranium, namely those corresponding to the atomic numbers 43, 61, 85, and 87, were still missing. None of these has a stable isotope. The most recent stable element identified before then was *hafnium* (Z = 72), discovered in 1923 by George Hevesy (1885–1966). This chemist was later to become one of the most famous radiochemists of the century.

The missing element with atomic number 43, toward the middle of the periodic table, is situated below manganese and between molybdenum and ruthenium. It had puzzled chemists for a long time. The pseudodiscovery of *masurium* in 1925 by the Noddacks (who had correctly identified rhenium in the same year) on the basis of its X-ray emission spectrum had never really been convincing.

In 1937, C. Perrier and E. Segrè observed a residual activity in a molybdenum target bombarded with deuterons and neutrons. This activity could possibly have arisen from an isotope of Mo, Zr, Nb, or from the elusive element with Z = 43. After dissolution of the target in aqua regia and addition of carriers, Mo was precipitated as the sulfide MoS_3, Nb as the oxide Nb_2O_5 and Zr as the hydroxide $Zr(OH)_4$. None of these precipitates carried the activity. However, most of it was found in a precipitate of Re complexed with 8-hydroxyquinoline. This behavior helped to identify the new activity as that of an isotope of element 43, formed in a reaction such as

$$^{94}_{42}\text{Mo} + {}^{2}_{1}\text{H} \rightarrow {}^{95}_{43}\text{Tc} + \text{n}$$

Element 43 was the first man-made component to be added to the periodic table.[44] This is recalled in the name *technetium* (Tc), derived from the Greek τεχνετοσ, artificial. Technetium is formed in the fission of uranium and is available in large amounts as the isotope ^{99}Tc, which has a half-life of over 200,000 years. This isotope is also produced by neutron irradiation of Mo in the sequence

$$^{98}_{42}\text{Mo} + \text{n} \rightarrow {}^{99}_{42}\text{Mo}\ \beta^-\ 66\ \text{h} \rightarrow {}^{99\text{m}}_{43}\text{Tc}\ \gamma\ 6\ \text{h} \rightarrow {}^{99}_{43}\text{Tc}$$

The short-lived isomer $^{99\text{m}}$Tc is the most frequently used radionuclide in nuclear medicine.

As indicated earlier, element 87, francium, was discovered in 1939 as the daughter of natural ^{227}Ac.[30]

The first isotope of element eka-iodine (Z = 85) was obtained in 1940 by D. R. Corson, K. R. MacKenzie, and E. Segrè by bombarding bismuth with accelerated α particles according to the reaction

$$^{209}_{83}\text{Bi} + {}^{4}_{2}\text{He} \rightarrow {}^{211}_{85}\text{At} + 2\ \text{n}$$

The name of the element, *astatine* (At), comes from the Greek αστατος, unstable, because it is the least stable of the artificial *cis*-uranium elements. The half-life of ^{211}At is 7.2 h, somewhat shorter than that of ^{210}At (8.3 h), which is the longest-lived isotope of the element.[45] The element of atomic number 85 was identified by its iodine-like properties and, curiously enough, by its concentration in the thyroid glands of guinea pigs. This was the first physiological confirmation of a new chemical element.

Three years after the synthesis of astatine, the element was identified in nature. Very short-lived isotopes are formed in small β^- branchings of ^{218}Po in the ^{238}U family and of ^{215}Po in the ^{235}U family. The latter includes also ^{219}At produced in the α branching of ^{223}Fr; it is the longest lived natural At isotope with a half-life of 0.9 min.

The last missing element, (Z = 61) belongs to the lanthanide series. It was discovered in 1947 by J. A. Marinski, L. E. Glendenin, and C. D. Coryell (1912–1971) among the fission products of uranium;[46] at that time four transuranium elements were already known. The name of this element, *promethium* (Pm) (initially, *prometheum*), is intended to recall the analogy between the liberation of nuclear energy leading to its discovery, and the theft of fire by Prometheus.

Promethium is also obtained by neutron irradiation of the neighboring lanthanide, neodymium, according to

$$^{146}_{60}\mathrm{Nd} + \mathrm{n} \rightarrow {}^{147}_{60}\mathrm{Nd}\ \beta^-\ 11.1\ \mathrm{d} \rightarrow {}^{147}_{61}\mathrm{Pm}\ \beta^-\ 2.62\ \mathrm{years}$$

The most stable isotope is ^{145}Pm, with a half-life of 17.7 years.

The periodic table was now complete. It had taken 78 years to fill the 29 blank spaces in Mendeleev's original classification. Eleven of the predicted elements have no stable isotopes and are accordingly termed *radioelements*. Among these, the trio Tc, Pm, and At is classified as *artificial*. However, the punctilious would rightly argue that the first two elements appear in the spontaneous fission of uranium, whereas the third, as said, occurs in two natural radioactive families.

On the other hand, radioactive isotopes of stable elements are produced naturally in nuclear reactions involving cosmic radiation. The year the periodic table was completed, W. F. Libby (1908–1980) discovered that ^{14}C, commonly designated *radiocarbon*, is continuously produced in the atmosphere by the reaction of neutrons of cosmic origin with nitrogen:

$$^{14}_{7}\mathrm{N} + \mathrm{n} \rightarrow {}^{14}_{6}\mathrm{C} + {}^{1}_{1}\mathrm{H}$$

This radionuclide has a half-life of 5736 years and is widely used for dating archeological samples.[47] A dozen cosmogenic radionuclides have since been discovered.

VII. THE CHEMISTRY OF NUCLEOGENIC ATOMS

Shortly after Rutherford had discovered emanation, one of his students, Hariett Brooks (1876–1933, later known as Mrs. Pitcher and the first woman, after Marie Curie, to devote her life to radioactivity), observed that atoms of *radium A* (^{218}Po) which had been deposited on a copper plate formed a volatile activity that was transferred to the walls of the ionization chamber. Brooks soon realized that this was not due to a radioactive emanation, but merely resulted from a mechanical effect. The ^{218}Po atoms decay to ^{214}Pb by emission of α particles. From simple momentum conservation,

$$M_\alpha V_\alpha = MV$$

where M and V pertain to the daughter atom. It follows that the ^{214}Pb atoms are formed with a kinetic energy of the order of 100 keV and are ejected from the decay site. Since they move in a direction opposite to that of the α particle, they are conveniently designated *recoil* atoms. This effect accompanies all spontaneous radioactive decays and nuclear reactions and has been widely exploited in the discovery of both natural and artificial radionuclides.

The first generation of radiochemists does not seem to have been concerned with the molecular aftermaths of transmutation. When Rutherford first announced the breakdown of a nitrogen nucleus by an α particle, he was asked what happens to the other nitrogen atom of the molecule when its partner is transmuted. It was obvious that the bond was broken by the recoil effect. However, Rutherford replied, "In terms of the size of the atomic nuclei, you have to imagine the other nitrogen atom miles and miles away."[48]

Thirty years after the recognition of radioactive recoil, R. A. Mortensen and P. A. Leighton published the first observation of a *chemical* effect of a nuclear transformation, "Molecular changes accompanying the radioactive transformations of radium D."[49] However, the birth of the chemistry of nascent atoms is generally credited to Leo Szilard (1898–1964) and T. A. Chalmers, who published at about the same time the note, "Chemical Separation of the Radioactive Element from its Bombarded Isotope in the Fermi Effect".[50]

This report on a simple experiment subsequently incited an immense field of research.[51,52] The authors observed that when ethyl iodide was irradiated with neutrons, about 50% of the radioactive ^{128}I formed in the nuclear reaction was no longer bound in ethyl iodide and could be easily separated by extraction with water. Thus, chemical bonds were severed in the course of neutron capture in the iodine nucleus, and, in particular, the radioactive atoms could be isolated from the stable isotope. For the first time, it became possible to separate isotopes in a very simple and efficacious way. The process was called a *Szilard-Chalmers reaction*. For a time, it was exploited in the preparation of radioisotopes having a high specific activity.

The phenomenon was explained by Fermi. The neutron capture reaction

$$^{127}_{53}I + n \rightarrow {}^{128}_{53}I^* \rightarrow {}^{128}_{53}I + \gamma$$

produces a radioactive ^{128}I nucleus in an excited state which rapidly decays to the ground state by emission of γ rays. As in the case of α decay noted by Brooks, the conservation of momentum must hold, i.e.,

$$\frac{E_\gamma}{c} = MV$$

The resulting recoil energy transferred to the nascent or *nucleogenic* radioactive atom is of the order of 100 eV, which far exceeds the energy of about 2 eV that is typical of a chemical bond.

The fast-moving recoil atoms are commonly termed *hot atoms*. The terminology is obvious since the thermal equivalent of 100 eV of kinetic energy is of the order of millions of degrees.

Nuclear recoil is the simplest way of transferring high translational energies to atoms and ions, together with simultaneous electronic excitation and intense ionization. Its discovery opened the new field of *hot atom chemistry*, which was initially the informal synonym for *chemical effects of nuclear transformations*. With time, expansion of the field led to the inclusion of non-nuclear techniques, such as crossed beams and lasers, for the production of atoms with controlled translational and excitation energies higher than those of chemical bonds. For the first time, it became possible to carry out chemical reactions that were otherwise not feasible. The progress that has since been made by the use of hot atoms in fundamental chemical physics is impressive. Hot atom reactions in a broad sense, including the consequences of radioactive decays, are operative in geochemistry, space chemistry, and in chemical and biological evolution. New pathways for labeling molecules with short-lived radionuclides such as those used in nuclear medicine have been elaborated by radiochemists on the basis of hot atom reactions.

VIII. THE INTRUSION OF RADIOACTIVITY IN CHEMISTRY

Initially, the scientific community was not particularly impressed by discoveries made by the pioneers of radioactivity. The new field was generally considered as a rather tedious sorting out of new radioactive elements by methods which relied more on ''kitchen recipes'' than on logical strategies. It was only after the discovery of the atomic nucleus by Rutherford in the 1920s that the interest of physicists in radioactivity became widespread. In fact, physics was undergoing a major revolution of its own, with Einstein's theory of relativity, the introduction of the quantum concept by Max Planck, the atomic model of Bohr, the discovery of the photoelectric effect, and the theoretical confirmation of the wave-particle dualism by Louis de Broglie. All these findings gave a decisive impulse to the birth of *nuclear physics*.

In all, the number of chemists involved in radioactivity during the period 1896 through 1915 did not exceed 40, with a dozen prominent leaders, working for the most part in Paris, Cambridge, Berlin, Vienna, and Montreal.

The investigation of radioactive elements for general applications in chemistry rather than for its own sake, probably began in 1913 when George Hevesy and Fritz Paneth (1887–1958) used radium D (^{210}Pb) to determine the solubilities of various lead compounds in water.[53] This was the first experiment in the field of tracer chemistry. It was followed by the use of natural radioactive tracers to demonstrate the exchange of atoms between solid and liquid phases. Subsequently, radioactive recoil was applied in diffusion experiments.

Hevesy also performed the first tracer experiment in life science, using radiolead to study uptake of this element by a broad bean, and radiobismuth to investigate distribution of bismuth in the rabbit.

Hevesy's invention in 1931 of isotopic dilution analysis for the determination of lead in 220 rock samples using ^{210}Pb was of particular interest to chemistry.[54] In 1936, the prodigious insight of this investigator led him to perform the first *neutron activation analysis* for the determination of rare earth elements.[55] This was just a few months after the discovery of artificial radioactivity.

The remarkable experiments of Hevesy were soon followed up and extended by many chemists. Within 5 years, several hundred papers were published; most of these were based on the use of new artificial radionuclides made available by neutron irradiation.

Chemists were becoming aware of the extreme sensitivity characteristic of radioactive measurements, which permitted studies of the behavior of matter at very low concentrations. Chemistry on the *tracer level* became increasingly important as radionuclides were applied to various special fields of chemistry.[56] The erratic behavior of matter at levels below 10^{-12} M in aqueous systems revealed the formation of *radiocolloids*.[57]

Radionuclides are presently being applied in most branches of chemistry as well as in other fields of science for *tracing* an atom, ion, or molecule in the course of a chemical reaction or transport process.

IX. ABOUT SEMANTICS

At the beginning of the century, the word *radioactivity* was used to describe all physical and chemical investigations dealing with phenomena which would now be considered as "nuclear". The title of Marie Curie's treatise published in 1910, as well as that of the second edition in 1934, was simply "Radioactivité".[58] It contains a full description of the physical and chemical properties of the natural radioelements.

Two books by Stefan Meyer and E. Schweidler and the book by Fajans that appeared between these two editions in 1925 had about the same content and shared the same title, "Radioaktivität".[59,60] The topic was covered similarly in the "Lehrbuch der Radioaktivität" published in 1931 by G. Hevesy and F. Paneth.[61]

Shortly after the discovery of radioactivity, Pierre Curie and Debierne noticed that aqueous solutions of radium salts released gaseous hydrogen and oxygen, and that hydrogen peroxide was progressively formed. Ozone appeared in air-sealed glass ampules containing radioactive substances. From these observations it was deduced that the radiations had a chemical action. By analogy with photochemistry, the word *radiochemistry* was introduced to designate the chemical effects due to all types of radiation emitted by radioelements. At first, this led to confusion, since other chemists had also used this term for the new branch dealing with the chemistry of radioelements.

Today the term *radiation chemistry* is universally accepted for designating the chemical effects of radiation, and this topic constitutes an extension of photochemistry. The main distinction between the two fields lies in the energy involved. Radiation chemistry is implicitly the realm of *ionizing* radiation, whereas photochemistry generally involves energies below the ionization potentials.

This point being clarified, endless debates have arisen with respect to the definitions of *radiochemistry* and *nuclear chemistry*. The former term became "official" with the publication in 1935 of Hahn's treatise, "Applied Radiochemistry", which was a collection of lectures presented by the author the year before at Cornell University in Ithaca.[62] Hahn had never been interested in radiation chemistry, and in fact the book does not contain a single reference to this topic. It was rather an advanced text on the chemistry of natural radioelements and on the behavior of radioactive matter at the tracer scale.

In 1937, F. Joliot received an appointment for a new chair entitled "Chimie Nucléaire" at the famous "Collège de France" in Paris. Ten years later, Moïse Haissinsky (1898–1976) published the first edition, in French, of his monumental *Nuclear Chemistry and its Applications*.[56] It was a truly ecumenical treatise because it covered the whole field of nuclear sciences, from particle physics to radiation biology.

From that time on, the choice of authors between the terms *radiochemistry* and *nuclear chemistry* seemed to be very subjective. A far from exhaustive list of textbooks is comprised of *Principles of Radiochemistry* (the English translation of Starik's *Osnovy Radiokhimii*, 1959), *Nuclear Chemistry* (M. Lefort, 1966), *Chimie Nucléaire Appliquée* (E. Roth, 1968), *Einführung in die Kernchemie* (K. H. Lieser, 1969), another *Principles of Radiochemistry* (H. A. C. McKay, 1971), *Nuclear Chemistry* (G. R. Choppin, J. Rydberg, 1980), *Essentials of Nuclear Chemistry* (H. J. Arnikar, 1982), and *Nuclear Chemistry* (A. Vertes, I. Kiss, 1987).[63-70] Friedlander and his co-authors preferred the solomonic title *Nuclear and Radiochemistry* (1981).[71]

Radioactivity Applied to Chemistry (A. C. Wahl, N. A. Bonner, 1951) and *Isotopes in Chemistry* (J. F. Duncan, G. B. Cook, 1968) are titles which elude the classification in either radiochemistry or nuclear chemistry.[72,73] Although the objectives of the authors differ widely, most of the books have parts in common that fit *radiochemistry* just as well as *nuclear chemistry*. The same lack of precision appears in the denominations of institutes and university chairs.

As mentioned before, Marie Curie and Marckwald had polemized for 5 years because both claimed to have discovered the true element with atomic number 84: it was polonium for Marie Curie and "radiotellurium" for Marckwald.[9] Although Marckwald probably remained convinced that Curie's original polonium was a mixture of several radioelements, he eventually capitulated in a most elegant fashion, quoting from Shakespeare's Romeo and Juliet:

> "What's in a name? that which we call a rose
> By any other name would smell as sweet."

These verses might also be addressed to polemists discussing the frontiers between radiochemistry and nuclear chemistry. The present author's standpoint on radiochemistry is described in Chapter 2.

REFERENCES

1. **Becquerel, H.,** Sur les radiations invisibles émises par les corps phosphorescents, *C. R. Acad. Sci. Paris,* 122, 501, 1896.
2. **Roentgen, W. K.,** Ueber eine neue Art von Strahlen, *Sitz. Ber. Med. Ges. Wuerzburg,* 132, 1895.
3. **Niepce de Saint Victor, A.,** Sur une nouvelle action de la lumière, *C. R. Acad. Sci. Paris,* 60, 505, 1867.
4. **Mostafavi, M., Marigner, J. L., Amblard, J., and Belloni, J.,** Nucleation dynamics of silver aggregates simulation of photographic development processes, *Int. J. Rad. Phys. Chem.,* 34, 605, 1989.
5. **Schmidt, G. C.,** Ueber die von Thorium und den Thorverbindungen ausgehende Strahlung, *Verh. Phys. Ges. Berlin,* 17, 4, 1898.
6. **Curie, M.,** Rayons émis par les composés de l'uranium et du thorium, *C. R. Acad. Sci. Paris,* 126, 1101, 1898.
7. **Curie, M.,** *Pierre Curie,* Payot, Paris, 1924.
8. **Curie, P. and Curie, S.,** Sur une nouvelle substance radio-active contenue dans la pechblende, *C. R. Acad. Sci. Paris,* 127, 175, 1898.
9. **Kirby, H. W.,** *The Early History of Radioactivity,* MLM-1960, TID-4500, Monsanto Research Corporation, Mound Laboratory, Miasmiburg, Ohio, 1972.
10. **Curie, P. and Debierne, A.,** Sur le polonium, *C. R. Acad. Sci. Paris,* 150, 386, 1910.
11. **Marckwald, W.,** Ueber das radioaktive Wismuth (Polonium), *Ber. Dtsch. Chem. Ges.,* 35, 2285, 1902.
12. **Curie, M.,** Sur la diminution de la radioactivité du polonium avec le temps, *C. R. Acad. Sci. Paris,* 142, 273, 1906.
13. **Curie, P., Curie, M., and Bémont, G.,** Sur une substance nouvelle radio-active contenue dans la pechblende, *C. R. Acad. Sci. Paris,* 127, 1215, 1898.
14. **Curie, M.,** Sur le poids atomique du radium, *C. R. Acad. Sci. Paris,* 135, 161, 1902.
15. **Wolke, R. L.,** Marie Curie's doctoral thesis: prelude to a Nobel prize, *J. Chem. Educ.,* 65, 56, 1988.
16. **Debierne, A.,** Sur une nouvelle matière radioactive, *C. R. Acad. Sci. Paris,* 129, 593, 1899.
17. **Kirby, H. W.,** Actinium, in *Gmelin Handbook of Inorganic Chemistry, Suppl. Vol. 1,* Springer-Verlag, Berlin, 1981.
18. **Rutherford, E. and Owens, R. B.,** Thorium and uranium radiation, *Trans. R. Soc. Canada Ser. 2, Sec. III,* 5, 9, 1899.
19. **Rutherford, E.,** A radio-active substance emitted from thorium compounds, *Philosophical Mag.,* 49, Ser. 5, 1, 1900.
20. **Rutherford, E. and Soddy, F.,** The radioactivity of thorium compounds. I. An investigation of the radioactive emanation, *J. Chem. Soc. Trans.,* 81, 321, 1902.
21. **Rutherford, E. and Soddy, F.,** The radioactivity of thorium compounds. II. The cause and nature of radioactivity, *J. Chem. Soc. Trans.,* 81, 837, 1902.
22. **Crookes, W.,** Radio-activity of uranium, *Proc. R. Soc.,* 66A, 409, 1899–1900.
23. **Fleck, A.,** The chemical nature of some radioactive disintegration products, *J. Chem. Soc.,* 103, 381, 1913.
24. **Kauffman, G. B.,** The atomic weight of lead of radioactive origin. A confirmation of the concept of isotopy and the group displacement law. Part I, *J. Chem. Educ.,* 59, 3, 1982.
25. **Kauffman, G. B.,** The atomic weight of lead of radioactive origin. A confirmation of the concept of isotopy and the group displacement law. Part II, *J. Chem. Educ.,* 59, 119, 1982.

26. **Soddy, F.,** Intra-atomic charge, *Nature (London),* 92, 399, 1913.
27. **Thomson, J. J.,** Rays of positive electricity, *Proc. R. Soc.,* 89A, 1, 1913.
28. **Fajans, K. and Göhring, O.,** Ueber das Uran X2 — das neue Element der Uranreihe, *Phys. Z.,* 14, 877, 1913.
29. **Brown, D. and Maddock, A. G.,** Protactinium, *Q. Rev.,* XVII, 289, 1963.
30. **Perey, M.,** Sur un élément 87, dérivé de l'actinium, *C. R. Acad. Sci. Paris,* 208, 97, 1939.
31. **Curie, I. and Joliot, F.,** Un nouveau type de radioactivité, *C. R. Acad. Sci. Paris,* 198, 254, 1934.
32. **Curie, I. and Joliot, F.,** Séparation chimique de nouveaux radioéléments émetteurs d'électrons positifs, *C. R. Acad. Sci. Paris,* 198, 559, 1934.
33. **Joliot, F. and Curie, I.,** Artificial production of a new kind of radioelements, *Nature (London),* 133, 201, 1934.
34. **Fermi, E., Amaldi, E., D'Agostino, O., Rasetti, F., and Segrè, E.,** Artificial radioactivity produced by neutron bombardment, *Proc. R. Soc. London,* 146, 483, 1934.
35. **Fermi, E.,** Possible production of elements of atomic number higher than 92, *Nature (London),* 133, 898, 1934.
36. **Herrmann, G.,** Fifty years ago: from the ''transuranics'' to ''nuclear fission'', *Angew. Chem. Int. Ed. Engl.,* 29, 481, 1990.
37. **Curie, I. and Savitch, P.,** Sur le radioélément de période 3,5 h formé dans l'uranium irradié par les neutrons, *C. R. Acad. Sci. Paris,* 206, 906, 1938.
38. **Hahn, O. and Strassmann, F.,** Ueber den Nachweis und das Verhalten der bei der Bestrahlung des Urans mittels Neutronen entstehenden Erdalkalimetalle, *Naturwissenchaften,* 27, 11, 1939.
39. **Hahn, O. and Strassmann, F.,** Nachweis der Entstehung aktiver Bariumisotope aus Uran und Thorium durch Neutronenbestrahlung, *Naturwissenchaften,* 27, 89, 1939.
40. **Meitner, L. and Frisch, O. R.,** Disintegration of uranium by neutrons: a new type of nuclear reaction, *Nature (London),* 143, 239, 1939.
41. **Frisch, O. R.,** Physical evidence for the division of heavy nuclei under neutron bombardment, *Nature (London),* 143, 276, 1939.
42. **McMillan, E. and Abelson, P. H.,** Radioactive element 93, *Phys. Rev.,* 57, 1185, 1940.
43. **Seaborg, G. T. and Loveland, W. D.,** *The Elements beyond Uranium,* John Wiley & Sons, New York, 1990.
44. **Perrier, C. and Segrè, E.,** Some chemical properties of element 43, *J. Chem. Phys.,* 5, 712, 1937.
45. **Corson, D. R., MacKenzie, K. R., and Segrè, E.,** Possible production of radioactive isotopes of element 85, *Phys. Rev.,* 57, 459, 1940.
46. **Marinsky, J. A., Glendenin, L. E., and Coryell, C. D.,** The chemical identification of radioisotopes of neodymium and of element 61, *J. Am. Chem. Soc.,* 69, 2781, 1947.
47. **Libby, W. F.,** *Radiocarbon Dating,* Chicago University Press, Chicago, 1951.
48. **Wolfenden, J. H.,** The dawn of hot atom chemistry, *J. Chem. Educ.,* 32, 276, 1955.
49. **Mortensen, R. A. and Leighton, P. A.,** Molecular changes accompanying the radioactive transformation of radium D, *J. Am. Chem. Soc.,* 56, 2392, 1934.
50. **Szilard, L. and Chalmers, T. A.,** Chemical separation of the radioactive element from its bombarded isotope in the Fermi effect, *Nature (London),* 134, 462, 1934.
51. **Matsuura, T., Ed.,** *Hot Atom Chemistry,* Kodansha, Tokyo, 1984.
52. **Adloff, J. P., Gaspar, P. P., Imamura, M., Maddock, A. G., Matsuura, T., Sano, H., and Yoshihara, K., Eds.,** *Handbook of Hot Atom Chemistry,* Kodansha, Tokyo, 1992.
53. **Hevesy, G. and Paneth, F.,** The solubility of lead sulphide and lead chromate, *Z. Anorg. Chem.,* 82, 322, 1913.
54. **Hevesy, G. and Hobbie, R.,** Lead content of rocks, *Nature (London),* 128, 1038, 1931.

55. **Hevesy, G. and Levi, H.,** The action of neutrons on the rare earth elements, *K. Dan. Vidensk. Selsk. Mat. Fys. Medd.,* 14, 5, 1936.
56. **Haissinsky, M.,** *La Chimie Nucléaire et ses Applications,* Masson, Paris, 1957; English translation, *Nuclear Chemistry and its Applications,* Addison-Wesley, Reading, 1964.
57. **Haissinsky, M.,** *Les Radiocolloides,* Hermann, Paris, 1934.
58. **Curie, P.,** *Traité de Radioactivité,* Gauthier-Villard, Paris, 1910; *Radioactivité,* Hermann, Paris, 1935.
59. **Meyer, St. and Schweidler, E., von,** *Radioaktivität,* Teubner, Leipzig, 1916.
60. **Fajans, K.,** *Radioaktivität,* Friedr. Vieweg & Sohn, Verlag, Braunschweig, 1921.
61. **Hevesy, G. and Paneth, F.,** *Lehrbuch der Radioaktivität,* 2nd ed., Leipzig, 1931.
62. **Hahn, O.,** *Applied Radiochemistry,* Cornell University Press, Ithaca, NY, 1935.
63. **Starik, I. J.,** Principles of Radiochemistry, AEC-tr-6314, Translation Series, U.S. Atomic Energy Commission, Washington, D.C., 1964.
64. **Lefort, M.,** *La Chimie Nucléaire,* Dunod, Paris, 1966; English translation, *Nuclear Chemistry,* Van Nostrand, London, 1968.
65. **Roth, E.,** *Chimie Nucléaire Appliquée,* Masson, Paris, 1968.
66. **Lieser, K. H.,** *Einführung in der Kernchemie,* 2nd ed., Verlag Chemie, Weinheim, 1980.
67. **McKay, H. A. C.,** *Principles of Radiochemistry,* Butterworths, London, 1971.
68. **Choppin, G. R. and Rydberg, J.,** *Nuclear Chemistry,* Pergamon Press, Oxford, 1980.
69. **Arnikar, H. J.,** *Essentials of Nuclear Chemistry,* 2nd ed., John Wiley & Sons, New York, 1987.
70. **Vertes, A. and Kiss, I.,** *Nuclear Chemistry,* Elsevier, Amsterdam, 1987.
71. **Friedlander, G., Kennedy, G. W., Macias, E. S., and Miller, J. M.,** *Nuclear and Radiochemistry,* 3rd ed., John Wiley & Sons, New York, 1981.
72. **Wahl, A. C. and Bonner, N. A., Eds.,** *Radioactivity Applied to Chemistry,* John Wiley & Sons, New York, 1951.
73. **Duncan, J. F. and Cook, G. B.,** *Isotopes in Chemistry,* Clarendon Press, Oxford, 1968.

FURTHER READING

Badash, L., Ed., *Rutherford and Boltwood, Letters on Radioactivity,* Yale University Press, New Haven, 1960.

Badash, L., Ed., *Radioactivity in America, Growth and Decay of a Science,* Johns Hopkins University Press, Baltimore, 1979.

Fermi, L., *Atoms in the Family: My Life with Enrico Fermi,* University of Chicago Press, Chicago, 1954.

Hahn, O., *Vom Radiothor zur Uranspaltung,* Friedr. Vieweg & Sohn, Verlag, Braunschweig, 1962.

Frisch, O. R., *What Little I Remember,* Cambridge University Press, Cambridge, 1979.

Ivimey, A., *Marie Curie, Pioneer of the Atomic Age,* Praeger, New York, 1980.

McKay, A., *The Making of the Atomic Age,* Oxford University Press, Oxford, 1984.

Romer, A., *The Discovery of Radioactivity and Transmutation, and Radiochemistry and the Discovery of Isotopes,* Dover Publication, New York, 1964.

Trenn, Th. J., Ed., *Radioactivity and Atomic Theory,* Taylor and Francis, London, 1975.

Weart, S., *Scientists in Power,* Harvard University Press, Cambridge, MA, 1979.

APPENDIX 1.1
NOBEL PRIZES RELATED TO RADIOCHEMISTRY

1903: H. A. Becquerel for his discovery of spontaneous radioactivity and P. and M. Curie for their joint researches on the radiation phenomena (physics prize).

1908: E. Rutherford, for his investigations into the disintegration of the elements and the chemistry of radioactive substances.

1911: M. Curie, for her services to the advancement of chemistry by the discovery of the elements radium and polonium, by the isolation of radium, and the study of the nature and compounds of this remarkable element.

1921 (awarded in 1922): F. Soddy, for his contributions to our knowledge of the chemistry of radioactive substances and his investigations into the origin and nature of isotopes.

1935: F. Joliot and I. Joliot-Curie in recognition of their synthesis of new radioactive elements.

1938: E. Fermi, for his demonstration of the existence of new radioactive elements produced by neutron irradiation and for his related discovery of nuclear reactions brought about by slow neutrons (physics prize).

1939: E. O. Lawrence, for the invention and development of the cyclotron and for results obtained with it, especially with regard to artificial radioactive elements (physics prize).

1943: (awarded in 1944): G. von Hevesy, for his work on the use of isotopes as tracers in the study of chemical processes.

1944: (awarded in 1945): O. Hahn, for his discovery of the fission of heavy nuclei.

1951: E. M. McMillan and G. T. Seaborg, for their discoveries in the chemistry of the transuranium elements.

1960: W. F. Libby, for his method to use carbon-14 for age determination in archaeology, geology, geophysics, and other branches of science.

CHAPTER 2

RADIOCHEMISTRY TODAY

I. FROM RADIOACTIVITY TO THE TOPICS OF NUCLEAR SCIENCE

Until the 1930s, the leit-motiv of everything dealing with the field opened by Becquerel in 1896 was ''radioactivity''. Scientists working with radionuclides did not necessarily consider themselves to be physicists or chemists. However, investigation of the physical and chemical properties of matter that emit ionizing radiations either spontaneously or after artificial transformation progressively led to the opening of new fundamental domains. Several frontiers began to emerge, for example, between the chemistry of the newly discovered elements and radionuclides and the physics of the nucleus and its radiations.

Presently, the field covered by nuclear science has grown to such an extent and complexity that it can hardly be assessed by a single discipline. It has split up into more or less loosely defined specialized topics. These have their own objectives and methodologies, but remain nevertheless interrelated by similar instrumentation and the common factor of radioactivity.

The realm of nuclear science can at present be subdivided into several main topics: radiation chemistry, radiochemistry, nuclear chemistry, nuclear physics, and particle physics. This order represents a progressive transition from a mostly chemical to an exclusively physical basis. Further, radioactivity has found plenty of applications in many other different branches of science and technology.

II. RADIATION CHEMISTRY

The objective of radiation chemistry is investigation of the chemical changes arising from the interaction of ionizing radiation with matter. It includes the identification of both transient and stable radiolytic species and the mechanism of their reactions and transformations.

The origin of this field dates back to the early period of radioactivity. The first radiation sources were natural radionuclides, such as ^{222}Rn and ^{210}Po, which were incorporated into the medium to be investigated or used as external irradiators, e.g., ^{226}Ra. Later, artificial radionuclides, and in particular ^{60}Co, were commonly used in radiation chemistry.

Important progress has been made with the substitution of radionuclide sources by accelerators which produce intense, monoenergetic, and controlled pulsed beams of electrons. With increasingly shorter pulses, early processes of short duration have become accessible, first on the microsecond time scale, and recently down to picoseconds and even femtoseconds. Laser spectroscopy is presently also widely used in fundamental radiation research.

Today, radiation chemistry is very close or complementary to photochemistry. A more recent application concerns the reactivity of free radicals and other species which can be produced selectively, and which are of major importance in radiobiology or in biology itself.

Although radiation chemists may now ignore the initial concepts of radioactivity, anyone working with radioactive matter requires a thorough knowledge of the chemical effects of radiations and must be aware of the physical and chemical changes induced in any system submitted to ionizing radiations.

III. RADIOCHEMISTRY

Semantically, the word *radiochemistry* indicates the association of *radioactivity* and *chemistry*. This does not mean that every chemical operation dealing with radioactive matter is relevant to radiochemistry.

Conversely, radiochemistry is concerned in all its aspects and objectives with radioactivity, including decay and the emission of radiations. On the basis of these phenomena, radiochemistry encompasses: (1) the investigation of the properties of chemical systems that cannot be studied on stable matter or on very long-lived radioactive matter, (2) the investigation of chemical systems consisting of stable or very long-lived matter which cannot be studied by conventional methods, and (3) the preparation of these systems.

As in chemistry, radiochemistry deals essentially with the *electronic* interactions between elements, but it also has the peculiarity of using *nuclear* phenomena for the investigation of chemical properties.

Several particular chemical systems investigated by radiochemists can be distinguished:

1. In a first example, radioactive matter may be available in nonponderable quantities, i.e., in amounts below, say 10^{-7} g. At this level, even the most sensitive classical methods do not provide adequate information on the chemistry or behavior of matter. Such systems may be gaseous, or consist of solid or liquid solutions or phases incorporating radionuclides. The smallest amount of matter which can be investigated is represented

by a few radioactive atoms, provided these can be detected. The concentration range which can be explored in radiochemistry has no equivalent in classical chemistry.

2. Other types of systems investigated in radiochemistry may be macroscopic. These include radioactive matter and stable matter, in both of which specific radionuclides may be incorporated in very small, nonponderable quantities. These radionuclides serve as *chemical* probes when they are used as tracers, or as *nuclear* probes, for example in perturbed angular correlation techniques and in Mössbauer emission spectroscopy.
3. A further type of system consists of weighable amounts of natural and artificial radioelements with high specific activity, either in solution or in the form of solid compounds. These require special precautions because of the chemical effects of the radiations and may present difficult safety problems. Experiments in this case are often performed at the microchemical level.

Main topics of radiochemistry include the following:

1. The physicochemical properties at very low concentration of all elements for which a suitable radioactive isotope with high specific activity is available
2. The physicochemical properties of radioelements for which the longest-lived isotope is available only at the tracer scale level
3. The chemical consequences, both atomic and molecular, of nuclear transformations which include change in the structure of solids and in the properties of the species formed in radioactive decays and in nuclear reactions; *hot atom chemistry* and the chemistry of *nucleogenic* atoms both deal with extremely small amounts of matter
4. The relation between nuclear and chemical properties, in particular the modification of certain nuclear properties in relation to the chemical environment
5. Properties of radioelements available in weighable quantities, but with very high specific activity

For these purposes, radiochemistry is highly dependent on two factors: the disposal of large amounts of radionuclides extracted from natural sources or produced with the aid of nuclear reactors and accelerators and the detection and measurement of radiation.

Radiochemistry has its own methodologies for the separation of one or more radionuclides from a mixture or from large amounts of nonradioactive matter and for the preparation of radionuclides in suitable form for measurement. With its particular methods and strategies, radiochemistry enables the study of thermodynamics, kinetics, and structural properties of matter at very high dilution or under conditions of high specific activity.

Radiochemistry has its constraints and these are set by radioactivity itself. Nevertheless, radioactivity remains the most simple, and often the only phenomenon which provides access to the smallest amount of matter.

IV. NUCLEAR CHEMISTRY

Nuclear chemistry unites the concepts of *nucleus* and *chemistry*, and hence basically involves the study of the nucleus.

Its aim concerns the structure of the nucleus, considered as a statistical assembly of nucleons in thermodynamic equilibrium. This includes nuclear excitation and decay modes, nucleon-nucleon interaction, production modes and cross sections, reaction mechanisms, nuclear masses, and stability.

In considering the spontaneous or stimulated transformations of nuclei as a reorganization of the nucleons, nuclear chemists, in a sense, operate in a manner similar to that of chemists who deal with atoms and molecules. However, in the case of subatomic systems, which are strictly defined by the number of neutrons and protons, the electronic core does not intervene.

It is an important distinction between radiochemistry and nuclear chemistry that the latter is not restricted to radioactive nuclei. From this point of view, the object of nuclear chemistry is considerably broader and includes all nuclei of the isotopes of all elements, particularly those with a large excess or deficiency of neutrons that are very far from the nuclear stability line, and the heaviest nuclei. These exotic nuclei are so short-lived that in the present state-of-the-art they are not amenable to radiochemical experiments. Their properties are investigated on-line and their production and detection require very heavy machinery.

Nuclear chemistry utilizes the full spectrum of modern technology, including physical and chemical techniques. The latter, however, were more frequently used in the past before the advent of solid state detectors and modern methods of data treatment and computation.

In summary, radiochemistry and nuclear chemistry are intimately related. Both require heavy tools for the production of artificial nuclides and use the same instrumentation. Such facilities are frequently located at the same site. However, radiochemists who prepare radionuclides for their own purposes are often little concerned by the mechanism of nuclear reactions or by the degree of refinement of the detectors used.

Radiation safety is, of course, equally important to both radiochemists and nuclear chemists, although the handling of high levels of radioactivity is accompanied by specific problems that are only encountered in radiochemistry.

The relationships between the various topics of nuclear science are shown in Figure 2.1.

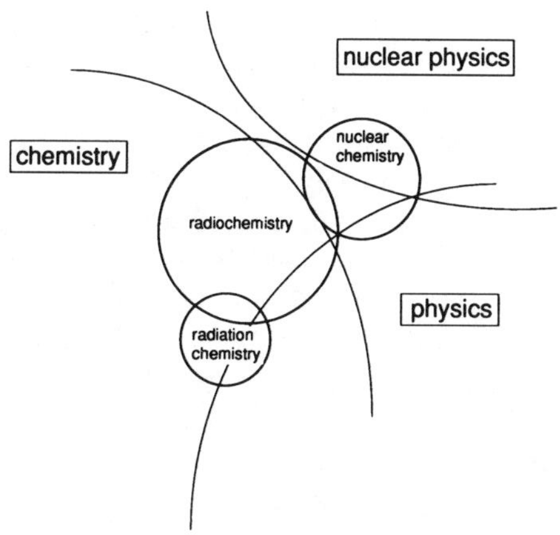

FIGURE 2.1. Diagram showing intersections of the domains of radiochemistry, radiation chemistry, nuclear chemistry, and nuclear physics, and their situation with respect to the fields of physical and chemical sciences. The four domains rely on the use of heavy equipment (reactors and accelerators) and nuclear instrumentation. The common objectives of the four domains can be assessed from their overlap.

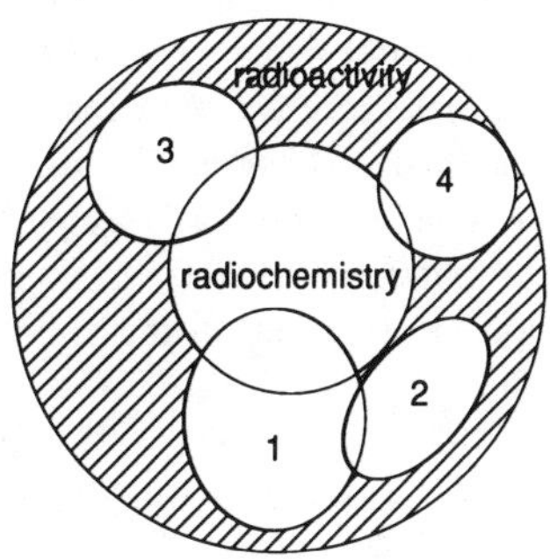

FIGURE 2.2. Illustration of the relative role of radiochemistry, within the field of radioactivity, with respect to (1) physics and chemistry of radioelements, (2) physics and chemistry of matter with very high levels of activity, (3) application of radionuclides as space and time probes, and (4) nuclear analytical methods.

V. FRONTIERS OF RADIOCHEMISTRY

A. *Radioactive Tracers*

Certain topics of chemistry which deal with radioactive matter and radionuclides are not strictly within the scope of radiochemistry (Figure 2.2). Such is the case for the multifarious uses of radioactive *tracers*. In this instance, an element is mixed with one of its radioactive isotopes such that the latter is present in very small proportion (typically, $1:10^{10}$). Alternatively, if no convenient radioisotope is available, an element can be labeled with a radioisotope of another chemically similar element.

Radionuclides as spatial and temporal tracers are used in widely differing fields of science and technology. Today, the most extensive applications are found in life sciences, and especially in nuclear medicine for diagnosis and, with intense sources, for therapy.

The utilization of radionuclides as tracers is relevant to tracer methodology rather than to radiochemistry. This is well supported by the fact that stable tracers or activable tracers are commonly used (an activable tracer is a stable nuclide which is introduced into the system and its ultimate location, at the end of the experiment, is revealed by activation analysis).

Radiochemists are frequently concerned with labeling of molecules if the synthesis involves reactions of hot nucleogenic atoms, such as the short-lived ^{11}C, ^{13}N, and ^{15}O used in life sciences. These chemists also intervene in the utilization of radionuclides like ^{99m}Tc, which has a rather complex behavior at the tracer level.

B. Radioanalytical Chemistry

Analytical chemistry is one of the branches which has profited most from radioactivty. The main advantage of activation analysis using neutrons and charged particles is sensitivity, which in many cases lies in the parts per billion ($1:10^9$) range. The development of solid state detectors for high resolution γ-ray spectroscopy has virtually obviated the necessity of chemical separation of radionuclides following irradiation. Modern activation analysis is fast, multielemental, instrumental, and highly computerized. However, nuclear analytical techniques generally do not provide information on the chemical state of the analyte.

Radioanalytical techniques, such as isotope dilution analysis, radioimmunoassay, and radiorelease, are mere tracer techniques and have little relevance to radiochemistry.

C. Nuclear Industry

Over the course of the last 50 years, the utilization of fission energy has required the production of impressive amounts of radioactive matter, which for the most part is confined to nuclear reactors and to the various units of the nuclear fuel cycle. The latter involve isotope separation plants, production of fuel enriched in ^{235}U or mixed with Pu, reprocessing of spent fuels, and waste disposal.

To give an idea of the amount of radioactive products involved, it may be stated that a typical 1-GWe nuclear power plant requires annually 25 tons of UO_2 enriched to 3.25% in ^{235}U. One ton of the combustible material, irradiated under routine conditions to produce an energy output of 33,000 MW/d, still has, after 3 years cooling, a β^- activity (due to fission and activation products) of 28 GBq and a 0.24 TBq α activity from actinides. The latter include about 9-kg plutonium, 0.5-kg neptunium, 0.3-kg americium, and a few grams of curium. In addition, about 2 kg of various fission radionuclides are formed.

The chemical industry of the nuclear fuel cycle uses classical methods, but these have to be adapted to the intense radiation and to the risk of criticality. Modern plants for the reprocessing of spent nuclear fuels, i.e., the separation of U and Pu from all other radioactive materials, are designed for the treatment of 800 tons/year.

The production of nuclear weapons has led to the accumulation of hundreds of tons of ^{235}U recovered from natural uranium and of ^{239}Pu, which has been separated from the irradiated fuel. Plutonium for military uses must be regularly purified in order to eliminate the accumulated ^{241}Am.

Nuclear industry is not limited to the fuel cycle. ^{90}Sr, ^{238}Pu, ^{244}Cm, and ^{242}Cm have been produced in kilogram amounts to serve as energy sources for thermoelectric devices used in satellites, (or *SNAP*, for Space Nuclear Auxiliary Power). ^{252}Cf, a powerful neutron source, is produced in amounts of several hundred milligrams per year.

Irradiators containing high activities of γ emitters, such as ^{60}Co, ^{137}Cs, and ^{192}Ir, have industrial applications, e.g., for sterilization or polymer grafting.

The preparation of all these radionuclides requires chemical operations which are not relevant to radiochemistry, but rather to the *chemistry of radioactive matter, possibly fissionable*. Many large-scale productions of radionuclides were set up on the basis of separation procedures established by radiochemists during the historical period of radioactivity.

However, radiochemistry is again concerned with some aspects of the terminal part of the nuclear fuel cycle, namely, the management of radioactive wastes and the dispersion of artificial radionuclides in the environment.

VI. RADIOCHEMISTRY AND PERIODIC TABLE OF THE ELEMENTS

Most of the elements in the periodic chart are of importance to radiochemistry. These include:

1. The numerous elements for which a short-lived radioisotope is available.
2. The radioactive fission products from binary fission, which are isotopes of 35 elements comprised between Zn and Tb, and tritium, the radioactive hydrogen isotope, resulting from ternary fission.
3. All *radioelements*, i.e., the elements which have no stable isotopes. Several of these have only short-lived isotopes and thus exist only in nonponderable amounts; examples are Po, At, Rn, Fr, and all radioelements from $Z = 99$ to $Z = 109$, the present limit. Some radioelements in macroscopic amounts have very high specific activities which perturb their chemical behavior and must be investigated at the tracer level. These include Pm, Cm, Bk, Cf, and Es.

Some radioelements may possess long-lived, weighable isotopes such as ^{99}Tc and ^{231}Pa, but which are very difficult to isolate. These are more conveniently investigated with short-lived isotopes that are readily available, for instance, ^{99m}Tc and ^{233}Pa. On the same lines, short-lived γ- or X-ray emitters may be substituted for pure α emitters (^{237}Pu for ^{239}Pu), for pure β^- emitters (^{99m}Tc for ^{99}Tc), or for highly radiotoxic nuclides (^{237}Pu for ^{239}Pu).

The majority of elements investigated in radiochemistry are those of the terminal part of the periodic chart, in particular the actinides. The latter are of special interest for several reasons. First, most of these elements belong to the 5f series and can be compared to the 4f series of the lanthanides. They possess numerous oxidation states, especially in the case of the lighter elements from Pa to Pu, and hence are characterized by an extensive range of redox chemistry. These light actinides are also formed in large amounts in the nuclear fuel cycle. In the oxidation states 3, 4, 5, and 6, the actinides form numerous solid compounds and complexes. For the heavier of these elements it is expected that there should be an influence of relativity on the chemical properties, such as oxidation state.

VII. RADIOCHEMISTRY AND CHEMISTRY

Although chemistry has played a determinant role in nuclear science, radiochemistry, in turn, has contributed to the development of several branches of chemistry. The greatest impact was undoubtedly the completion and an extension of over 10% of Mendeleev's table of the elements. The discovery of transuranium elements has brought a renewal to the whole field of chemistry, particularly in the inorganic sector. The element plutonium, although unknown 50 years ago, has since been, for various reasons, the object of more research than has been devoted to many other elements in the periodic table.

Investigations of the heaviest radioactive elements have provided a wealth of data, especially on the oxidation states, structure and complexation of aqua ions, and on the complexation and stability of solid compounds. This information has completed our knowledge of the trends of chemical properties of the elements within the periods and groups of the periodic table. Previous concepts have been reconsidered and frequently revised.

Radiochemistry has opened the new field of chemistry at very low concentration (below 10^{-10} *M*) that is becoming increasingly important not only for knowledge of the role and behavior of trace elements, but also in space chemistry. Unsuspected phenomena have been revealed, such as the existence of radiocolloids, the peculiar first steps of hydrolysis, adsorption, and exchange processes on solid surfaces, diffusion in solids, and the presence of radioactive aerosols of radon and its progeny. Competition between thermodynamics and kinetics for the advancement of certain reactions has been observed at the tracer level. At the ultimate limits of concentration, radiochemistry has revealed the problem of chemistry at the level of only a few atoms.

The properties of solutions at the other extreme of very high, i.e., multimolar concentrations, can in some cases be explored more conveniently with the positron probe than by classical means. Hence, radiochemistry can furnish valuable tests of solution models both at infinite dilution and at very high concentration.

The positron may be considered as a "radioactive" tracer of the electron, and in this respect has an immense potential application in all fields of chemistry. Positron chemistry proceeds at an extremely high rate and presently provides access to early events in the interaction of radiation with matter.

The chemistry of hot nucleogenic atoms has brought considerable insight into the relaxation behavior of highly excited atoms, ions, and molecular fragments with kinetic energies well above those of thermal levels. This field includes the reactions of species which are not in thermal equilibrium with their environment, and is of particular importance in the study of biomolecules in life sciences.

Finally, the dynamic aspects of chemical equilibria can be conveniently studied with the use of radionuclides for visualizing isotopic exchange reactions and chemical isotope effects. These various contributions of radiochemistry to the field of physical chemistry are concomitant with new incentives in furthering knowledge of the chemistry of the elements.

CHAPTER 3

ESSENTIALS OF RADIOACTIVITY

I. INTRODUCTION

In introducing the topic, it is expedient to start with a chemist's point of view in dealing with the phenomenon of radioactivity and various properties of radioactive matter. The concepts useful for the radiochemist are not necessarily those required by a nuclear physicist or a reactor engineer. Hence, the definitions and facts related to radioactivity will be treated in a comprehensive but partially unconventional approach within the objective of this book. Some aspects that are overlooked or that receive minor attention in classical treatises will require more detailed consideration when radioactivity is discussed from the chemist's standpoint.

A *nuclide* is any atomic species defined by specific values of the atomic number Z and the mass number A and is represented by the symbol

$$^{A}_{Z}X \quad \text{or} \quad ^{A}X$$

which represents one of the isotopes of an element X. The mass number A is the total number of nucleons, i.e., of Z protons and N neutrons:

$$A = Z + N$$

Nuclear masses are represented by a small letter m and atomic (or nuclidic) masses by a capital M. They are expressed in atomic mass units (amu or u)

$$1\ u = 1.660\ 565 \times 10^{-27}\ kg$$

or

$$1\ u = 931.5016\ MeV/c^2$$

Here the abusive but common practice 1 u = 931.5016 MeV will be followed. The rest masses of the three atomic constituents are given below:

Proton	1.007 276 u	$1.672\ 648 \times 10^{-27}$ kg	935.279 MeV
Neutron	1.008 665 u	$1.674\ 954 \times 10^{-27}$ kg	939.673 MeV
Electron	$5.485\ 802 \times 10^{-4}$ u	$0.910\ 953 \times 10^{-30}$ kg	0.511 003 MeV

Most illustrative examples will refer to isotopes of elements that have dominated the history of radiochemistry and to those that are of particular present-day importance. This is the case for isotopes of the elements U and Pu, the cornerstones of nuclear fission energy. The examples of radioactive matter will concern mainly aqueous solutions of these elements because of their dominance in chemistry and radiochemistry.

II. RADIOACTIVITY AND RADIOACTIVE DECAY MODES

Of the many possible ways of introducing radioactivity, a phenomenological description will be used, based on the observation of the emission of particles or radiations by matter in the absence of external stimuli. Chemists deal with atoms rather than with nuclei deprived of the electronic cloud. Accordingly, radioactivity is defined as the *spontaneous* emission of corpuscular or electromagnetic radiations by a species consisting of a nucleus surrounded by electrons with a total charge z. To begin with, the species will be considered as monoatomic, i.e., either an atom or an ion. More complex radioactive species will be treated in later sections. The species is said to be radioactive, meaning the nucleus is radioactive. This definition fails only on very rare occasions.

The spontaneous transformation of the species is called a *disintegration*. The term *radiation* embraces all types of emission that carry momentum and energy. Particles, also designated as *corpuscular* radiations, transport matter, as opposed to *electromagnetic* radiation. Today, all radiations with an energy greater than 50 eV are called *ionizing radiations*, regardless of their nature. They can generally be easily detected with very high degrees of sensitivity. The only exception concerns neutrinos that accompany the emission of electrons; neutrinos are uncharged particles with a rest mass equivalent to a few electronvolts (eV). Neutrinos can traverse the earth without suffering any interaction and travel billions of kilometers in space.

The characteristics of the radiations are fingerprints of the combination of atomic and mass numbers of the emitters.

Chemistry depends essentially on Z and, with the exception of hydrogen and other light elements, shows little variation with A for a given Z. On the other hand, radioactivity is very sensitive to both Z and A.

A. RADIOACTIVE DECAY MODES

There are six very common radioactive decay processes; a few other modes are much less frequent. It is customary to distinguish between the emissions proceeding from the nucleus alone and those that result from the interaction of the nucleus with its orbital electrons.

1. Emissions from the Nucleus

The primary emissions of a radioactive nucleus include particles and electromagnetic radiations which are designated as γ photons or γ rays. Particles emitted by the nucleus are

- The electron (e^-) associated with the electron antineutrino ($\overline{\nu}$) or β^- radioactivity, as in ^{210}Bi
- The positron (e^+) associated with the electron neutrino (ν), or β^+ radioactivity, as in ^{30}P
- The α particle, consisting of a helium nucleus $^4He^{2+}$, or α radioactivity, as in ^{210}Po
- Two heavy particles (fission fragments), together with γ photons, neutrons, and neutrinos in spontaneous fission (SF) radioactivity, as in ^{238}U

The four cited nuclides are those for which the various decay modes were first observed.

Disintegrations by emission of electrons, positrons, and α particles are observed in hundreds of radionuclides. In most cases, emission of the primary particle is followed very rapidly by γ photons, which are emitted in the transition to the ground state of the excited entities formed in the primary step. For this reason they are called de-excitation photons.

The primary emission of a γ photon by a radioactive nucleus represents decay by *isomeric transition* (IT). As before, it corresponds to the de-excitation of the nucleus to the ground state. The distinction may appear arbitrary. It is customary to qualify nuclear de-excitation as an isomeric transition if the lifetime of the excited species is long enough to permit a chemical experiment to be performed, say, a few seconds. In contrast, "true" de-excitation may be as fast as 10^{-12} s. The excited and ground states which have the same A and Z, but differ in their internal energy, are called *isomers*. The excited state is marked with the superscript m for metastable, e.g., ^{234m}Pa.

Less-common nuclear decay modes include the delayed emission of particles immediately following β^- or β^+ decay; for example, ^{87}Br decays by β^- radioactivity, followed by emission of a neutron; ^{21}Mg decays by β^+ radioactivity, followed by emission of a proton.

Still rarer events are represented by proton radioactivity (e.g., ^{53m}Co) and by the emission of ^{14}C (e.g., ^{223}Ra), which is a case of radioactivity by heavy charged particle emission.

Decay by emission of two β^- particles in cascade as would correspond to *double β^- decay* is predicted by theory but is still being searched. Direct neutron radioactivity has never been observed.

Electrons and positrons emitted in radioactive decay behave like relativistic particles. The rest mass m_0 increases to 1.1 m_0 for electrons with 50-keV kinetic energy and to 3 m_0 at 1 MeV.

2. Emission Involving the Nucleus and the Orbital Electrons

a. *Electron Capture*

In the *electron capture* (EC) decay process, the nucleus seizes a bound electron, usually from the K or L shell as in ^{48}V. In this decay mode, the primary emitted particle is a neutrino.

EC decay is a rather peculiar mode, because when the nucleus is deprived of its surrounding electrons, the process is no longer possible. Such would be the case for ^7Be in the Be^{4+} configuration which may occur in cosmic radiation or in the sun. The nuclide ceases to be radioactive under these conditions.

A reverse EC, the orbital electron *creation*, in which a very slow β^- particle is trapped in the electron cloud of the radioactive atom has been considered, but never really proven.

b. *Internal Conversion*

The transfer of nuclear energy to a core electron represents the process of *internal conversion* (IC). The electron is ejected and becomes a *conversion* electron. This process competes in most cases with γ de-excitation; the probability of conversion may take any value between 0 and *nearly* 100%. Full conversion is known for a single nuclide, ^{72}Ge.

Isomeric transitions may be highly, but never totally, converted. Thus, IC cannot be considered as a mode of radioactive decay, since no example of totally converted transition is known.

3. Competing Modes

In a group of identical radionuclides, several decay modes may be observed. The competition is expressed by the *branching ratios* which correspond to the relative probability of occurrence of a decay mode. Three modes are known for ^{40}K [β^- 89%, β^+ 11%, EC 0.2%] and for ^{152}Eu [EC 72%, β^- 28%, β^+ 0.021%]. Competition is the rule rather than an exception. Examples of competing modes are β^+ with EC, β^+ with EC and α, α with EC, α with β^-, IT with EC and β^-, α with SF. Low branching ratios are masked by the more intense decay modes. When examined in full detail, the decay of a radionuclide is a very complex process.

As an example, the 15 isotopes of Pu with mass number from 232 to 246 have very different properties, depending on A:

- ^{232}Pu to ^{235}Pu and ^{237}Pu decay by EC with a very weak α branching.
- ^{236}Pu, ^{238}Pu, ^{239}Pu, ^{240}Pu, ^{242}Pu, and ^{244}Pu are α emitters.
- ^{241}Pu is a β^- emitter with a very weak α branching.
- ^{243}Pu, ^{245}Pu, and ^{246}Pu are pure β^- emitters.
- ^{238}Pu, ^{239}Pu, and ^{241}Pu also emit γ rays with intensities below 4%.
- ^{238}Pu to ^{240}Pu, as well as ^{242}Pu and ^{244}Pu, also decay by spontaneous fission. For ^{238}Pu, ^{242}Pu, and ^{244}Pu, the fission rate is about 10^3 g^{-1} s^{-1}.

B. Secondary Emissions

The primary emission of a radioactive nucleus is followed by secondary processes in which the orbital electrons participate.

1. Shake-Off

With the exception of γ emission, all decay modes are characterized by a sudden change ΔZ of the proton number in the parent nucleus. The electric field exerted by the nucleus on the orbital electrons is abruptly modified and the electron cloud expands or contracts, depending on the sign of ΔZ. Usually, the electronic cloud adapts adiabatically to the field of the daughter nucleus, but sometimes excited electronic states or ionization will result. The term *shake-off* denotes the loss of electrons following the transmutation. In the case of β decay, the excitation energy ΔE of the orbital electrons can be calculated from the expression

$$\Delta E = 22.8\ Z^{0.4}\ \text{eV}$$

where Z is the proton number of the parent nucleus. For the β^- decay of ^{41}Ar or ^{39}Ar to the respective potassium daughters, the excitation energy amounts to 72.5 eV; in comparison, the first four ionization potentials of the K atom lie between 4.34 eV and 60.91 eV and electrons can be lost by shake-off ionization, but the atomic charge increase does not exceed a few units.

2. Inner-Shell Ionization

Electron capture and internal conversion leave the daughter atom with a hole in an inner electronic shell, most often in the K or L shell. The consequences are well known from atomic physics and appear as competitive emission of fluorescence radiation and of Auger electrons.

In the first instance, characteristic X-rays are produced. The most frequently observed are those belonging to the K series: $K_{\alpha 1}$ ($K \rightarrow L_{III}$ transition), $K_{\alpha 2}$

($K \rightarrow L_{II}$), $K_{\beta'1}$ ($K \rightarrow M_{IV}, M_{III}, M_{II}$), and $K_{\beta'2}$ ($K \rightarrow N_{III}, N_{II}$). The energy of X-rays emitted in these transitions lie below 150 keV. X-rays of the L series are recorded in the case of heavier atoms ($Z > 50$) and especially in the actinides.

The relative probability of X-ray emission is expressed by the fluorescence yield ω. For the K shell,

$$\omega(K) = \frac{\text{Number of K X-rays}}{\text{Number of initial ionizations in the K shell}}$$

The value of ω(K) increases with Z; it is very small for the light elements (0.035 for carbon) and approaches unity for $Z > 60$. The ratio of K X-ray intensities to the intensity of the $K_{\alpha 1}$ ray is slightly Z-dependent.

From the recorded values of the X-ray energies, of the relative intensities, and of the fluorescence yields, the chemical nature of a radioactive nuclide can be established unambiguously *without* the preliminary excitation which is required in X-ray spectroscopy. This is particularly important in EC decay, since the primary radiation released in this process is an undetectable neutrino. The emitted X-ray is characteristic of the daughter nuclide. ^{55}Fe, which decays only by EC, is identified by the 5.9 keV Mn K_α ray. Pu isotopes decaying by EC emit the X-rays of Np.

Internal conversion is a very common source of X-rays, lacking only in the rare cases of pure α and β decays, i.e., direct transitions to the ground state of the daughter. X-rays emitted from internal conversion following emission of particles are also characteristic of the daughter nuclide, but in an isomeric transition they correspond to the element involved in the transition. Thus, ^{80m}Br emits the X-rays of bromine.

An alternative event following inner-shell ionization is the *Auger* process. The electronic excitation energy is no longer released as a fluorescence radiation, but serves to eject a second electron called *Auger electron*. The inner shell now has two vacancies. These holes will in turn be filled by external electrons. If the Auger process is repeated, the charge of the ion becomes +4. Once triggered, the electron *vacancy cascade* proceeds very rapidly (10^{-14} s) to the valence shell with an increased efficiency after each step. The resulting loss of electrons can become very high. The limit of the charge spectrum of Br^{z+} ions from the IT of ^{80m}Br in CH_3Br is +11; the charge spectrum of xenon ions following IT in ^{131m}Xe atoms extends up to +22.

In a few cases, EC is followed by IC. The ensuing double vacancy cascade amplifies the mechanism of electron loss. The limit of the charge spectrum of the Te^{z+} ions from the EC and IT decays of ^{125}I in CH_3I is Te^{20+}.

The occurrence of Auger emission is complementary to that of fluorescence, and thus more frequent in the light elements. Transitions between subshells (Coster-Kronig transitions), which are prohibited for the radiative relaxation of orbital electrons, become possible but are of minor importance.

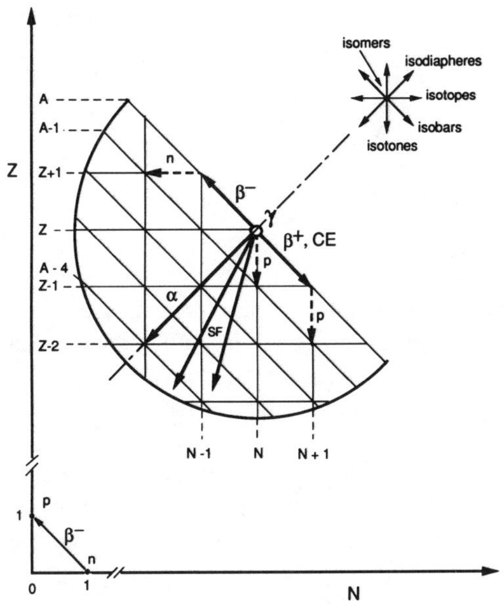

FIGURE 3.1. Representation in the N, Z plane of the daughter nuclides resulting from the decay modes of a parent with Z, protons; N, neutrons; and mass number, A. Dashed lines indicate rare decay modes.

C. The Immediate Products of Radioactive Decay

The total number of nucleons is invariant in any mode of decay, as well as the total charge, except in the case of EC.

Isobaric emissions are those in which the daughter nuclide conserves the mass number of the parent. They include

- β^- decay, whereby Z increases by one unit
- β^+ and electron capture decay, whereby Z decreases by one unit
- Isomeric transitions, for which Z remains constant

In all other types of decay, in which the parent emits a proton, an α particle, a heavy ion or splits into fragments as in spontaneous fission, Z and A of the resulting nuclide are changed. The relation between atomic number and mass number of parent and daughter nuclides, together with the associated definitions, is shown in Figure 3.1.

The formal balance of nucleon conservation does not provide information about the actual *chemical* state of the species formed immediately after decay. If it is assumed that the transformation is adiabatic, the orbital electrons will not be excited and the balance of the charge z of the parent species is easily described, for example

$$^{212}_{82}Pb^{2+} \rightarrow ^{212}_{83}Bi^{3+} + \beta^- + \bar{\nu}$$

and

$$^{239}_{94}Pu^{6+} \rightarrow ^{235}_{92}U^{4+} + ^{4}_{2}He^{2+}$$

On the other hand, if the products are formed with a high initial velocity they may ''leave behind'' a few electrons among those which are less firmly bound. This effect is very pronounced in fission fragments which have velocities of the order of 10^9 cm s^{-1} and as many as 10 to 20 electrons can be lost.

Secondary processes, such as shake-off, can also change the expected formal oxidation states. As a result, the exact nature of *decay products* is often poorly defined and from a chemical point of view many transient situations can arise between the end of the disintegration and the complete relaxation of the electron cloud.

The most evident chemical effect in a decay with a change of Z is the appearance of a new element. In this manner, a free neutron decays by β^- emission to the simplest element, hydrogen.

III. THE ORIGIN OF PRIMARY EMISSIONS

A. *Nuclear Properties*

A nucleus represents a very complex point in space at which, three (electromagnetic, weak, and strong) out of the four fundamental interactions intervene. The observed properties of nuclei cannot be explained satisfactorily with a single nuclear model. Some of the common representations are familiar to the chemist; these include analogy with the atomic structure, rotational and vibrational motions similar to those of molecules, or parameters taken from macroscopic matter. Most nuclear properties are interpreted in terms of semitheoretical (or semiempirical) expressions which comprise many parameters denoted by values determined from experimental data.

The quantum mechanical description of the nucleus is based on a Hamiltonian that covers all of the nucleons. The derived models are either of the single-nucleon type in which all nucleons behave independently, or they take into account the collective motions of the nucleons.

In both models, the nucleons, like the electrons in an atom, can occupy either a ground level or various excited levels depending on their arrangement inside the nucleus. The result is the occurrence of nuclei in the fundamental or ground state and in many possible stationary excited states. The term *state* refers to experimentally identified *levels*.

The energy E of the various states is expressed with respect to that of the ground state. The most usual values of E are in the kiloelectronvolt (keV) or megaelectronvolt (MeV) range rather than on the electronvolt (eV) scale, reflecting the relative sizes of the nucleus and the atom and the relative intensities of atomic and nuclear forces. The values of the excitation energies represent the differences between the nuclear mass in the excited states and that in the ground state.

Among the nuclear properties, only those necessary for a description of the phenomenon of radioactivity will be considered.

Nuclei with even values of Z and N are well represented by a spherical assembly of nucleons, whereas the heavy odd-A nuclei are *deformed* with respect to this simple shape.

Each level is characterized by its energy E and its mean lifetime τ. These are related by the uncertainty principle

$$\Delta E \Delta t = h/2\pi$$

which takes the form

$$\Gamma \tau = h/2\pi = 6.56 \times 10^{-16} \text{ eV s}$$

Here τ is defined for a single nucleus and represents the mean time required for a transition from one state to another. The statistical significance of τ for an assembly of nuclei will be indicated in a later section.

The uncertainty in the energy E of any level is

$$E \pm \Gamma/2$$

The energy is sharply defined only for the ground state, which has an infinite lifetime. The width Γ of the excited states increases with the energy since the lifetime of the levels generally decreases with higher excitation energies.

When dealing with radioactivity, it is important to consider the energy E of states and the separation ΔE between successive levels, in particular for the first few. The inter-state separation decreases with increasing values of the energy. Above 10 to 15 MeV, the state density is so high and the separation so narrow that the width Γ of a level exceeds the separation ΔE. The resulting energy *continuum* is not important in the description of radioactivity.

It is found experimentally that the energies E and ΔE are strongly dependent on the values of A, N, and Z. Specific values of the proton number and neutron number, referred to as *magic numbers*, affect particularly the state distribution. Magic proton and neutron numbers are 2, 8, 20, 50, and 82. Two additional magic neutron numbers are 126 and 152. These figures are interpreted theoretically by the *nuclear shell model*, which assigns the nucleons to specific shells in the same manner as for relativistic electrons in n, *l*, j shells of the atom.

For an even value of A and identical magic numbers Z and N, the first excited state always lies very high above the ground state, e.g., about 6 MeV in the case of ^{16}O. In semi-magic odd-A nuclei, with N or Z one unit removed from a magic number, the energy of the first state is much lower and is less than 1 MeV in ^{17}O. In both cases, the state separation is in the MeV range for low Z values, and of the order of 50 keV for the heavy nuclei. These nuclei are spherical, and the states are identified as *nucleonic* levels, in which the nucleons behave independently.

The description is quite different for deformed nuclei with $160 < A < 180$ and $A > 220$. For even-A, the first state still has a high energy, but now the nuclei possess numerous collective rotational levels within vibrational levels. The coupling of nucleonic and collective levels in odd-A nuclei leads to a still more complicated picture with an extremely large number of close lying states.

Each nuclear level is characterized by the *nuclear spin I* related to the total angular momentum. Ground states of even-A nuclei have integer I values ranging from 0 to 6; if Z and N are even, the spin is 0. The spin values are half-integral for odd-A nuclei, and lie between 1/2 and 9/2. Excited states may have much higher angular momenta.

The various parameters describing a state are interrelated. As an example, for transitions involving the same spin change ΔI, and for a given Z value, the lifetime decreases with increasing energy; for a fixed energy value, τ also decreases with increasing Z.

The *parity* Π is associated with each state. The symbols Π^+ or Π^- indicate that the nuclear wave function is, respectively, symmetric or antisymmetric when the signs of the space coordinates of each nucleon are changed. The pattern of energy states for a few nuclei are shown in Figure 3.2.

B. Nuclear Decay

The emission of particles and photons in the radioactive process results from a reorganization of the nucleons inside the nucleus. The nucleus of the radioactive species is changed from an initial state (from the ground state or one of the excited states) *i* to a final state *f* of the same nucleus (γ emission) or of a different nucleus. In spontaneous fission decay, the final state comprises two different nuclei.

The *i* to *f* transitions are representative of different types of interactions. The *strong* or nuclear interaction is at the origin of decay by emission of α particles and nucleons, as well as spontaneous fission. The *weak* interaction is responsible for β^+, β^-, and EC decays, and the *electromagnetic* interaction for the emission of photons in nuclear de-excitation and in isomeric transitions, as well as in the process of IC.

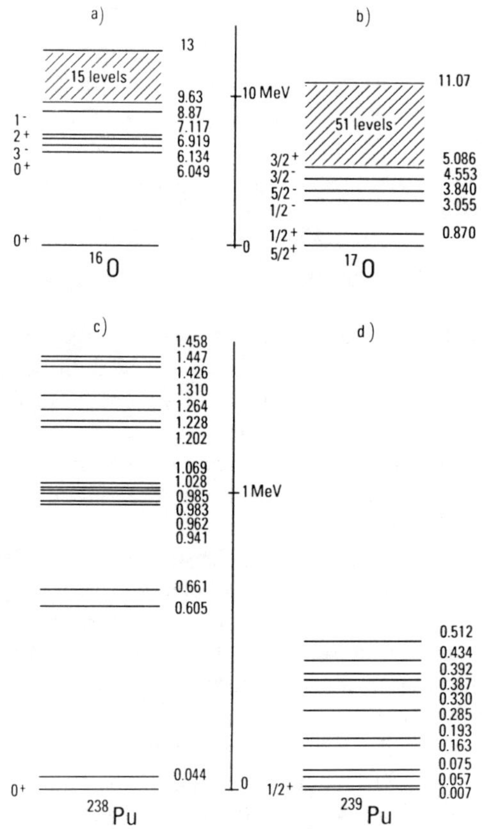

FIGURE 3.2. Partial schemes with spin, parity, and energy (in MeV) of nuclear states. (a) and (b): Light spherical nuclei, stable in their ground state. The first excited states are well separated from the ground state; at higher energies the inter-state separation decreases. These states are populated by various nuclear reactions. (c) and (d): Heavy radioactive nuclei, α-emitters in the ground state. The first excited states are close to the ground state and the inter-state separation decreases rapidly with increasing energy. For clarity, spin, and parity are given only for ground states. The excited states of ^{238}Pu with energies up to 1.426 MeV are rotational or vibrational and the upper three are nucleonic states. ^{239}Pu has nucleonic states at 0.285, 0.392, and 0.512 MeV. The states of ^{238}Pu are populated (with intensities $\geq 0.1\%$) by β^- decay of ^{238}Np, α decay of ^{242}Cm, and EC decay of ^{238}Am, and those of ^{239}Pu by β^- decay of ^{239}Np, α decay of ^{243}Cm and EC decay of ^{239}Am.

1. Nuclear and Atomic States

Calculations of transition probabilities require knowledge of the wave functions of the initial and final systems. In principle, the description of the states should include the nucleus and its electronic environment. The necessary information is most readily available for the initial state, since the chemical properties and the oxidation state of the radioactive species are known in many cases. On the other hand, as explained previously, the electronic configuration of the daughter species is initially poorly defined and the best one can do in choosing appropriate wave functions is to assume that the relaxation of the orbital electrons is fast and complete.

However, it is a common practice to consider that the *i* and *f* states are purely nuclear. This assumption is most readily verified in the transitions not involving the orbital electrons. Indeed, the hyperfine interactions, i.e., the coupling of nuclear parameters with the electromagnetic field produced by the electrons at the nuclear site, are extremely weak.

A sizable hyperfine effect would be expected for the electric monopole interaction. The s-type electrons have a finite probability of penetrating into the nuclear volume. The ensuing change in the energy of the nuclear states is of the order of 10^{-8} eV, which is the magnitude of the energy width Γ of the states. Although this effect is commonly observed in Mössbauer spectroscopy and is widely used in studies of chemical structure, it is overlooked in the general practice of radiochemistry. Similarily, the lifetime of the levels is practically unaffected by the atomic configuration (see Chapter 8).

Under these conditions, the presence of electrons may be ignored and only the nuclear wave functions need be taken into consideration. In consequence, radioactive decay modes and the properties of the emitted radiations do not provide information on chemical properties. This is a serious disadvantage as far as speciation of radioelements is concerned. In EC and IC, the role of orbital electrons, particularly of the s-electrons, is preponderant. However, even for this nucleus-electron interaction, the effect on the lifetime of excited states or radioactive ground states is extremely small. For example, the relative changes in lifetime of EC-decaying ^{7}Be (half-life 53 d) in various beryllium compounds do not exceed a few per million.

For these processes the radioactive properties thus depend on chemistry, but, as will be shown, it is still a difficult task to glean pertinent information from experiments.

Finally, from the chemists' standpoint, radioactivity is a purely nuclear phenomenon, although this point of view is not necessarily shared by radiochemists.

2. Energy Balance of the Transition

A radioactive decay is a spontaneous exoenergetic process which releases an energy Q. The amount of this energy is expressed as the difference between the respective rest-mass energies of the initial and final nucleus (or nuclei in case of fission), both in the ground state, diminished by the mass of the emitted particles:

$$Q(\text{MeV}) = 931.5\ [m(N_i, Z_i) - \sum m(M_f, Z_f) - \sum m(\text{free particles})]$$

In practice, atomic masses M rather than the nuclear masses m are used for the calculation. Both are related by

$$M(N,Z) = m(N,Z) + Z\, m_o - \frac{B_{el}}{c^2}$$

which includes the nuclear mass, the mass of the Z electrons and the total electron binding energy B_{el}, which is a function of Z. This last term is generally approximated using expressions such as

$$B_{el}(Z) = 15.73\ Z^{7/3}\ \text{eV} \quad \text{or} \quad B_{el}(Z) = 13.5\ Z^{12/5}\ \text{eV}$$

although precise computed values for atoms or ions are now available. For the heaviest atoms, B_{el} is of the order of 1 MeV, to be compared with the 1600 MeV total binding energy of the nucleons in a nucleus for which A = 200.

The error introduced by replacing nuclear masses with atomic masses and *decay products* with atoms is inconsiderable because of the small values of B_{el} and their compensation in the definition of Q. In a few cases, however, the value of B_{el} may be decisive; the calculated Q of the observed β^- decay of ^{176}Lu to ^{176}Hf has a positive value when the atomic, instead of the nuclear, masses are used in the energy balance.

Q values per nucleus are in the MeV range and exceed by a factor of 10^6 the energy released in the most exothermic chemical reactions, which involve only a few eV per molecule. In terms of thermodynamics, the energy released in a radioactive decay is expressed solely by the expression of Q, which is the equivalent of ΔU or ΔH in chemical reactions. Although any radioactive decay represents an increase in entropy, the corresponding term $T\Delta S > 0$ is not considered. Its value is very small with respect to Q and, for example, would amount to only a few eV in α decay.

Q values for each decay mode can be estimated with semiempirical expressions as a function of A, Z, and a few additional parameters. The limiting values of A and Z which make $Q > 0$ can be deduced.

The energy Q is distributed among all reaction products in accordance with the laws of conservation of energy and momentum. The kinetic energy acquired by the heavy products is called *recoil energy*. The energy of the released particles and photons is easily measured. Its experimental determination, together with the calculated recoil energy, provides a means of determining Q values.

The emission of particles and radiations is generally isotropic; however, with respect to the direction of a quantization axis, this is no longer true owing to the conservation of the total angular momentum in the disintegration. Thus the direction of emission of a photon or particle in a disintegration may influence the orientation of a further radioactive event (see Chapter 8).

Frequently, the final nucleus formed in the decay is itself radioactive and in turn engenders a new nuclide. The pair of radionuclides is straightforwardly termed a *radioactive filiation* in which the *parent* nuclide decays to its *daughter*. A filiation is not necessarily restricted to two radioactive members; the ^{238}U family in itself comprises over 15 radioactive members. Tens of filiations with up to six or eight members occur among fission products.

In a few cases, the daughter nuclides are produced in highly excited states, which relax by the emission of particles. This is the origin of *delayed* proton, neutron or α emissions following β^-, β^+, and spontaneous fission decays, respectively.

C. Stable Atoms

A nucleus (and its corresponding atom) is reputed to be stable as long as no spontaneous emission is observed. The definition of a stable nucleus may be reconsidered in the future when improved, more sensitive detection methods become available.

274 atoms with $Z < 82$ have lifetimes greater than 10^{18} years, the present limit of detection of a radioactive emission, and are listed as stable. This limit corresponds to 50 disintegrations per hour in 1 mole of nuclides. This definition of stability does not necessarily mean that all these atoms are effectively eternal. The longest-lived radioactive atoms include ^{50}V, which decays by EC and β^- emission with a mean lifetime over 4×10^{17} years and the α-emitter ^{148}Sm, with a mean lifetime of 10^{16} years.

At present it is considered that the lifetime of the proton with respect to decay into subprotonic particles exceeds 2×10^{30} years.

In the N, Z plane, the stable atoms are represented to a first approximation by a curve which satisfies the relation

$$N = Z\left[1 + \frac{\sqrt[3]{A^2}}{60}\right]$$

Up to $Z = 20$, the number of neutrons and protons is approximately equal. For higher Z, the stable atoms deviate increasingly from the $N = Z$ line; for a given value of Z, the number of neutrons becomes progressively larger. The curves ends with $Z = 82$, but is interrupted twice, for $Z = 43$ and 61.

All possible atoms which are not located on the curve possess an inherent trend to undergo spontaneously transformation in one or several steps until they reach one of the 274 stable configurations. This behavior may serve as another definition for radioactivity.

Figure 3.3 is a representation in the N, Z plane of all stable atoms, together with the known radioactive atoms for which the mean lifetimes τ lie between 10^{-3} s and 10^{15} years.

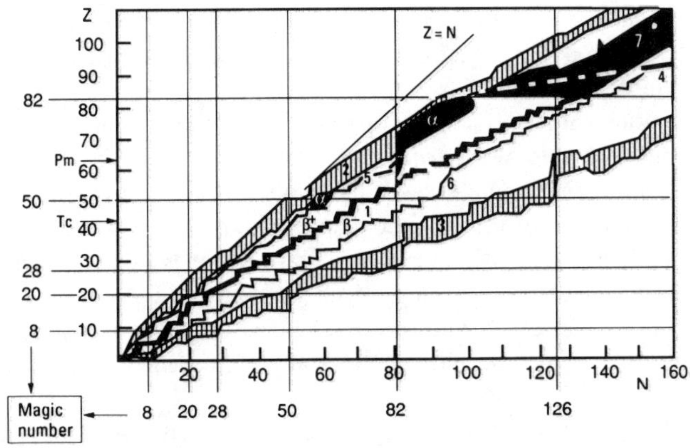

FIGURE 3.3. Stable and radioactive nuclides in the N-Z plane. The regions and lines numbered 1 to 7 refer to the following nuclides: (1) location of stable nuclides. Interruptions occur in the stability line for the elements Tc ($Z = 43$) and Pm ($Z = 61$). (2) Nuclides with proton excess. The limits of the region are odd-Z and even-Z nuclides with a lifetime of 10^{-3} s with respect to spontaneous proton emission. (3) Nuclides with neutron excess. The frontiers correspond to even and odd values of N. (4) Nuclides decaying by spontaneous fission with a detection limit of 10^5 fissions per second and per mole. (5) Limit of identified β^+ or EC emitters with $\tau < 10^{-3}$ s. (6) Limit of identified β^- emitters with $\tau < 10^{-3}$ s. (7) Region in which spontaneous fission competes progressively with α, β^-, and EC decays.

IV. PROBABILITY OF RADIOACTIVE DECAY

A. The Parameter λ

A chemical reaction that is thermodynamically possible may not occur, or may progress only at a very slow rate, because of kinetic hindrance. Similarily, a positive Q value is a necessary although insufficient condition in itself for a radioactive decay to be observed. For practical purposes, the probability of the transition must be such that the emission of particles or radiation is measurable with the presently available detection devices.

The fundamental parameter is the probability λ, also known as *radioactive* or *decay constant*, that the transition occurs per unit of time. Like other characteristics of radioactive decay, such as the energy of the radiations or the direction of emission with respect to a reference axis, the probability depends on the properties of the initial and final states of the transition. It is related to the mean lifetime τ of the nucleus in the initial state by

$$\lambda \tau = 1$$

The probabilistic nature of λ is easily understandable when the emitted particles are known to preexist inside the nucleus (proton or neutron), or at least supposedly preexist (α particle, spontaneous fission). In such cases λ represents the probability that the particle will escape from the nucleus, notwithstanding the interaction between nucleons.

In transitions with emission of β particles or γ photons, and in EC decays, λ represents the value of the *transition moment* between the *i* and *f* states. It can be calculated in simple cases if the wave functions of the states together with the hamiltonian of the transition perturbation are known. Selection rules can be established which indicate the situations for which λ is zero or very small. The calculated λ will depend on the values of A and Z of the two states, on the value of Q, and on the changes in parity and spin following the transition, i.e., on the momentum and parity carried by the emitted radiations.

In practice, λ is determined experimentally for all transitions, including spontaneous fission. The transition probability may be very sensitive to magic nucleon numbers. Such nuclei are either stable or have very long lifetimes.

An initial state may decay by competing modes in attaining various final states. Each pathway is characterized by a specific transition probability λ_i such that

$$\lambda = \sum_{i=0}^{i} \lambda_i$$

The observation of a unique radioactive atom represents a special event, since at the moment of the decay the atom has ceased to exist. A population of several atoms also presents problems and these will be considered in Chapter 9. In the majority of cases, however, the number of atoms is larger by many orders of magnitude and the evolution of the population can be described statistically.

At ordinary temperatures, all nuclei in the initial state are in the ground state; even in the case of very low-lying excited levels, the thermal energy kT is largely insufficient to raise the nuclei above the ground state.

B. The Radioactive Decay Law

First, a single radioactive atom will be considered. Two basic assumptions are made for λ (or λ_i as the case may be): (1) the probability of decay per unit time is the same, whatever the time elapsed since the beginning of the observation. This latter time may, or may not, coincide with the time of formation of the atom. In other words, the *age* of a radioactive atom is a constant; and (2) the probability λ is the same for all atoms with equal A and Z, whatever the chemical state. This assertion is not strictly true only in a few cases which will be considered in Chapter 8.

The probability that the atom will decay during the time interval Δt at any given time t is

$$P(1)_{t,t+\Delta t} = \lambda \, \Delta t$$

and the probability of survival is

$$P(0)_{t,t+\Delta t} = 1 - \lambda \, \Delta t$$

In these two expressions, the time interval Δt can be replaced by its limit dt when Δt becomes very small with respect to t. The probabilities are now replaced by the differential probabilities

$$dP(1)_{t,t+dt} = \lambda \, dt$$

and

$$dP(0)_{t,t+dt} = 1 - \lambda \, dt$$

This will now be applied to an assembly of atoms. In order to simplify the reasoning, three atoms, numbered 1, 2, and 3 are considered at time t. The probability $\mathcal{P}(1)_{t,t+dt}$, in short $\mathcal{P}(1)$, of observing one disintegration during dt is the sum of the probability of decay of one of the atoms and the probability that the two others, numbered 2 and 3, 1 and 3, or 1 and 2 do not decay, whatever the order in which these situations occur. Clearly, the number of such situations is $C_3^1 = 3$. It follows that

$$\mathcal{P}(1) = C_3^1 \, P(1)_{t,t+dt} \, [1 - P(1)_{t,t+dt}]^2$$

or

$$\mathcal{P}(1) = C_3^1 \, \lambda \, dt(1 - \lambda \, dt)^2$$

Similarly,

$\mathcal{P}(2) = C_3^2(\lambda \, dt)^2(1-\lambda \, dt)$, $\mathcal{P}(3) = C_3^3(\lambda dt)^3$ and $\mathcal{P}(0) = C_3^0(1 - \lambda \, dt)^3$ with $C_3^2 = 3$, $C_3^3 = 1$ and $C_3^0 = 1$.

Continuing by recurrence for 4, 5, 6 . . . , and finally N atoms present at time t, the probability that N_i atoms will decay during the interval dt becomes

$$\mathcal{P}(N_i) = \frac{N!}{N_i! \, (N - N_i)!} (\lambda \, dt)^{N_i} \, (1 - \lambda \, dt)^{N-N_i}$$

This expression corresponds to the binomial distribution law, for which the average value is

$$\overline{N}_i = n \lambda dt$$

Even at very small concentrations, the number of radioactive atoms is so large that N can be considered as a continuous variable. Hence, the number of atoms, among the population of N atoms present at time t, which will decay during the time interval dt is

$$|dN| = \overline{N}_i$$

As long as the diminution of the number of atoms does not significantly alter the overall population N, i.e., $dN << N$, it follows that

$$-dN = N \lambda dt$$

This expression is integrated with the initial condition $N = N_0$ at time zero to yield the well-known radioactive decay law

$$N = N_0 e^{-\lambda t}$$

The latter is usually presented without further justification as the decrease in number of radioactive atoms according to a first order kinetic reaction.

The *half-life* T of a population of radioactive atoms is by definition the time required for the initial number N_0 of atoms to be reduced to half that number. Hence, after one half-life,

$$N = \frac{N_0}{2} = N_0 e^{-\lambda T}$$

from which

$$2 e^{-\lambda T} = 1$$

The numerical relation between the half-life and the radioactive constant is

$$\lambda T = \ln 2 = 0.693147$$

which for most purposes can be simplified without much loss of precision by

$$\lambda T = 0.7$$

The fraction of atoms which remains after n half-lives is

$$\frac{N}{N_0} = 2^{-n} \approx 10^{-0.3n}$$

As a useful rule of thumb, the population of radioactive atoms decreases to about 1% after 7 half-lives ($2^{-7} = 7.8 \times 10^{-3}$) and after 10 half-lives it is 1‰ ($2^{-10} = 9.8 \times 10^{-4}$) of its initial value.

For a time t short with respect to the half-life T,

$$N = N_0\left(1 - \lambda t + \frac{\lambda^2 t^2}{2} + \ldots\right)$$

which for $t < 0.2\ T$ simplifies to the linear decay

$$N = N_0\ (1 - \lambda t)$$

In a population of N atoms, the lifetime of any single atom may lie between zero and infinity. The mean lifetime θ of a radioactive atom is obtained by dividing the total lifetime of all atoms between zero and infinity by the initial number N_0 of atoms. This value θ represents the average lifetime of an atom or is defined mathematically on the basis of the time t and its statistical weight $-dN/N_0$:

$$\theta = \int_0^\infty t\left(-\frac{dN}{N_0}\right) dt = \frac{1}{N_0}\int_0^\infty t N \lambda\ dt$$

The value θ also represents the mean time t_m between the birth and the disappearance of this atom:

$$t_m = \frac{\int_0^t t N \lambda\ dt}{\int_0^t N \lambda\ dt}$$

On the other hand, $\lambda\ \tau = 1$ so that

$$\theta = t_m = \tau$$

These three definitions relate the elemental and statistical aspects of the phenomenon of radioactivity.

After one mean lifetime, the initial population has decreased to

$$N_0 = e^{-1} = 0.368\ N_0$$

The half-life is an important property of a radioactive species and must be known with precision since its value determines the radioactive constant and the mean lifetime. The initial population N_0 is computed from the measured number of residual atoms at time t

$$N_0 = Ne^{\lambda t}$$

The degree of precision for N_0 is highest when the measurement is performed at time $t = \tau$, for which the derivative

$$\frac{dN_0}{d\lambda} = Ne^{\lambda t}\left[t - \frac{1}{\lambda}\right]$$

is zero.

The exponential decay law is valid providing two conditions be fulfilled: the differential observation time dt must be sufficiently short with respect to the decay rate, i.e., $dt << \lambda^{-1}$ or, equivalently, $dt << T$ and the number N of atoms must be treated as a continuous variable.

Exceptional cases in which these conditions are not met will be encountered later. From the exponential decay law, it is obvious that after a certain length of time, the number of residual atoms will have decreased to the level of a few units, and then N can no longer be treated as a continuous variable.

A further use of the exponential decay law is the calculation of the average number N_i of disintegrations occurring during a given time t,

$$\mathcal{N}_i = \overline{N}_i\,(0,t) = \int_0^t N\lambda\,dt = N_0[1 - e^{-\lambda t}]$$

Considering the binomial distribution law which gives the probability P(m) for observing an event characterized by the value m and its probability p of occurrence among n trials

$$P(m) = \frac{n!}{(n - m)!\,m!}\,p^m\,(1 - p)^{n-m}$$

for which the mean value is

$$\overline{m} = np$$

and setting $m \equiv \mathcal{N}_i$, $n \equiv N_0$ and $p \equiv 1 - e^{-\lambda t}$ it follows that

$$P(\mathcal{N}_i) = \frac{N_0!}{(N_0 - \mathcal{N}_i)!\,\mathcal{N}_i!}\,(1 - e^{-\lambda t})^{\mathcal{N}_i}\,(e^{-\lambda t})^{N_0 - \mathcal{N}_i}$$

This relation holds for any observation time t. Clearly, if t = dt and $\lambda dt << 1$ (i.e., $dt << T$) the exponential $e^{-\lambda dt} \approx 1 - \lambda t$ and $\mathcal{P}(\mathcal{N}_i) = P(N_i)$.

C. Activity

The activity A of a population of N radioactive atoms is defined as the number of decays occurring per unit time and is expressed as

$$A = -\frac{dN}{dt} = \lambda N$$

Since

$$-\frac{dN}{dt} = N_0 \lambda\, e^{-\lambda t} = \left(\frac{dN}{dt}\right)_{t=0} e^{-\lambda t}$$

the time change of the activity is

$$A = A_0\, e^{-\lambda t}$$

in which A_0 is the initial activity at time zero. The decrease in activity as a function of time is represented in Figure 3.4. The area below the decay curve represents the initial number of atoms:

$$\int_0^\infty A\, dt = N_0 \int_0^\infty e^{-\lambda t}\, dt = N_0$$

The SI unit of activity is the *becquerel* (symbol Bq), which denotes the activity of a radionuclide decaying at the rate of one spontaneous decay per second. It has replaced the unit *curie* (symbol Ci) which was initially defined as the number of decays occurring per second in one gram of ^{226}Ra,

$$1\ \text{Ci} = \frac{0.7}{T_{Ra}} \times \frac{6.02 \times 10^{23}}{226}$$

Because this expression was subject to change with the refinement of the experimental value of the half-life, the curie was defined exactly and independently of ^{226}Ra as

$$1\ \text{Ci} = 3.7 \times 10^{10}\ \text{decays per second} = 3.7 \times 10^{10}\ \text{Bq}$$

This activity corresponds to 1.025 g of radium.

The curie is still used when dealing with large amounts of radioactive matter, rather than, say, megabecquerels (MBq, 10^6 Bq) or terabecquerels

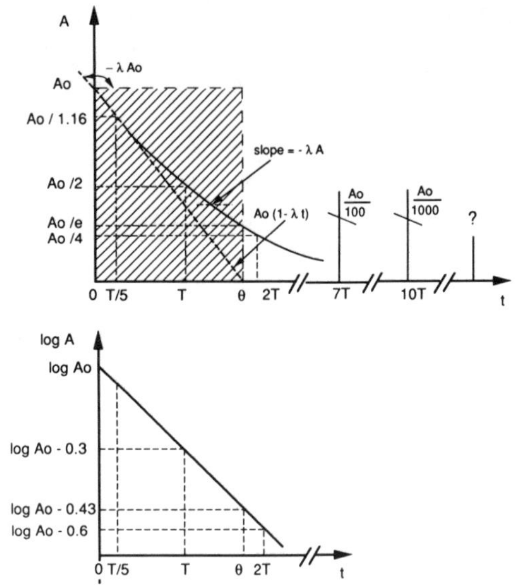

FIGURE 3.4. The radioactive decay law. The number of radioactive atoms is assumed to be larger than 20, whatever the time t. The area of the dashed rectangle and that below the decay curve are equal to the number N_0 of atoms present at time zero. The decay curve practically coincides with its tangent up to T/5. The equation of the straight line in the semi-log representation is: $\log A = \log A_0 - 0.303\ t/T = \log A_0 - 0.434\ \lambda t$.

(Tbq, 10^{12}Bq). The official Bq unit is well adapted to the expression of very low activities, such as those encountered in the environment. Useful conversion factors are

$$1\ \text{Bq} = 2.7027 \times 10^{-11}\ \text{Ci} = 27.027\ \text{pCi};$$

$$1\ \text{Ci} = 0.037\ \text{TBq} = 37\ \text{GBq};\ 1\ \text{mCi} = 37\ \text{MBq};\ 1\ \mu\text{Ci} = 37\ \text{kBq}$$

Measurement of the absolute activity of a radionuclide is a very difficult task, since it requires a precise recording of all decay events in 4π steradians, and must take into account all decay modes and all types of radiations emitted.

In the vast majority of cases, only relative activities are measured. An appropriate detector delivers a number of pulses proportional to the activity, and hence to the number of disintegrations per unit time:

$$N(\text{pulses per s}) = \eta k N(\text{disintegrations per s})$$

The proportionality factor η represents the overall detection efficiency, which includes the intensity of the detected radiations with respect to all possible decay modes and the geometry of the source-detector assembly. The factor k is related to the intrinsic properties of the detector (see Chapter 5). The constancy of the factor η is a prerequisite in any radiochemical experiment and must be checked regularly.

V. DECAY SCHEMES

The properties of radionuclides and those of the emitted radiations can be summarized in an illustrative representation, the *decay scheme* of the nuclide. The schemes of the lightest elements are relatively simple, but may be very complex for heavy nuclides.

The scheme comprises the ground state of a stable or radioactive nuclide and all known excited states with their characteristics including energy or mass with respect to the ground state, spin, parity, lifetime, and frequently the electric quadrupolar and magnetic dipolar momenta. The various observed transitions between the levels are indicated.

The nature and properties of the radiations emitted in the transition from an initial to a final level are derived from the properties of the states. The total decay energy Q, in MeV, from the ground state of the parent nuclide to the ground state of the daughter is indicated in the scheme, together with all pathways which populate the various levels.

A. γ Emission and Internal Conversion

With very few exceptions, all radionuclides undergo de-excitation with the emission of γ radiations. The transition to a lower state by emission of a nucleon is possible if the excitation energy exceeds the binding energy of a neutron or a proton, which amounts to about 8 MeV. This is very common in nuclear reactions, but is observed only rarely in radioactive decays.

1. Energy of γ Photons

The energy released in the transition is

$$Q_\gamma = E_i - E_f = \Delta E_{if}$$

If the nuclide is at rest in the initial level, it acquires after the emission a momentum MV such that

$$\vec{p}_\gamma = \vec{p}_R$$

$$\frac{h\nu}{c} = MV$$

and a recoil energy

$$E_R \approx \frac{E_\gamma^2}{2Mc^2}$$

If E_γ is expressed in MeV and M in atomic mass units (amu), the recoil energy in eV is obtained from the approximation

$$E_R \approx 536 \frac{E_\gamma^2}{M}$$

It follows that the energy of the γ photon is

$$E_\gamma = Q_\gamma - E_R$$

For a γ transition of 100 keV, the recoil energy of a nuclide with M = 100 is only 0.05 eV. Hence, the correction due to recoil can be neglected in most cases and $E_\gamma = Q_\gamma$.

The energy of a photon emitted by a nucleus in motion with a velocity v is slightly changed by the Doppler effect. The energy shift of a photon emitted in the direction of displacement of the nucleus is E_γ v/c and in most cases does not exceed a few eV. Doppler broadening of γ lines must sometimes be taken into account in radiochemistry (see Chapter 8).

2. Character of γ Radiation

In electronic transitions, the angular momentum l of the photon is equal to its own spin s = 1 and the radiation is dipolar, electric, or magnetic. On the other hand, a nucleus may emit radiations with higher l values, and thus with higher *multipolarities*, represented by 2^l. Atoms may possess many fast de-excitation channels, which is not the case for isolated nuclei; the higher the order l, the slower the transition. If

$$\Delta I = I_i - I_f$$

denotes the difference of spin between the two levels in the transition, the allowed values for the photon angular momentum are given by

$$\Delta I = |I_i - I_f| < l < I_i + I_f$$

The observed photons have an angular momentum $l = \Delta I$, and sometimes $l = \Delta I + 1$. The most common l values are 1, 2, and 3. The transition $I_i = I_f = 0$ is prohibited since the smallest l value must be $l = 1$.

The γ photons are further distinguished according to their electric or magnetic character, which is related to the parity Π_i and Π_f of the levels. The conservation of parity requires that

$$\Pi_i = \Pi_f \, \Pi_\gamma$$

The γ radiation with angular momentum l is electric E(l) if

$$\frac{\Pi_i}{\Pi_f} \equiv \Pi_\gamma = (-1)^l$$

and magnetic M(l) if

$$\frac{\Pi_i}{\Pi_f} \equiv \Pi_\gamma = (-1)^{l+1}$$

It follows that the parity is positive for the radiations M1, E2, M3, E4, and negative for the radiations E1, M2, E3, and M4.

3. De-Excitation Rate

The lifetime τ of a level and the decay rate λ depend on ΔI, ΔE_{if} and the mass number A. For a given Q_γ, the emission of E(l) radiation is more probable than that of M(l). An approximate expression for the relative emission probabilities is

$$\lambda_{M(l)} = \frac{\lambda_{E(l)}}{4.4\ A^{2/3}}$$

in which $\lambda_{E(l)}$ is a function of E_γ^{2l+1}. For a nuclide with A = 100, the probability of emission of an E(l) radiation is 100-fold that of an M(l) radiation.

The duration of a transition depends on l and on the energy E_γ. For a 1-MeV dipolar γ photon, the emission time is less than 10^{-15} s, but a 0.1-MeV photon requires 10^{-10} s, which is also the transition time of a 1-MeV quadrupolar γ ray. For 1-MeV octupolar and 0.1-MeV quadrupolar photons the emission time is about 10^{-4} s.

The lifetime of excited states decreases with increasing values of the excitation energy, but may become important for the combination of a high l value and a low transition energy. When the lifetime exceeds a few seconds, the nucleus in the excited state together with the nucleus in the ground state constitute an *isomeric pair*. The two isomers may be in filiation if the excited (parent) state decays by γ emission to the ground (daughter) state: $^{80m}Br \rightarrow {}^{80}Br$. Alternatively, the excited state may decay by a different mode and the isomers are then no longer in filiation (^{209m}Bi, ^{116m}In).

A partial detailed decay scheme of ^{60}Ni is represented in Figure 3.5a.

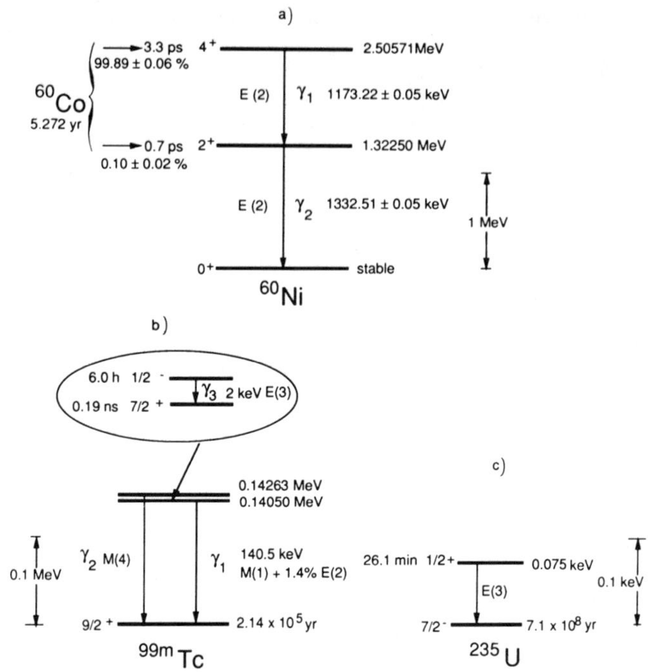

FIGURE 3.5. Examples of decay schemes. (a) Partial decay scheme of excited ^{60}Ni states populated in β^- decay of ^{60}Co. The picture shows the origin of the two photons γ_1 and γ_2 which have multifarious applications in science and technology. The allowed l values for the photon γ_1 extend from two to six and the radiation comprises a mixture of E(2), M(3), E(4), M(5), and E(6) with a preponderance of E(2). IC coefficients are: $\alpha_K = (1.51 \pm 0.08) \times 10^{-4}$ and $\alpha_{total} = (1.67 \pm 0.12) \times 10^{-4}$. The photon γ_2 is E(2) and converted with $\alpha_K = (1.15 \pm 0.06) \times 10^{-4}$ and $\alpha_{total} = (1.07 \pm 0.05) \times 10^{-4}$. The atomic parameters of Ni are: $\omega_K = 0.414 \pm 0.0028$; $K_{\alpha_2} = 7.46089$ keV; $K\alpha_1 = 7.477815$ keV; $K\beta_1 = 8.26466$ keV; intensity ratio $\beta_1/\alpha_1 + \alpha_2 = 13.9/100$. With such low α values, the emission of conversion and Auger electrons and X-rays is practically excluded. The values indicated give an idea of the precision attainable in modern nuclear and atomic spectroscopy. (b) ^{99m}Tc isomeric transition. The main transition corresponds to the emission of γ_1. The $\gamma_3-\gamma_1$ cascade accounts for only 0.02%. All γ photons and particularly γ_3 are highly converted, with K shell conversion coefficients: $\alpha_K(\gamma_1) = 0.1$, $\alpha_K(\gamma_2) = 23$; $\alpha_K(\gamma_3)$ is too high to be measured. The energy of the conversion electrons is 119.5 keV (8.7%) and 137.5 keV (1.1%); X-rays include $K\alpha_2 = 18.3$ keV (6%) and $K\alpha_1 = 20.6$ keV (1%). The spectroscopy of the electrons from the very intense conversion of γ_3 is used for speciation of Tc (see Chapter 8). (c) ^{235m}U isomeric transition. The very soft E(3) γ photon (75 eV) is not converted.

4. Internal Conversion

The competition between radiative de-excitation and internal conversion (IC), i.e., transfer of the nuclear excitation energy to the electron core, is expressed by the total IC coefficient α:

$$\alpha = \frac{N_e}{N_\gamma}$$

N_e and N_γ are, respectively, the number of conversion electrons and γ photons emitted in the same time interval. The energy E_e of the conversion electrons is

$$E_e = E_\gamma - E_b$$

in which E_b is the binding energy of the electrons in the K shell and in the L, M, and N subshells involved in the process. Conversion electrons are emitted as several groups (up to eight) of monokinetic particles. The coefficient α is the sum of the partial conversion coefficients α_K, α_{LI}, α_{LII} etc. in the various atomic states. Conversion is more pronounced in the K shell since the 1s electrons have the highest probability of penetrating into the nuclear volume.

The conversion coefficients vary with l, Z and E_γ according to

$$\alpha_{KE(l)} \approx \frac{1}{(1 + l)} \frac{Z^3}{E_\gamma^{(l+5/2)}}$$

and for $E_\gamma < 0.5$ MeV and $E_K < E_\gamma$,

$$\frac{\alpha_{KE(l)}}{\alpha_{KM(l)}} \sim \frac{1}{1 + l}$$

As a result of the Z^3 dependence isomeric transitions in heavy atoms are highly converted.

The competition between IC and γ emission is also expressed by partial decay constants λ_i. The overall decay probability is

$$\lambda = \lambda_\gamma + \lambda_{IC}$$

and the total probability of IC is

$$\lambda_{IC} = \lambda_K + \lambda_{LI} + \lambda_{LII} + \ldots$$

from which it follows that

$$\lambda = \lambda_\gamma (1 + \alpha)$$

The IC process decreases the lifetime of the level with respect to the de-excitation by emission of E(l) or M(l) radiation.

In the case of a transition between two levels with $I = 0$, IC is the only decay mode possible and α is infinite. This results in an exceptional E(0) transition $0^+ \rightarrow 0^+$ which necessarily has a positive parity. It is observed in ^{72}Ge; its very occurrence proves that IC results from a *direct* interaction between the electric nuclear field and an orbital electron.

IC is followed by X-ray emission. The conversion coefficient and the fluorescence yield are related. For instance, for the K shell

$$\alpha_K \omega_K = \frac{N_{eK}}{N_\gamma}\frac{N_{KX}}{N_{eK}} = \frac{\text{Number of K X-rays}}{\text{Number of } \gamma}$$

Examples of the 142-keV isomeric transition of ^{99m}Tc and of the 75-eV transition of ^{235m}U are presented in Figures 3.5B and 3.5C.

B. β Emission and Electron Capture

1. Decay Energy

The emission of a β^- particle results from the process

$$\text{Neutron} \rightarrow \text{proton} + \beta^- + \bar{\nu}$$

in nuclides with an excess of neutrons with respect to a stable isobar.

The energy released in the decay is

$$Q_{\beta^-} = 931.5\,[M(A,Z) - M(A,Z+1)] \quad \text{MeV}$$

The recoil energy following emission of an electron is calculated from the momentum conservation

$$\vec{p}_e = \vec{p}_R$$

Since the velocity of the electron may be a considerable fraction of the velocity of light, it is necessary to use relativistic expressions for momentum and energy. The result is

$$E_R = \frac{E_e^2}{2\ Mc^2} + E_e \frac{m_0}{M}$$

in which m_0 is the electron rest mass.

For practical use, the energy E_e of the electron is expressed in MeV and M in amu. The recoil energy, in eV, is then given by

$$E_R = 536 \frac{E_e^2}{M} + 541 \frac{E_e}{M}$$

The recoil energy following emission of a 1-MeV electron by a nuclide of mass 100 is about 10 eV, which is negligible in comparison to nuclear energies.

The energy Q_{β^-} is shared between the electron and the antineutrino. The energy spectrum of the electrons is continuous between zero and a maximum value E_{max} (usually quoted as the *β-ray energy*). The corresponding recoil energy spectrum extends up to a limit given by the previous expression with $E_e = E_{max}$.

Neutron-deficient nuclides have two decay modes. It is convenient to begin with the EC

$$\text{Proton} + e^- \rightarrow \text{neutron} + \nu$$

for which the energy balance is

$$Q_{EC} = 931.5\ [M(A,Z) - M(A,Z-1)] - 10^{-6}\ E_b \quad \text{MeV}$$

E_b is the binding energy, in eV, of the electron captured in the nucleus. It is quite negligible with respect to the nuclear energy, as is also the recoil energy due to the emission of the neutrino.

An alternative mode is decay by positron emission:

$$\text{Proton} \rightarrow \text{neutron} + e^+ + \nu$$

The energy balance is

$$Q_{\beta^+} = 931.5\ [M(A,Z) - M(A,Z-1) - 2m_0] = Q_{EC} - 1.022 \quad \text{MeV}$$

from which it appears that the process is possible only when the decay energy Q_{EC} exceeds two electron rest masses, i.e., $2 \times 0.511 = 1.022$ MeV. If this condition is not fulfilled, the nuclide decays only by electron capture.

Like β^- rays, positrons have a continuous energy spectrum. The positron rapidly loses its kinetic energy and finally disappears in the annihilation process

$$e^- + \text{positron} \rightarrow 2\gamma$$

in which photons have an energy of 0.511 MeV. The observation of these photons among the radiations emitted by a radionuclide is the fingerprint of decay by positron emission.

When positron emission is allowed, it is always in competition with electron capture. For $Z > 80$, EC is more probable than β^+ decay. Like IC, EC involves mainly 1s electrons, i.e., K-capture. If Q_{EC} is insufficient for K-capture, the process occurs in the L-shell.

2. Decay Rate

Transitions due to the weak interaction are achieved in less than 10^{-6} s. The decay rate is expressed by the product log fT in which T stands for the half-life of the radionuclide, and f is a parameter which depends on the atomic number Z of the daughter nuclide, on E_{max} (for β^- and β^+ decays) and on E_ν (for EC). Representative f values for $Z < 100$ and $0.1 < E_{max} < 10$ MeV, are given below:

β^- decay: $\log f = 4 \log E_{max} + 0.60 + 0.02\, Z - 0.005\, Z \log E_{max}$
β^+ decay: $\log f = 4 \log E_{max} + 0.60 - 0.07\, Z - 0.009\, Z (\log E_{max}/3)^2$
EC decay: $\log f = 2 \log E_\nu + 3.5 \log Z - 5.6$

Log fT values (T in s) lie between 3 and 13, depending on the changes ΔI and $\Delta\Pi$ in the decay. The transitions may be allowed or forbidden; the more probable ones, with log fT = 3, are those for which $\Delta I = 0$ and $\Delta\Pi$ is positive. Half-lives always exceed 10^{-3} s and may be as long as 10^{17} years.

Partial decay schemes for ^{11}C (β^+, EC), ^{22}Na (β^+, EC), ^{237}Pu (EC, α), and ^{241}Pu (β^-, α) are represented in Figure 3.6.

C. α Decay

1. Decay Energy

The energy balance in decay by emission of an α particle is

$$Q_\alpha = 931.5\,[M(A,Z) - M(A-4, Z-2) - M(^4_2He^{2+})] \quad \text{MeV}$$

Since the decay involves emission of a heavy particle, the recoil energy cannot be neglected. From the momentum conservation

$$M_\alpha V_\alpha = MV$$

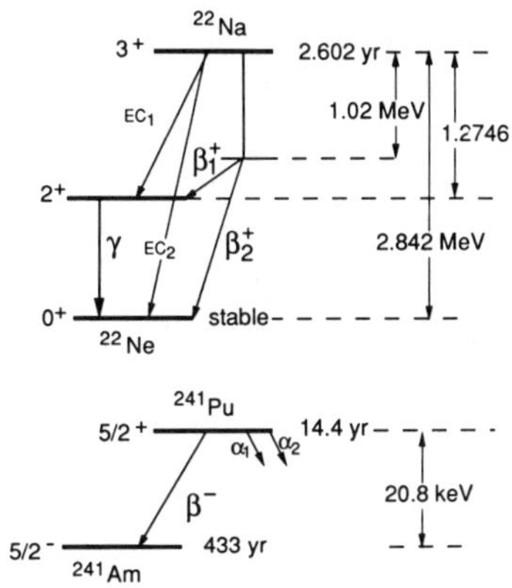

FIGURE 3.6. Partial decay schemes of ^{22}Na and ^{241}Pu. ^{22}Na: β_1^+: 90.50 ± 0.15%, E_{max} = 0.5454 ± 0.0013 MeV, log ft = 7.4. β_2^+: 0.05 ± 0.15%, E_{max} = 1.820 ± 0.0013 MeV, log ft = 14.9. EC_1: 9.45 ± 0.15%, 1.5647 ± 0.0013 MeV, log ft = 7.4. EC_2: <2 × 10^{-4}%, 2.8420 ± 0.0013 MeV, log ft = 14.9. K capture: 92.3%; L capture: 7.7%. The energy of the γ photon is 1.27454 ± 0.00004 MeV. ^{241}Pu: This nuclide decays to the extent of 99.99759% by emission of a β^- radiation of E_{max} = 0.0208 MeV, log ft = 5.8. α-Emission accounts for 2.4 × 10^{-3}% of the decays with total decay energies α_1 = 4.897 MeV (2.005 × 10^{-3}%) and α_2 = 4.853 MeV (0.292 × 10^{-3}%).

in which M is the mass of the *daughter* atom, the recoil energy is calculated as

$$E_R = E_\alpha \frac{M_\alpha}{M} = 4 \frac{E_\alpha}{M}$$

which for a typical case E_α = 5 MeV and M = 200, yields E_R = 0.1 MeV. It results that

$$Q_\alpha = E_\alpha + E_R$$

and for the energy of the α particle,

$$E_\alpha = Q_\alpha \left[1 + \frac{M_\alpha}{M}\right]^{-1} = Q_\alpha \frac{M(\text{daughter})}{M(\text{parent})}$$

This expression holds for the ground-state to ground-state transition in even-even nuclei between two levels with spin I = 0, which is considered as a *reference* transition. In most cases, however, the radionuclide decays to an excited level with non-zero spin of the daughter. Several transitions become possible, which constitute the *fine structure* of α emission. The Q_α and E_α values of these transitions lie below those of the reference.

In all nuclides other than the even-even ones, the most probable transition leads to an excited state for which ΔI = 0.

Exceptionally, α emissions from excited states of the parent nuclide have been observed. Here, Q_α and E_α exceed the reference values, and the anomalous α particles are called *long range* α. The only known examples of these are found in ^{212}Po and ^{214}Po, the β^- decay daughters of ^{212}Bi and ^{214}Bi.

2. Decay Rate

The emission of α particles is an extremely fast process, characteristic of the strong interaction, and is achieved in less than 10^{-21} s. The half-life of the reference transition is given by

$$\log T = \frac{a}{\sqrt{E_\alpha} - b} \equiv A$$

in which a and b depend on Z. Other transitions are generally slower than the computed value. In this simplified model, the difference is accounted for by a factor f such that

$$\log f = A + \log T$$

Observed half-lives lie between 10^{-6} s and 10^{15} years.

Decay schemes of several α-emitting plutonium isotopes are shown in Figure 3.7.

D. Spontaneous Fission

Owing to its complexity, spontaneous fission cannot be represented on the decay scheme. There exist at least 30 possible pathways for the splitting of a nucleus into two *fission fragments* with mass number A_x and A_y, and atomic numbers Z_x and Z_y. The fission is accompanied by emission, on the average, of two to three neutrons and up to eight γ photons, as well as neutrinos. Over 99% of the neutrons are emitted at the instant of the scission; the remainder, together with most of the γ's, are released by excited fission fragments.

Alternative, but very rare, modes are also possible with emission of α particles, protons, deuterons, or tritons. Tripartition can also occur.

The energy balance in the bipartition with emission of ν neutrons

$$Q_{SF} = [M(A,Z) - M(A_x,Z_x) - M(A_y,Z_y) - \nu m_n]$$

is such that to 200 MeV are released. Most of this energy (about 160 MeV) appears as kinetic energy of the fission fragments. This energy is dissipated as heat during deceleration of the fragments and constitutes the essence of nuclear energy. The remainder is shared between the γ photons (about 8 MeV), the neutrons ($\approx$ 5 MeV) and neutrinos ($\approx$ 10 MeV).

Once the fission fragments have been slowed down, they are considered as *primary* fission products. Most of them are radioactive and possess a large

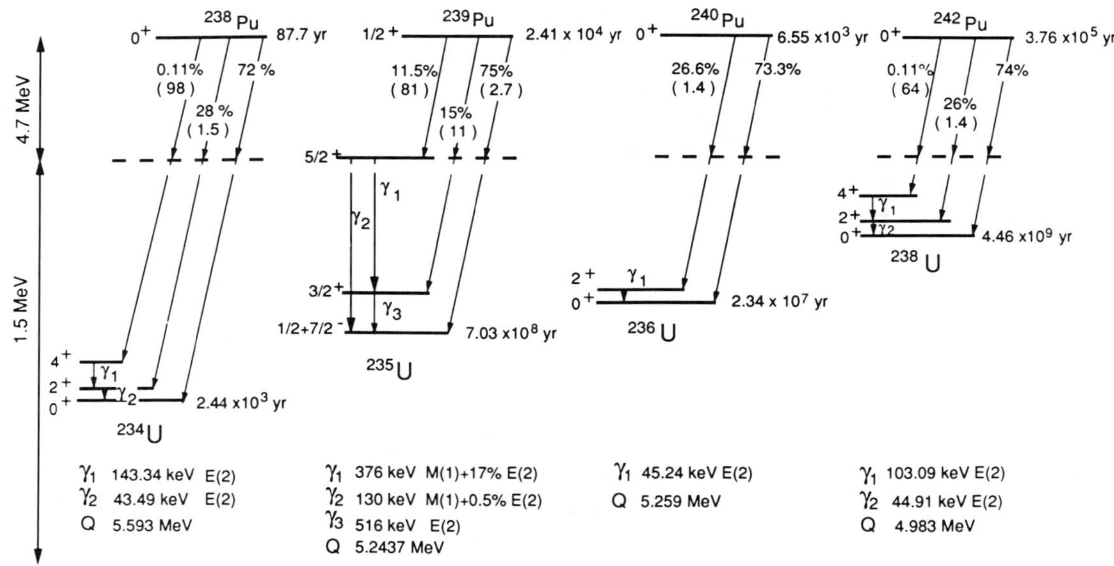

FIGURE 3.7. Decay schemes of ^{238}Pu, ^{239}Pu, ^{240}Pu, and ^{242}Pu. Note the different energy scales.

excess of neutrons. They decay by successive β^- emissions to form the *secondary* fission products. The fission decay chains may have up to six or eight members, and include isomers. They release an additional energy of about 8 MeV in β^- decays and again as much in γ emissions. The half-lives usually increase from one term to the next until the final stable isobar is reached.

The total number of nuclides produced in fission is quite large: more than 250 have been identified with atomic numbers comprised between 30 (Zn) and 66 (Dy). The mass numbers are distributed between 72 and 161, with peaks at A = 90, characteristic of the *light* fission products, and A = 161 for the group of *heavy* products.

Decay by spontaneous fission does not necessarily involve nuclides in the ground states. The process is also possible for low-lying (E < 2.5 MeV) and very short-lived (τ from 10^{-12} to 10^{-3} s) excited states, which for this reason are called *fission isomers*, although they cannot be isolated as radionuclides. Such cases are known for some isotopes of plutonium.

The spontaneous fission process is achieved in 10^{-8} s. The half-life of this mode of decay depends on the ratio Z^2/A, but the range of values is very large and is determined by Z and N.

The half-lives of even-even nuclides of a given element with $90 < Z < 102$ can be represented by a parabola whose maximum corresponds to N = 152 neutrons. Half-lives for other nuclides are much smaller (Figure 3.8).

Spontaneous fission competes with α emission in nuclides with Z > 90 and A > 230. It may be the dominant decay mode, e.g., in ^{248}Cm and ^{250}Cm. For Z > 104, α emission is again more probable than spontaneous fission.

VI. NUCLEAR REACTIONS

A. Introduction

During the 40 years which followed the discovery of radioactivity, all knowledge in nuclear science was based on observations made on radionuclides found in nature. However, the immense majority of the 2500 known radionuclides is artificial and is produced in nuclear reactions. A great number of these nuclides have descendants which are themselves radioactive, and constitute artificial radioactive families, sometimes called *radioactive chains*. These chains are headed mainly by transuranium nuclides and fission products.

The starting material for a nuclear reaction is a target consisting of a stable or radioactive long-lived nuclide. In order to modify the distribution of the nucleons in the target nucleus, the latter is bombarded with an appropriate projectile, which may be either a charged particle, a neutron or a γ photon. The reaction is written

$$a + X \rightarrow b + Y$$

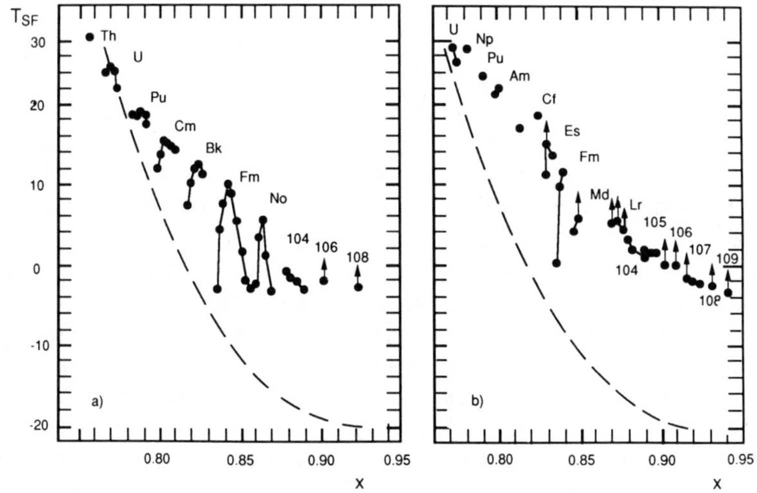

FIGURE 3.8. Trends of spontaneous fission half-lives T_{SF} as a function of the fissionability parameter.

$$x = \frac{Z^2/A}{50.8\,[1 - 1.78\,(N-Z)/A^2]}$$

(a) Even-even nuclei. (b) Even-odd or odd-even nuclei. The dashed line represents the trend expected from the liquid drop model. Arrows indicate that T_{SF} may be larger. (Redrawn from Oganessian, Yu. Ts. et al., *Radiochim. Acta*, 37, 113, 1984.)

or in short, X(a,b)Y. The entrance channel comprises the target X and the projectile a, and the exit channel contains the produced radioactive nuclide Y (in the ground or in an excited state) and the emitted particle or radiation b.

Charged projectiles, when approaching the target, suffer a coulombic repulsion. The height B_c of the coulomb barrier is satisfactorily approximated by the expression

$$B_c(\text{MeV}) \approx \frac{Z_a Z_X}{A_a^{1/3} + A_X^{1/3}}$$

where terms in Z and A refer to the two entities in the entrance channel. Even for the case of protons impinging on the lightest elements, the barrier exceeds several MeV. Thermal excitation of the projectile is precluded, since it would require 10^8 to 10^9 K, so the only alternative is that the particles must be accelerated.

The α particles emitted by, say, ^{210}Po are naturally accelerated to 5.3 MeV and these were the first projectiles used for artificial transmutation. Today, the most recent accelerators provide a great variety of projectiles with a wide range of available energy and intensity of the particle beams.

Neutrons can readily penetrate into any nucleus. *Thermal* neutrons with an energy of 0.025 eV are particularly suitable projectiles. *Photonuclear* reactions result from the absorption of γ rays in the target nucleus.

Most nuclear reactions for the preparation of radionuclides require projectiles with energies below a few ten MeV. The mechanism of these reactions is satisfactorily described by the *compound nucleus* model.

In a first step, the target and the projectile merge to form an excited amalgam of all nucleons. This compound nucleus, which recalls the activated complex of a chemical reaction, is very ephemeral, with a lifetime of 10^{-16} to 10^{-18} s. In a second reaction step, it undergoes transformation by one or several decay modes typical of spontaneous disintegrations: emission of charged particles, neutrons, γ radiation, or splitting into fragments. However, the number of possible exit channels is larger than in radioactive decay.

The identity of the products in the exit channel is deduced from the conservation of the number Z of protons and N of neutrons. As in spontaneous decays, the chemical nature of the *species* which appear in the reaction is largely unknown.

The characteristics of the nuclear reaction are determined by the conservation of all unvariants in the entrance channel: mass-energy, spin, momentum, and parity.

B. Energetics of a Nuclear Reaction

A very important parameter is the energy balance of a nuclear reaction. The total mass-energy conservation is written

$$m_a c^2 + E_a + m_X c^2 + E_X = m_b c^2 + E_b + m_Y c^2 + E_Y$$

In this expression the rest mass-energy term mc^2 and kinetic energy E of each partner in its ground state are explicit. If, as in the case of spontaneous decay, atomic rather than nuclear masses are introduced, the reaction heat becomes

$$Q = 931.5\,[M_a + M_X - M_b - M_Y] = E_b + E_Y - E_a \quad \text{MeV}$$

assuming that the target nucleus is at rest. A positive Q value characterizes an *exoenergetic* reaction in which part of the masses in the entrance channel is converted into kinetic and excitation energy of the reaction products. If the latter are in the ground state,

$$E_b + E_Y = Q + E_a \frac{M_X}{M_a + M_X}$$

The kinetic energy of the product nuclide Y may be much higher than the recoil energy following spontaneous decays.

An *endoenergetic* reaction has a negative Q value and the masses in the exit channel will exceed those at the entrance. The reaction proceeds only if sufficient energy is brought into the system as kinetic energy from the projectile. The *reaction threshold* is defined by the condition

$$E_a > -Q\,\frac{M_a + M_X}{M_X}$$

which in most cases reduces to $E_a > -Q$. In contrast to the coulomb barrier, which can be traversed by tunneling, the energy condition is an unsurmountable obstacle to the occurrence of the reaction.

The excitation energy E_{ex} of the compound nucleus is given by the binding energy B of the projectile in the target nucleus and, owing to momentum conservation, by a fractional part of E_a:

$$E_{ex} = B + E_a\,\frac{M_X}{M_a + M_X}$$

For a heavy target nucleus, this reduces to $B + E_a$, and for thermal neutrons this energy corresponds simply to the binding energy of the particle in the target, which amounts to a few MeV.

C. Cross Section

A nuclear reaction is a comparatively rare event. It is characterized for a given energy E_a of the projectile by the value of the *cross section* σ defined as

$$\sigma = \frac{\text{Number of nuclei Y produced per target nucleus X and per unit time}}{\text{Number of projectiles per unit surface and per unit time}}$$

or

$$\sigma = \frac{N_Y}{N_X\,\Phi}$$

Φ is the fluence rate or fluence density of the particles, often called *flux*, defined as the number of incoming particles per unit surface and unit time. For thermal neutrons it may also be expressed by the product of the particle density times the velocity.

As indicated by its definition, the cross section is proportional to a surface area. The product

$$P = \sigma\Phi = \frac{N_Y}{N_X}$$

is the probability of occurrence of the reaction per unit time. It is customary to express the cross section in units of barn, symbol b, which is defined as

$$1 \text{ b} = 10^{-24} \text{ cm}^2$$

To a good approximation, 1 b represents the geometrical section of a nucleus of mass number A = 100.

Cross sections vary within very large limits and depend critically on Z and N of the target nucleus.

Radiative neutron capture or (n,γ) reactions, are very common for thermal neutrons and have cross sections extending from 0.5 mb for ^{2}H (but 0.33 b for ^{1}H) to 2.65 × 10^6 b for ^{135}Xe. Nuclides with high capture cross sections include ^{10}B (3837 b), ^{113}Cd (19,910 b), ^{157}Gd (254,000 b), and ^{174}Hf (390 b).

The fission cross sections for thermal neutrons have very high values for odd-A nuclides, such as ^{239}Pu, ^{241}Pu, ^{233}U, and ^{235}U (724 b, 1000 b, 530 b, and 583 b, respectively), while even-A isotopes of Pu and U do not undergo fission with thermal neutrons. Cross sections of reactions induced by charged particles may vary from about a tenth of a barn to 10^{-12} b.

D. Excitation Function

The variation of the cross section with the energy E_a of the projectile is the *excitation function* of the nuclear reaction. It depends on the values of Q, on the angular momenta and on other parameters, and in general it cannot be calculated.

In many cases, provided the excitation energy of the compound nucleus is not too high, the function exhibits a peak each time E_{ex} coincides with the energy of a level of the compound nucleus. The successive peaks appear as *resonances*; they are well separated as long as the interstate distance is large enough. At high excitation energies, the peaks overlap and σ varies in a monotonic manner.

Excitation functions of neutron-induced reactions used for the preparation of radionuclides, mainly (n,γ), (n,particle) and (n,fission) reactions may have hundreds of resonances within a few tenths of eV. The resonances disappear when the neutron energy E_n is below ≈ 10 eV and the cross section varies monotonically with $E_n^{-1/2}$, a behavior also known as the *1/ν law* (Figure 3.9).

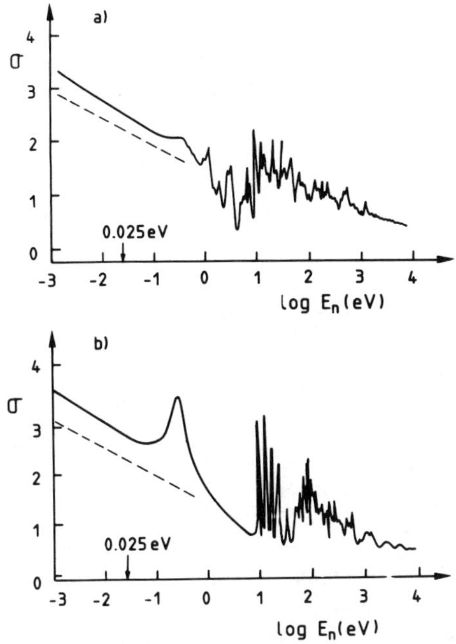

FIGURE 3.9. Fission cross section of (a) ^{235}U and (b) ^{239}Pu as a function of neutron energy. The cross section follows the 1/v law at low energy (dashed line).

The concept of cross section may be used for quantifying any type of interaction which produces dN events during a time interval dt when a number N of entities is bathed in a flux of Φ particles. In this case,

$$dN = \sigma\Phi N \, dt$$

which recalls the radioactive decay law

$$-dN = \lambda N \, dt$$

and which results in the equivalence of λ and σΦ, the kinetic parameters of nuclear transformations.

VII. THE KINETICS OF NUCLEAR TRANSFORMATIONS

The decay law of a parent radionuclide into a stable daughter is the most simple kinetic expression of a radioactive decay. In most cases, however, a radionuclide with a large excess or deficiency of neutrons converts to a stable nuclide in several steps, forming a radioactive chain or filiation. The successive members build up and decay by one or more pathways.

In a filiation, such as a natural or artificial radioactive family or a fission chain, in which the members are related by spontaneous radioactive decays, the kinetics depend solely on the radioactive constants. For radionuclides prepared by irradiation of a target in a particle flux, the kinetic equations involve both the radioactive constant λ and the build-up factor $\sigma\Phi$. These two cases will be considered separately.

A. Kinetics in a Filiation

For a filiation of the type

$$1 \rightarrow 2 \rightarrow 3 \rightarrow \cdots i \rightarrow n - 1 \rightarrow n$$

the number N_n of the atoms of the nth member can be computed from n differential equations of the form

$$\frac{dN_n}{dt} = \lambda_{n-1} N_{n-1} - \lambda_n N_n$$

which include the formation and decay channels of this member. The n^2 integration constants are determined by initial values of N_n and dN_n/dt. The corresponding system of equations has no general solution, but can be resolved by computer methods.

A relatively simple case prevails when the initial conditions at time $t = 0$ are

$$N_1 = N_1^0 \qquad N_i^0 = 0 \qquad \frac{dN_i}{dt} = 0 \text{ except for } i = 1$$

for which the number N_n of atoms of the nth member is calculated with the Bateman formula

$$N_n = N_1^0 \prod_{i=1}^{n-1} \lambda_i \sum_{i=1}^{n} \frac{e^{-\lambda_i t}}{\prod_{j=1\ j \neq i}^{n} (\lambda_j - \lambda_i)}$$

A member i of the chain may have a dual decay mode, with radioactive constants λ_i' and λ_i'', of which only one, say λ_i', leads to the nth member. In this case the expression is modified by substitution of λ_i' for λ_i in the first product and $\lambda_i' + \lambda_i''$ for λ_i in the second product.

For the most simple case of a filiation with only two members, the parent and daughter, and assuming the initial conditions $N_1 = N_1^0$ and $N_2^0 = 0$, the respective numbers of atoms at time t are

$$N_1 = N_1^0 e^{-\lambda_1 t}$$

$$N_2 = N_1^0 \frac{\lambda_1}{\lambda_2 - \lambda_1} (e^{-\lambda_1 t} - e^{-\lambda_2 t})$$

The corresponding activities as a function of the half-lives are given by

$$A_1 = \lambda_1 N_1 = A_1^0 e^{-0.7(t/T_1)}$$

$$A_2 = \lambda_2 N_2 = A_1^0 \frac{T_1}{T_1 - T_2} (e^{-0.7(t/T_1)} - e^{-0.7(t/T_2)})$$

where A_1^0 is the initial parent activity.

The ratio of the daughter-to-parent activity at any time is

$$\frac{A_2}{A_1} = \frac{T_1}{T_1 - T_2} \left[1 - e^{\frac{-0.7(T_1 - T_2)t}{T_1 T_2}}\right]$$

The development of the activities as a function of time depends on the relative values of T_1 and T_2.

1. The Half-Life of the Parent Exceeds that of the Daughter: $T_1 > T_2$

As long as the time t is one-fifth or less of T_1 and T_2, the activities vary linearly with time:

$$A_1 = A_1^0 \left(1 - 0.7 \frac{t}{T_1}\right) \qquad A_2 = 0.7\, A_1^0 \frac{t}{T_2}$$

For greater values of t, the daughter activity will eventually attain a maximum, for which

$$\frac{dN_2}{dt} = \lambda_1 N_1 - \lambda_2 N_2 = 0$$

The parent and daughter activities become equal. This situation is transient and is called *ideal equilibrium*; it is reached at time t_m such that

$$t_m = \frac{T_1 T_2}{0.7\,(T_1 - T_2)} \ln \frac{T_1}{T_2}$$

For $t > t_m$, the daughter activity eventually decreases with the half-life of the parent:

$$A_2 = A_1^0 \frac{T_1}{T_1 - T_2} e^{-0.7(t/T_1)}$$

and the limiting value of the parent-to-daughter activity ratio is

$$\frac{A_2}{A_1} = \frac{T_1}{T_1 - T_2}$$

Until decay is total, the two nuclides coexist with a constant activity ratio $A_2/A_1 > 1$, characteristic of a *transient equilibrium*.

When $T_1 >> T_2$, the activity ratio is close to unity: parent and daughter have practically the same activity and are in *radioactive equilibrium*. For times t very short with respect to T_1, the parent activity may be considered constant and equal to the initial activity A_1^0. In this case, the daughter activity increases according to

$$A_2 = A_1^0 (1 - e^{-0.7(t/T_2)})$$

until it reaches the constant parent activity. The growth equation of the daughter is complementary to the decay law. Once equilibrium is established and the daughter is separated from its parent, its activity decreases again according to

$$A_2 = A_1^0 e^{-0.7(t/T_2)}$$

in which t is the time elapsed since the separation.

This situation was first encountered for the very long-lived natural radioelements such as ^{226}Ra (T_1 = 1600 years) which decays to ^{222}Rn (T_2 = 3.8 d). Owing to the fact that the radioactive equilibrium persists throughout the lifetime of the parent, it is also designated as *secular* equilibrium.

Figure 3.10 represents the development of the activities of the parent and daughter for the filiation ^{227}Th (18.7 d) (α) ^{223}Ra (11.4 d), which leads to a transient equilibrium, and for ^{243}Am (7370 years) (α) ^{239}Np (2.35 d), typical of a secular equilibrium.

2. The Half-Life of the Daughter Exceeds that of the Parent: $T_2 > T_1$

Here the activity of the daughter increases up to a maximum value; afterwards it decreases with its own half-life according to

$$A_2 = A_1^0 \frac{T_1}{T_2 - T_1} e^{-0.7(t/T_2)}$$

A

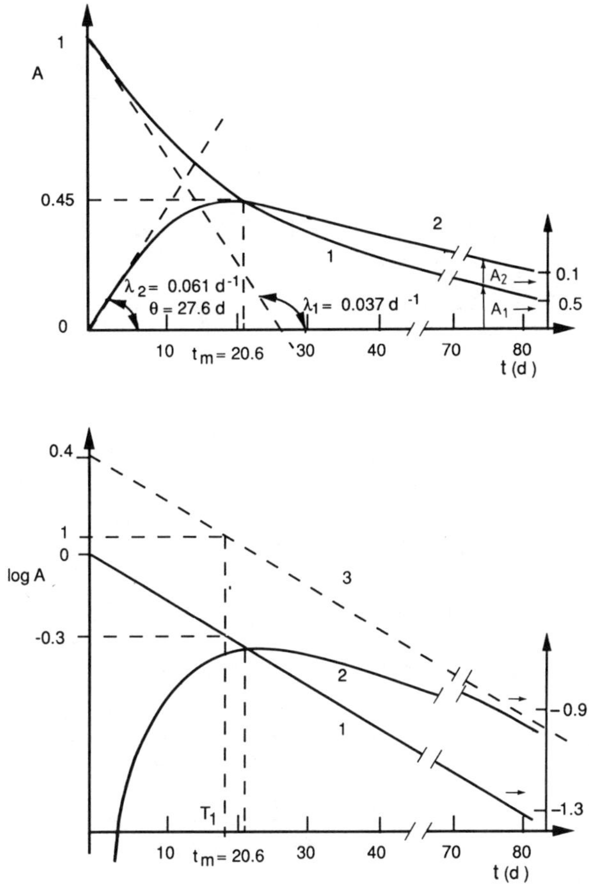

FIGURE 3.10. Examples of the development of activities in radioactive filiations. The initial activity of the daughter is assumed equal to zero. (A) ^{227}Th (α emitter, T_1 = 18.7 d) → ^{223}Ra (α emitter, T_2 = 11.4 d). Activities of the parent (curve 1) and the daughter (curve 2). In the semi-log representation curve 3 is the plot of $T_1/(T_1 - T_2)$ $\exp(-\lambda_1 t)$ which shows that A_2 decreases with the half-life T_1. Since the half-life values are close, the time of maximum activity of the daughter is approximately given by $t_m = 1.4\sqrt{T_1T_2} = 20.4$ d. The activity of ^{223}Ra reaches 90% of its maximum value after 13 d and remains above this value during the following 20 d. For $t \approx 7\,T_2$, a *transient* equilibrium is established, for which A(^{223}Ra)/A(^{227}Th) = 2.5. (B) ^{243}Am (T_1 = 7370 years) → ^{239}Np (T_2 = 2.35 d). The parent is a pure α emitter and ^{239}Np a convenient β^- emitting tracer. This is an example of a radiogenerator in which the short-lived ^{239}Np is periodically *milked* from its parent. Curve 2 indicates decay in activity of the separated daughter; curves 1 and 3 show the ingrowths of ^{239}Np in the parent at time zero and after milking. Curves 2 and 3 are complementary: BC = DE. The slope of curve 1 is the sum of the slopes of curves 2 and 3. *Radioactive* equilibrium (for which $A_1 = A_2$) is established in about 3 weeks.

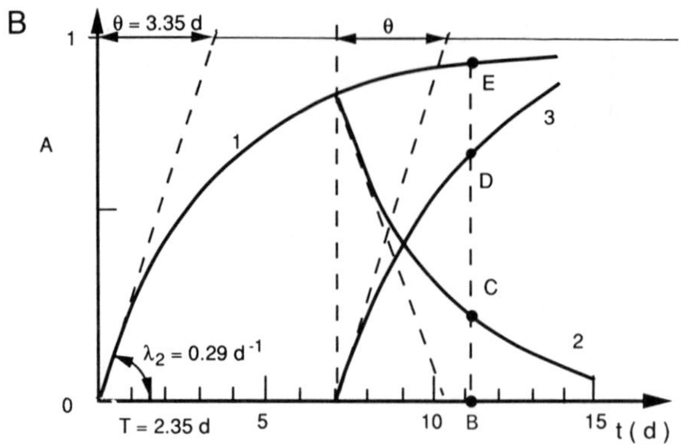

FIGURE 3.10 (continued).

and equilibrium conditions no longer hold. This is illustrated in Figure 3.11 for the pair ^{241}Pu (14.4 years) (β^-) ^{241} Am (432 years).

3. The Parent and the Daughter have Similar Half-Lives

When T_1 is close to T_2, one can write

$$T_1 = T_2 (1 + \delta)$$

and the time at which the daughter attains maximum activity is given by

$$t_m = T_2 \frac{(1 + \delta)}{0.7\,\delta} \ln(1 + \delta)$$

which after expansion of the logarithmic term yields

$$t_m \approx 1.43 \sqrt{T_1 T_2}$$

Finally, in the theoretical case of two identical half-lives, $T_1 = T_2 = T$, and $\lambda_1 = \lambda_2 = \lambda$:

$$A_2 = A_1^0 \lambda t e^{-0.7(t/T)}$$

The maximum of daughter activity is then reached for $t_m = 1.43$ T and the activity ratio becomes

$$\frac{A_2}{A_1} = 0.7 \frac{t}{T}$$

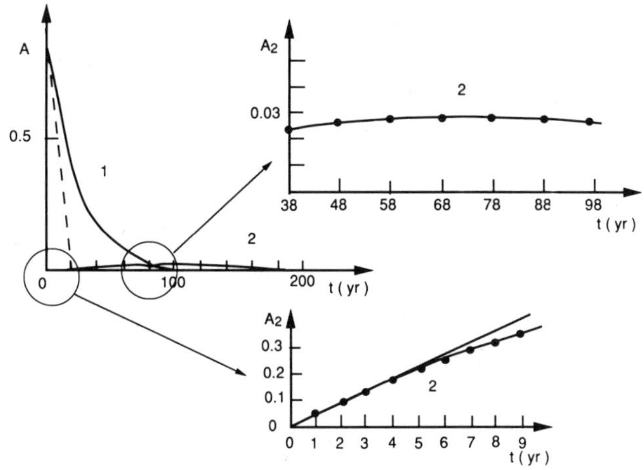

FIGURE 3.11. Development of activities in the filiation ^{241}Pu (T_1 = 14.4 years) → ^{241}Am (T_2 = 432 years). The parent is a β^- emitter, the daughter decays by α emission. The activity of ^{241}Am (curve 2) increases by about 5% per year during the first years and reaches a maximum value after 72 years. Equilibrium can never be established, since the half-life of the daughter exceeds that of the parent.

The preceding relations can be extended to filiations with more than 2 members. For example, in the case of times t much shorter than all half-lives in the filiation and provided the Bateman equation applies, one has

$$A_3 = (0.7)^2 A_1^0 \frac{t^2}{2T_2T_3} \qquad A_4 = (0.7)^3 A_1^0 \frac{t^3}{6T_2T_3T_4}$$

When the half-life T_1 of the first parent largely exceeds all other half-lives and $t >> T_1$, secular equilibria prevail between all members in a filiation. In the natural radioactive families secular equilibrium is established and all members have the same activity. Disruptures of the equilibrium state are sometimes observed owing to geochemical processes.

B. Build-Up Kinetics

A target containing a stable or long-lived isotope is irradiated in a flux Φ of incoming particles. Each radionuclide may be formed by two pathways: radioactive decay of the parent or a nuclear reaction, e.g.,

$$\begin{array}{r} {}^{243}Cm \\ \downarrow\alpha \\ {}^{238}Pu\ (n,\gamma){}^{239}Pu \end{array}$$

Similarly, two disappearance channels are open corresponding to radioactive decay and transmutation in the nuclear reaction. The scheme of the build-up matrix for (n,γ) reactions and β^- decays is

$$
\begin{array}{ccc}
{}^{i-1}_{j-1}X(n,\gamma) & {}^{i}_{j-1}X(n,\gamma) & {}^{i+1}_{j-1}X \\
\downarrow \lambda^{i-1}_{j-1} & \downarrow \lambda^{i}_{j-1} & \downarrow \lambda^{i+1}_{j-1} \\
{}^{i-1}_{j}X(n,\gamma) & {}^{i}_{j}X(n,\gamma) & {}^{i+1}_{j}X \\
\downarrow \lambda^{i-1}_{j} & \downarrow \lambda^{i}_{j} & \downarrow \lambda^{i+1}_{j} \\
{}^{i-1}_{j+1}X(n,\gamma) & {}^{i}_{j+1}X(n,\gamma) & {}^{i+1}_{j+1}X
\end{array}
$$

The rate equation of the atoms of nuclide ${}^{i}_{j}X$ with mass number A_i and atomic number Z_j is

$$\frac{dN({}^{i}_{j}X)}{dt} = \sigma^{i-1}_{j}\,\Phi\,N({}^{i-1}_{j}X) + \lambda^{i}_{j-1}\,N({}^{i}_{j-1}X) - \sigma^{i}_{j}\,\Phi\,N({}^{i}_{j}X) - \lambda^{i}_{j}\,N({}^{i}_{j}X)$$

in which the terms in N represent the number of atoms present at time t and the σ terms the cross sections of the (n,γ) reactions.

The first step in the sequence is the nuclear reaction (1)(a,b)(2), whose kinetics are determined by the cross section σ_1 of the target and the half-life T_2 of the produced nuclide. The disappearance rate of the target atoms is

$$-\frac{dN_1}{dt} = (\lambda_1 + \Phi\,\sigma_1)$$

Quite often, the first term is negligible and the population decreases with an apparent half-life

$$T^*_1 = \frac{0.7}{\sigma_1 \Phi}$$

and the equations of the filiation kinetics can be used after substituting $\sigma_1\Phi$ for λ_1.

This is illustrated by the preparation of ^{230}Pa (T_2 = 17.7 d) from ^{232}Th (T_1 = 4.1 × 10^{10} years) in the reaction ^{232}Th(p,3n)^{230}Pa. In a typical process, the target is irradiated with a 20 μA proton beam over an area of 0.5 cm^2. The beam intensity is equivalent to 6 × 10^{13} particles s^{-1}. With the cross section value of 0.28 b, the calculated apparent half-life is 1.3 × 10^3 years.

A simple and very frequent case is the formation of a radionuclide from a stable nuclide under conditions for which consumption of the target atoms and higher-order capture reactions are negligible. The activity A_2 of the nuclide increases with the irradiation time until it reaches a plateau corresponding to the *saturation activity*, which represents the maximum activity that can be obtained in the particle flux,

$$A_2 = \sigma_1 \Phi N_1^0 (1 - e^{-0.7(t/T_2)}) = A_{max} (1 - e^{-0.7(t/T_2)})$$

The build-up law is complementary to that of radioactive decay; after a time $t = T_2$, the activity is one half of the saturation value. The same expression holds for a radioactive target, provided $T_1 >> T_2$.

Thus far, the cross section and flux have been assumed to be constant. This approximation holds only for thin targets. Otherwise, the intensity and energy of the particles diminish progressively with the penetration x in the target and $\sigma_1 \Phi$ must be replaced by the integral

$$\int_0^t \sigma_1(x) \Phi(x) \, dx$$

in which t is the target thickness. To solving this expression, $\sigma_1(x)$ should be known, and this depends on the energy loss of the beam in the target and on the excitation function and flux $\Phi(x)$, which is obtained from the integration of

$$\frac{d\Phi}{\Phi} = -\sigma_1(x) \, dx$$

The variation of cross sections in thick targets is not important for neutron-induced reactions in a nuclear reactor, in particular for thermal neutrons. The solutions of build-up equations of radionuclides in a neutron flux are of considerable importance in nuclear reactor technology and in the preparation of isotopes of transuranium elements.

GENERAL REFERENCES

Arnikar, H. J., *Essentials of Nuclear Chemistry,* 2nd ed., John Wiley & Sons, New York, 1987.

Choppin, G. R. and Rydberg, J., *Nuclear Chemistry,* Pergamon Press, Oxford, 1980.

Evans, R. D., *The Atomic Nucleus,* McGraw-Hill, New York, 1955.

Friedlander, G., Kennedy, G. W., Macias, E. S., and Miller, J. M., *Nuclear and Radiochemistry,* 3rd ed., John Wiley & Sons, New York, 1981.

Lefort, M., *Nuclear Chemistry,* Van Nostrand, London, 1968.

Lieser, K. H., *Introduction to Nuclear Chemistry,* 2nd ed., Verlag Chemie, Weinheim, 1980 (in German).

Malcolme-Lawes, D. J., *Introduction to Radiochemistry,* The MacMillan Press, London, 1978.

Segrè, E., *Nuclei and Particles,* Benjamin, New York, 1964.

Valentin, L., *Subatomic Physics. Nuclei and Particles,* Vol. 1 and 2, Hermann, Paris, 1982 (in French).

Vertes, A. and Kiss, I., *Nuclear Chemistry,* Elsevier, Amsterdam, 1987.

CHAPTER 4

RADIOACTIVE MATTER

I. ACTIVITY OF RADIOACTIVE MATTER

The existence of an isolated, purely monoatomic and mononuclidic radioactive species such as ^{239}Pu is only conceptual. In reality, after the very first decay, the radioactive species are mixed with the stable or radioactive (isotopic or nonisotopic) daughter nuclide, in this case, ^{235}U. The relative proportion of daughters increases steadily with time and eventually the original population of radionuclides is replaced by an assembly of stable atoms. In a very remote but calculable future, all uranium present on earth will be replaced by the same number of stable lead atoms and the intermediate radioelements will be extinguished. The same will apply after a still longer period for thorium. All stable nuclides in the universe are the end products of the decay of radionuclides produced during the nucleogenesis of the elements.

Radioactive species are usually incorporated in some other type of other matter. Metallic ^{239}Pu, for instance, would be an exception. The radiations emitted by the radionuclides, as well as the decay itself, trigger processes that lead to the emission of new, additional types of radiations. Thus, radioactive matter may spontaneously emit radiations that are different from the truly specific ones.

Most radioactive matter is a mixture of radioactive and stable species. Here the term *species* is taken in a very general sense and includes ions, both simple and complex, and molecules in any host, such as an ore or a biosystem. Physical and chemical alteration of matter containing radionuclides results from the action of radiation on the medium.

A sample of radioactive matter may contain only isotopes of a single element. In this case, the radioactive species acts as a tracer for the isotopic matter and, apart from isotope effects, all species, whether radioactive or not, have the same chemical behavior. In many cases, especially at very low concentration, only the properties of the radioactive species can be readily determined. Of course, no stable isotopic species is available for radioelements. For the latter, very long-lived radioactive species play the role of stable matter, for instance ^{238}U (half-life T = 4.45 × 10^9 years) for uranium. Shorter-lived uranium isotopes behave like radioactive tracers of the long-lived ones. ^{236}Pu (T = 2.9 years) and ^{237}Pu (T = 45.17 d) serve as tracers for the long-lived ^{239}Pu (T = 24,110 years).

Strictly speaking most, if not all, matter whether living or inert, incorporates radionuclides, either because of the unavoidable presence of cosmogenic nuclides such as tritium or radiocarbon, or from contamination by natural or anthropogenic radionuclides.

A. Total and Partial Activities

A given radionuclide may be found in radioactive matter distributed among various chemical species E_i, e.g., radiophosphorus in $^{32}PO_4H_2^-$ and $^{32}PO_4H^{2-}$. In a simple case, only one type of radionuclide, X* (e.g., ^{32}P), isotope of the elements X (e.g., P) contained in the matter, is present. The number of radioactive atoms is N and the number of species E_i (e.g., PO_4H^{2-}), labeled by one atom X*, is N^*_{Ei} [e.g., $N(^{32}PO_4H^{2-})$].

The total number of the species E_i, whether labeled or not, i.e., containing one atom X or X*, is N_{Ei} [e.g., $N(^{32}PO_4H^{2-}) + N(^{31}PO_4H^{2-})$] and that of all entities, labeled or not, is N_E. It results from these definitions that

$$N = \sum N^*_{Ei} \quad N_E = \sum N_{Ei}$$

The total activity of the radioactive matter is

$$A = \lambda N = \lambda \sum N^*_{Ei} = \sum A_{Ei}$$

A_{Ei} is the partial activity of the species E_i, which may be written in the form

$$A_{Ei} = \lambda \frac{N^*_{Ei}}{N_{Ei}} N_{Ei} = \lambda \, \sigma_{Ei} \, N_{Ei}$$

In most cases, the factor σ_{Ei} is much less than unity. It decreases according to the radioactive decay law, but otherwise remains constant in the absence of isotope effects.

B. Specific Activity

The partial specific activity of the species E_i is

$$s_{Ei} = \lambda \frac{N^*_{Ei}}{N_{Ei}} = \lambda \, \sigma_{Ei}$$

Its highest possible value is obtained when σ_{Ei} is equal to unity; in this case all species are labeled. The maximum specific activity can be attained practically only for radioelements, i.e., when the elements have no stable isotopes, such as with Po or Pu.

The radionuclide ^{32}P is conveniently obtained by neutron irradiation of a target containing Cl according to the reaction $^{35}Cl(n,\alpha)^{32}P$. The theoretical specific activity of ^{32}P, which has a half-life of 14.28 d is 1.05×10^{16} Bq g^{-1} or 2.85×10^{5} Ci g^{-1}. Such a sample is isotopically pure and *carrier-free*. In practice, it is extremely difficult to avoid the presence of isotopic impurities (in this case the stable ^{31}P). The addition of small amounts of isotopic carrier may even be necessary in order to avoid the erratic behavior of extremely small amounts of radioactive matter and to ensure an acceptable separation yield from a target. *No-carrier added* synthesis of radiolabeled molecules is important for application of radionuclides in medicine (see Chapter 10).

The specific activity of the radionuclide distributed amongst all species is

$$s = \frac{\lambda N}{N_E} = \lambda \frac{\sum N^*_{Ei}}{\sum N_{Ei}}$$

It should be noted here that specific activities are not additive.

II. NON-GENUINE RADIATION EMITTED BY RADIOACTIVE MATTER

The emission of radiation from a radionuclide embedded in matter should be compared carefully with that from the free radionuclide. Differences may be significant in relation to radiochemical purity, interferences in spectroscopy, and radiation safety. The inventory of these radiations points to several origins, which are summarized in Figure 4.1.

A. Nuclear Reactions

Alpha particles emitted by radionuclides and neutrons produced in (α,n) reactions, as well as recoil protons released by elastic collisions of fast neutrons with hydrogen atoms during their slowing down, may in turn act as projectiles for nuclear reactions. In a similar manner, γ photons may participate in photonuclear reactions. The result is the appearance of new radionuclides in the host of the primary radionuclides.

At the low specific activities commonly encountered, the flux of these various secondary projectiles is rather small, and even for reactions with the highest cross sections the yield of induced new radionuclides or radioelements is negligible, as is also the emission of new radiations.

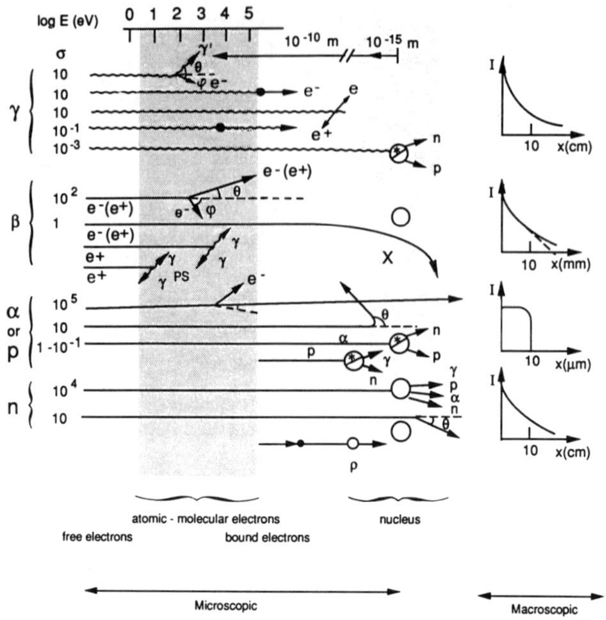

FIGURE 4.1. Interaction of radiation emitted by radionuclides with electrons and nuclei leading to the emission of non-genuine secondary radiation. The upper scale represents the bonding energy of electrons involved in the interaction. The approximate cross sections σ (in barns per atom) indicate the relative importance of the interactions. Stars indicate excited nuclei. Successive types of interaction from top to bottom are γ rays: Compton scattering, photoelectric effect, pair creation, elastic scattering, photonuclear reaction; β particles: ionization, scattering, annihilation, formation of Ps; α particles and protons: ionization, backscattering, nuclear reactions; neutrons: nuclear reactions, scattering. At the left, interactions are indicated at the microscopic (atomic) scale; the response to the interaction at the macroscopic level is represented at right hand as absorption curves in water, i.e., the diminution in intensity I of the incoming beam with the absorber thickness x.

However, the situation may be quite different in radioactive matter containing macroscopic amounts of radioelements which emit α particles with energies up to 8 MeV or more. Many reactions induced by α particles are exoenergetic. They are favored for the lighter elements owing to the lower coulombic barrier. The most frequent are (α,n) reactions, e.g., $^{12}C(\alpha,n)^{15}O$, $^{16}O(\alpha,n)^{19}Ne$, $^{23}Na(\alpha,n)^{26}Al$, $^{27}Al(\alpha,n)^{30}P$, because neutrons are readily evaporated from the compound nucleus. The released neutrons, whether fast or thermalized, are in turn captured by many nuclei and produce new radionuclides.

The type of reactions observed depends on the neutron energy E_n and on the Z of the atoms present in the sample. Radiative capture reactions (n,γ) occur with all elements and are the most probable for $E_n < 0.5$ MeV. With increasing values of E_n, the threshold of (n, particle) reactions is attained.

Well-known reactions observed in radioactive matter include $^{23}Na(n,\gamma)^{24}Na$ and $^{27}Al(n,\alpha)\ ^{24}Na$. Neutron-induced fission occurs in uraniferous hosts such as ores and minerals.

With increasing Z, the reactions (n,n′) become more probable because of the decrease in energy of the first excited level of the target nucleus.

The (α,n) reactions are deliberately exploited in isotopic neutron sources consisting of a mixture of an α emitter (Ra, Am) and light elements such as Be or F. Emission rates exceeding 10^7 neutrons s^{-1} per Ci are readily obtained.

(α,p) reactions occur with α-energies above 5 MeV. They are more probable with light nuclei, e.g., $^{26}Mg(\alpha,p)^{29}Al$ or $^{30}Si(\alpha,p)^{33}P$.

Photonuclear reactions such as (γ,n) are observed in radioactive matter. The reactions are endoenergetic, with a threshold close to the Q value. The latter is -2.22 MeV for deuterium and -1.67 MeV for Be, which makes the $^9Be(\gamma,n)$ reaction a convenient means of obtaining monoenergetic neutrons. For this purpose, Be is mixed with ^{124}Sb (T = 60.20 d, E_γ = 1.69 MeV). For other elements the threshold is above 6 MeV. The possible release of neutrons from radioactive matter containing energetic γ-emitters should never be overlooked, for reasons of radiation safety requirements.

Energetic recoil protons are produced in the slowing down of fast neutrons by elastic collisions in hydrogenous materials. Nuclear reactions such as $^{13}C(p,n)^{13}N$ lead to additional radionuclides in aqueous solutions or in hydrocarbon media.

B. Emission of Electromagnetic Radiation

During the slowing down of charged particles, kinetic energy is converted into electromagnetic radiation known as *Bremsstrahlung*, literally, *braking radiation*. The emission depends on $(z\ Z/m)^2$ in which z and m refer, respectively, to the charge and mass of the particle and Z is determined by the medium. Owing to their light mass, electrons, i.e., β^- and β^+ particles, are mainly the source of the emission, which is more intense in heavy materials. The radiation produced exhibits a continuous spectrum extending from the upper limit of the β spectrum, which may correspond to several MeV, down to about 150 keV. A good example is provided by ^{90}Sr (T = 28.5 years, $E_{\beta-}$ = 0.546 MeV) in radioactive equilibrium with ^{90}Y (T = 64.1 h, $E_{\beta-}$ = 2.27 MeV). A material containing this pair of nuclides behaves like a source of electromagnetic radiation with an energy up to 2.27 MeV.

C. Emission Involving Positrons

In the majority of cases, the annihilation of a positron and an electron releases *two* 0.511-MeV γ photons. This process occurs mostly after thermalization of the positron, which requires about 1 ps in the condensed state. Quite often, before annihilation, the positron forms with an electron an ephemeral bound state, the *positronium* atom, Ps.

Conversely, γ photons with an energy above two electron rest masses, (1.022 MeV) interact with the nuclear field and disappear in a materialization process which creates an electron-positron pair. The energy in excess of the 1.022-MeV threshold is shared by the two particles as kinetic energy:

$$E_{e^+} = E_{e^-} = \frac{E_\gamma - 1.022}{2} \text{ MeV}$$

The electron and positron behave as in the preceding case. With the γ rays encountered in radiochemistry the probability of pair production depends on Z^2 and the process is more frequently observed in radioactive matter containing heavy elements.

III. INTERACTIONS OF RADIATION IN RADIOACTIVE MATTER

All radiation emitted primarily or as a result of secondary processes and having energies above 50 eV is described as *ionizing* radiation. Charged species are directly ionizing, whereas electromagnetic radiations and neutrons act as such by virtue of a transducer effect which converts their energy into kinetic energy of electrons or charged particles.

These radiations transfer their energy to matter mainly by coulombic interaction, stripping electrons from the atomic constituents, and thereby forming ions. If the energy transfer to the electrons lies below the ionization potential, the atoms and molecules in the absorbing medium are excited. Ionization and excitation always occur concomitantly.

The direction of emission of electromagnetic radiations, as well as that of charged particles and neutrons, is modified by interaction with the constituents of the medium. This *scattering* may be *elastic* and proceed without energy loss, or can be *inelastic*. In the latter instance, part of the initial energy is lost and is generally absorbed in the medium.

The dominant diffusion interactions occurring in the energy range of radiation emitted by radionuclides include the inelastic scattering of γ photons and charged particles by electrons and the elastic scattering of neutrons by nuclei. The most important absorption processes are those of photons in atoms and neutrons in nuclei.

Elastic and inelastic scattering of charged particles by interaction with nuclei are less probable than absorption processes by several orders of magnitude.

The mechanism, kinematics, and cross sections of the various interactions of radiation with matter are described in numerous textbooks. Here, only the relevant facts of importance in experiments with radioactive matter are summarized.

A. Interaction of Electromagnetic Radiation

The main process of photon absorption is the *photoelectric* effect. In contrast with internal conversion (IC), which involves the atom containing the radionuclide, the photons are absorbed in the orbital electrons of any atom in the medium, mainly in the K (80%) and L (20%) shells. The photon disappears in the interaction and is replaced by a photoelectron with an energy

$$E_e = E_\gamma - E_b$$

in which E_b is the binding energy of the electron. This energy balance, which neglects the small recoil energy of the atom, expresses also the threshold condition for the process, $E_\gamma > E_b$. The photoelectron may be a further source of Bremsstrahlung.

The photoelectric effect is the dominant interaction for heavy elements and γ rays with $E_\gamma < 50$ keV. In the K shell, the probability of the photoelectric effect depends on Z^5 and on $E_\gamma^{-3.5}$. The ionization of the K or L shell is followed by secondary emissions similar to those which succeed internal conversion or electron capture.

The main scattering process is the *Compton* effect, which is the inelastic interaction of electromagnetic radiation with electrons. The energy lost by the photon and acquired as kinetic energy by an electron depends on the scattering angle θ. The energy E'_γ of the scattered photon is given by the expression

$$E'_\gamma = \frac{E\gamma}{1 + \alpha(1 - \cos\theta)}\ \text{MeV} \qquad \alpha = \frac{E_\gamma}{0.511}$$

Monoenergetic γ photons are replaced by a γ spectrum extending between the limits fixed by the values $\theta = 0$ for which $E'_\gamma = E_\gamma$ (equivalent to elastic or Rayleigh scattering) and $\theta = \pi$. The latter corresponds to *backscattering* for which E'_γ is the smallest energy that a photon can have after a Compton interaction,

$$E'_\gamma = \frac{E_\gamma}{1 + 2\alpha} \approx \frac{E_\gamma}{1 + 4E_\gamma}$$

As the initial photon energy E_γ increases beyond 1 MeV, the energy of the backscattered photons tends toward 0.250 MeV, and that of the photons scattered at $\theta = \pi/2$ is 0.511 MeV, i.e., the same as for γ rays appearing in the annihilation process. The energy of the Compton electrons is complementary to that of the scattered photons: it is zero when $\theta = \pi$ and its maximum value corresponds to the backscattering of the photons. It amounts

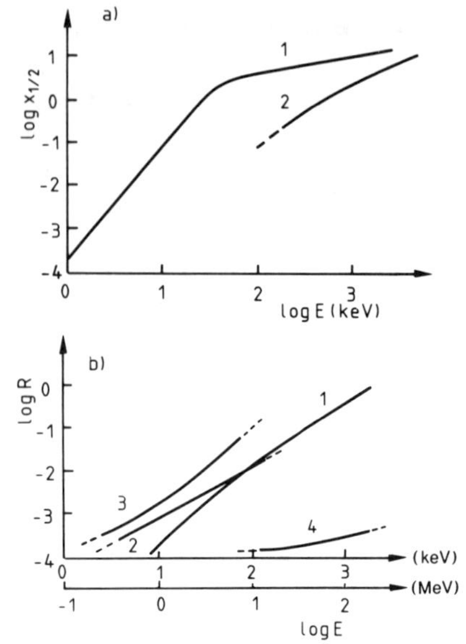

FIGURE 4.2. Absorption of radiation in water. (a) Half-thickness values $x_{1/2}$ for water (cm or, equivalently, g cm^{-2}) in the energy range of primary radiations emitted by radionuclides. The mass absorption coefficient μ_m ($cm^2\ g^{-1}$) is equal to 0.693 $x_{1/2}$; (1) γ rays up to several MeV; (2) neutrons up to several MeV. (b) Ranges in water; (1) electrons (keV scale); (2) α particles; (3) protons; (4) fission fragments.

to 0.8 MeV for 1-MeV photons. Similarly to photoelectrons, Compton electrons constitute an additional source of Bremsstrahlung.

Compton scattering occurs with loosely bound, nearly *free* electrons. The cross section is a complicated function of E_γ and proportional to the mass density of the electrons. The latter itself is proportional to Z/A, which is nearly constant for light and medium-mass elements. The scattering has about the same probability for water, dilute aqueous solutions, and organic compounds.

Compton scattering is the dominant mode of interaction of 50-keV to 1.5-MeV photons with light atoms and intervenes to about the same extent whatever the composition of the medium. At higher energies, pair production becomes important.

Conjunction of the photoelectric effect, Compton scattering, and pair creation results in an exponential attenuation law of photon intensity (Figure 4.2). A thickness x of any medium of mass density ρ reduces the initial intensity I_0 (in number of photons $cm^{-2}\ s^{-1}$) to the value I according to

$$I = I_0\, e^{-\mu x} = I_0\, e^{-\mu_m \rho x}$$

where μ (cm^{-1}) is the *linear attenuation* coefficient and $\mu_m = \mu/\rho$ ($cm^2\ g^{-1}$) is the *mass* attenuation coefficient. Attenuation results from scattering and absorption. Only the fraction of photon energy that is *absorbed* in radioactive matter is of concern. It includes the energy of both photoelectrons and Compton electrons.

B. Interaction of Charged Particles

Charged particles lose energy very rapidly by inelastic coulombic interaction between their own electrostatic field and that of electrons. In each step, a heavy particle (with energy E, mass m, charge ze, and velocity v) after interaction with an electron assumed to be at rest is deviated from its initial trajectory by an angle θ, while the electron is ejected in a direction ϕ with respect to the trajectory. The angular relation is

$$\tan\theta = \frac{m_0 \sin 2\phi}{m - m_0 \cos 2\phi}$$

where m_0 is the electron rest mass. The energy q transferred to the electron is

$$q = E\,\frac{4\,m\,m_0}{(m + m_0)^2}\cos^2\phi$$

with a probability inversely proportional to q^2. The further fate of the electron is highly dependent on the value of q.

The most common charged particle encountered in radiochemistry is the α particle. The maximum energy which it transfers to the electron in a head-on collision with $\theta = \pi$, $\phi = 0$ is given by

$$q_{max} = E\,\frac{4\,m_0}{m}$$

It amounts to 2.5 keV for a 5-MeV particle. The kinetic energy of the electron is equal to q_{max} minus the binding energy and may be relatively small for interaction of the particle with K- or L-shell electrons in heavy atoms.

The most probable values of q lie close to or below the ionization potentials of the orbital electrons. The interactions lead either to ions with simultaneous release of electrons (that in turn act as ionizing species), or to electronically excited atomic and molecular states; in both cases the angular deviation remains quite small.

The maximum deviation of the particle occurs for $\phi = \pi/4$ and is equal to m_0/m rad, corresponding to the extremely small value of 0.1 mrad for an α particle. The interactions proceed until the particles eventually come to rest. Because the number of interactions is nearly the same for all particles with a given energy, the latter have the same well-defined and quasi-linear short range.

A practical and useful kind of information is given by the range of a particle (subscript 1) in comparison with the known range of another particle (subscript 2) of the same velocity traversing the same medium. The range ratio is

$$\frac{R_1}{R_2} = \frac{m_1}{m_2}\frac{Z_2^2}{Z_1^2}$$

A definition for the range of a particle is the *linear energy transfer* (LET), which is the energy loss per unit path length and can be expressed as $(-\,dE/dx)$. An analogous parameter for describing the rate of energy loss is the specific ionization, i.e., the number of ion pairs produced per unit path length. Both parameters are energy-dependent and increase toward the end of the trajectory. The mean value of the LET of charged particles is very high. For 5-MeV α particles in water, it is 40 keV μm^{-1}. It is even higher for recoil nuclei and fission fragments. Alternatively, the interaction of charged particles can be expressed by the *mass stopping power* of the medium, $S = (dE/dx)\,\rho^{-1}$ in units of MeV $cm^2\ g^{-1}$.

In radiochemistry, the range R of a particle is an important property. For a particle of initial energy E_0 it is given by

$$R = \int_0^{E_0} \frac{dE}{dE/dx}$$

The range is expressed in centimeters, or more often in units of superficial mass ($g\ cm^{-2}$), i.e., by the product $R\rho$ of the range and the mass density. The values of R as a function of energy, for many materials, can be graphically represented or calculated by approximate relations. For α particles,

$$R(g\ cm^{-2}) = 4.134 \times 10^{-4}\ E_\alpha^{3/2}\ (\text{MeV})$$

This expression yields $R = 5 \times 10^{-3}\ g\ cm^{-2}$, equivalent to 3.9 cm of air for the 5.3-MeV α particles of ^{210}Po. On the average, 2700 ion pairs are produced per micron of air. The mean range of these particles in water is 38.9 μm and the average LET in water 136 keV μm^{-1}. For any medium, the relative stopping power S is defined by the ratio

$$S = \frac{R(\text{air})}{R(\text{medium})}$$

which varies only slightly with energy. For 5.3-MeV α particles in water it is about 1000. The range of α particles in solids is very short, e.g., 0.04 mm in Al (see Figure 4.2b).

C. Interaction of Electrons

Both electrons and positrons lose energy mainly by coulombic interaction with the electrons of the medium. Primary electrons release secondary ones, which in turn may displace tertiaries, and so on, as long as the transferred energy q is sufficient. The trajectories of electrons are much longer than those of heavy charged particles of the same energy, but they are ill-defined and erratic, owing to a significant deflection at each step. It follows from the angular relation

$$\tan \theta = \frac{\sin 2\phi}{1 - \cos 2\phi}$$

that $\theta + \phi = \pi/2$ and the maximum deviation occurs for $\phi = \pi/4$, $\theta = \pi/4$. The energy transfer $q = E \cos^2\phi$ is important and may be total for a head-on collision ($\phi = 0$). The extent of energy dispersion is also high.

In radioactive matter, β particles with a continuous energy spectrum are more common than monokinetic electrons, and the concept of range has little meaning. However, the absorption of β particles follows a simple empirical law, at least for the first stage of deceleration and is of the form

$$N = N_0 e^{-\mu x}$$

Here, N_0 is the initial number of β rays and N the remaining rays after passing through a thickness x.

The coefficient μ is similar to an absorption coefficient. On the basis of this expression, a range for β rays can be defined in terms of the half-thickness $x_{1/2}$ (in cm or in g cm^{-2}) for which $N = N_0/2$ (see Figure 4.2a). The maximum range, R_{max}, is then obtained by extrapolation to $7x_{1/2}$. Graphs are available that give R_{max} as a function of E_{max} for various materials. In air, and with β rays such as $0.010 < E_{max}$ (MeV) < 2.5 one uses

$$R_{max}\ (\mathrm{g\ cm^{-2}}) = 0.412\ E^n_{max}\ (\mathrm{MeV}) \quad n = 1.265 - 0.0954 \ln E_{max}$$

for which $E_{max} = 1$ MeV, yields $R_{max} = 300$ cm. The stopping power in water is $S = 730$, and the maximum range of the β rays in this medium is 4.1 mm.

D. Absorption of Neutrons

The ionizing action of neutrons is indirect, arising from the energy released by the charged particles produced in (n,p) and (n,α) nuclear reactions. If these reactions and inelastic scattering are excluded, the neutrons are slowed down by elastic collisions with the nuclei in the medium. This type of interaction is the most common for neutrons in radioactive matter. The energy E of the neutron of initial energy E_n, after one collision step with a nucleus of mass number A, is

$$E = \frac{E_n}{(A + 1)^2} \left(\cos\theta + \sqrt{\cos^2\theta + A^2 - 1} \right)$$

and the recoil energy E_r of the struck atom is

$$E_r = E_n \frac{4\,A \cos^2\phi}{(A + 1)^2}$$

θ and ϕ are the scattering angles of the neutron and the recoil atom, respectively. In a head-on collision ($\theta = \pi$, $\phi = 0$) on a hydrogen atom, the energy transfer to the struck atom is total. The latter is expelled from its molecule and becomes a *recoil proton* that may also cause ionization of the medium.

The neutrons are eventually thermalized after a number N of collisions calculated with the expression

$$N = \frac{1}{\xi} \ln \frac{E_n}{0.025}$$

in which the neutron energy is expressed in eV and ξ is the logarithmic energy loss parameter defined as

$$\xi = 1 + \frac{(A - 1)^2}{2\,A} \ln \frac{A - 1}{A + 1}$$

With hydrogen, about 20 steps are sufficient to decrease the energy of a 1-MeV neutron down to the thermal energy of 0.025 eV.

The diminution in intensity of neutrons in radioactive matter depends on the cross sections of all possible interactions which are a function of E_n. It varies markedly with the isotopic composition of the medium; the intensity decrease is approximately an exponential function of the thickness (see Figure 4.2).

Half-thickness values and thermalization ranges can be evaluated with the help of computed graphs. As for electrons, the range is ill defined. The mean square range of neutrons in water is estimated from the relation

$$\overline{R}^2\ (\mathrm{cm}^2) = 0.635 \ln \frac{E_n\ (\mathrm{eV})}{0.025}$$

Thermalized neutrons, if not captured, escape from the medium or decay into a proton and a β^- particle.

E. Spatial Distribution of Electrons

Eventually, any type of radiation absorbed in radioactive matter is converted into electrons. The spatial distribution of the latter depends on the energy q acquired by the electrons in the ionization process.

When $q < 100$ eV, the successive electron generations are located in a heterogeneous zone called *spur*. The size of the spur in an aqueous medium is 1 to 2 nm. For values of q between 100 and 500 eV, the events triggered by the primary electrons are situated in *blobs*, analogous to a drop, of about 12 nm in length. *Short tracks* refer to the trajectories of 0.5- to 5-keV electrons; they are characterized by a very high density of ionizations, and the preceding description holds again for the released electrons. Finally, electrons with energies above 5 keV are considered as a primary radiation (Figure 4.3). LET values for electrons in water are represented in Figure 4.4.

The energy spectrum of the electrons depends on the properties of the primary radiations. Interactions with small energy transfer have the highest probability; most electrons have energies from 50 to 200 eV and are preferentially found in spurs.

Primary radiations with LET > 100 keV μm^{-1} produce many close-lying spurs which may overlap, forming a cylindrical zone of high ion density.

The ionization and excitation interactions are achieved very rapidly, i.e., in 10^{-15} to 10^{-18} s. There then follows a progressive diffusion of the electrons, ions, and excited species into the irradiated medium; simultaneously, various chemical reactions occur and build up free radicals and other reactive entities. After 10^{-6} s, all chemical species are in equilibrium.

F. Quantification of the Absorbed Energy

The amount of energy transferred by the various radiations to radioactive matter is called the radiation *dose*. The SI unit of dose is the *gray* (symbol Gy), equivalent to 1 J/kg.

The energy E_{rel} released by a radionuclide of activity A during a time t in radioactive matter is

$$E_{rel} = Q_d \int_0^t A \, dt = \frac{Q_d A_0}{\lambda} (1 - e^{-\lambda t})$$

where A_0 is the activity in Bq at time $t = 0$, and Q_d is the decay energy per nuclide.

However, only that part of the decay energy that is transferred to ionizing particles contributes to the energy E_{abs} *absorbed* in matter. The energy carried by neutrinos and antineutrinos escapes entirely, together with a part from the γ-photon energy. The absorbed energy depends on the mode of decay.

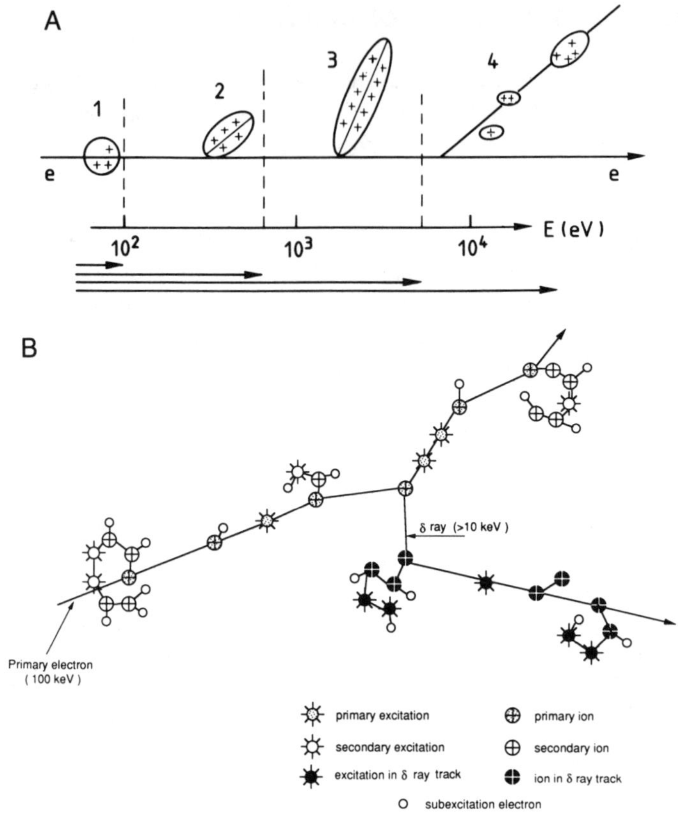

FIGURE 4.3. Topology of ionization and excitation events. (A) Schematic representation of processes along the track of an electron as a function of the initial energy. (1) Spur, about 2-nm size, containing 1 to 3 ion pairs and 3 to 10 excited molecules (not shown); (2) blob; (3) short track; (4) delta ray. Successive ionizations are separated by 1 nm for E < 100 eV and by 100 to 500 nm for fast electrons. The distance between spurs is initially 500 nm and decreases to less than 100 nm toward the end of the trajectory. (B) Ionization and excitation processes along the trajectory of a 1-MeV electron. The energy of subexcitation electrons is below that of the first excitation potential of the medium. (C) Events along the trajectory of a high LET particle such as an α ray or fission fragment. The central cylindric zone A with a high density of ions and excited species results from the coalescence of spurs. Short tracks are localized in the concentric zone B. The trajectory of δ rays extends into zone C. Dots represent ions; excitations are not shown.

In a pure α decay, Q_α is shared between the α particle and the recoil atom. The sample is generally large compared to the range of these ionizing species, which are entirely absorbed and $E_{abs} = E_{rel}$.

In a pure β^- decay, the energy spectrum of the particles is continuous and the released energy, which is equal to that absorbed, is determined by the average E_β value.

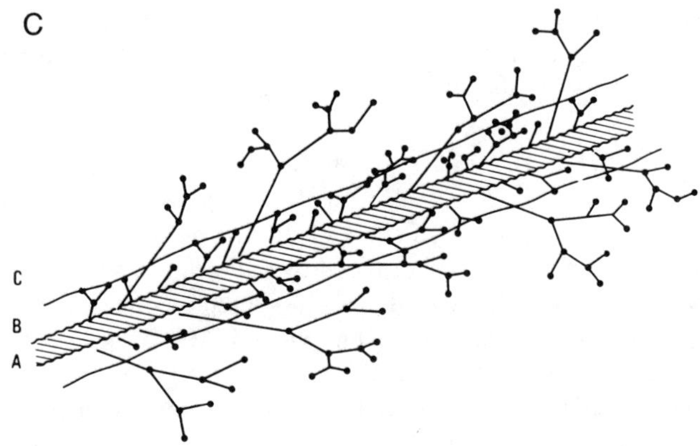

FIGURE 4.3 (continued).

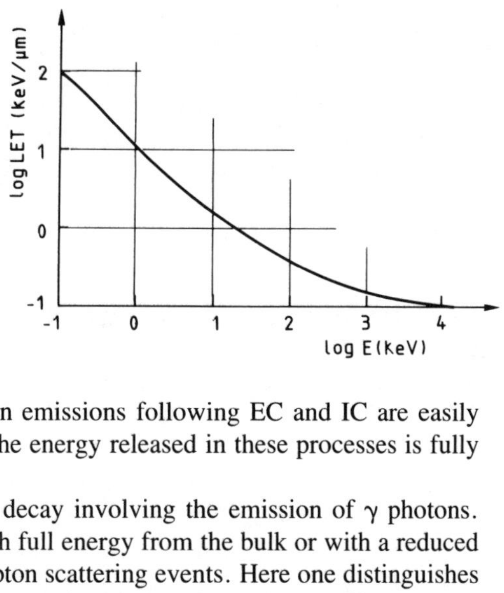

FIGURE 4.4. Linear energy transfer (LET) of electrons in water as a function of energy.

The X-ray and Auger electron emissions following EC and IC are easily stopped in a bulk material and the energy released in these processes is fully absorbed.

The situation is different for decay involving the emission of γ photons. The latter may either escape with full energy from the bulk or with a reduced energy following multiple Compton scattering events. Here one distinguishes the total attenuation coefficient and the energy absorption coefficient. For 1-MeV photons in water, the former is 0.070 $cm^2\ g^{-1}$ and the latter only 0.031 $cm^2\ g^{-1}$.

Spontaneous fission releases fragments which possess very high LET and are fully retained in the material. About 93% of the energy released in the fission reaction is thus absorbed in radioactive matter; the remainder is carried by energetic γ rays, neutrinos, and neutrons that may escape partially or totally from the bulk. Further energy is released in the medium from the radioactive fission fragments and fission products.

These considerations give only a very rough idea of the amount of energy released in radioactive matter. Actual quantities may vary within wide limits depending on the size, shape, density, and composition of the sample.

G. Heat from Radioactive Matter

The ultimate form of radiation energy is heat. Volcanic eruptions are large scale demonstrations of the heat released by radioelements and radionuclides inside the earth. Uranium, with its present isotopic composition, generates 3.05 J/year and per gram and potassium (via ^{40}K) 100 μJ/year and per gram. Nuclear power plants use the heat which originates from the kinetic energy of the fission fragments and the radioactivity of the fission products. The energy available per fission event is about 200 MeV. The consumption of 1 g of a fissile material (^{235}U, ^{239}Pu), i.e., of about 10^{21} atoms, produces 8.2×10^{10} J. Containers of liquid high-level radioactive wastes from the nuclear fuel cycle must be artificially cooled in order to evacuate the heat from the decay of fission products.

The release of heat from the currently handled radioactive sources is usually very small, e.g., 54.6 J/h for 10 Gbq ($\approx$ 1 Ci) of ^{60}Co. A dose of 10 kGy would be required to increase the temperature of 1 g of water by 2°C.

Radionuclides that emit α particles produce more heat. The equilibrium temperature of a test tube containing 0.5 TBq of ^{210}Po is 75°C. The specific power of ^{252}Cf, which decays mostly by α emission (96.91%) and partly by spontaneous fission (3.09%) is 38 W g^{-1}.

IV. CHEMICAL CHANGES IN RADIOACTIVE MATTER

Radioactive matter is *chemically* unstable owing to modifications produced by the radioactive process. These changes include radiolytic and transmutation effects. In addition, isotope exchange reactions, which are not specific to radionuclides but remain unnoticed for stable atoms, may change the isotopic composition of the system with time. In consequence, the radiochemical purity, which is the proportion of radionuclides present in specified chemical forms, is altered.

A. Radiolysis and Transmutation Effects

Radioactive decay implies the simultaneous emission of radiation and the birth of a nucleogenetic atom, which, with the exception of IT, differs chemically from its parent. The chemical and physical action of the radiation, which is referred to as *radiolysis*, or more precisely, as *autoradiolysis*, has two characteristics: it is operative only in so far as the radiation energy is absorbed in the radioactive matter, and it occurs throughout the bulk of the material. On the other hand, the transmutation effects are essentially localized around the decay site. The decay of a single ^{14}C atom creates one ^{14}N recoil atom, but the emitted β^- particle may destroy or alter up to 5000 organic molecules.

1. Autoradiolysis

Radiation chemistry is more often concerned with the description of effects of external radiation on matter, and mostly with respect to γ rays. Conversely, autoradiolysis is generally due to the strongly absorbed α and β particles emitted inside radioactive matter, but it may also be governed by the nucleogenic recoil atoms.

The extent of chemical transformations due to radiation is expressed by the *radiochemical yield* or *G value*. It is defined as the number of molecules (or other species) decomposed or formed per 100 eV of absorbed energy. A high G value means that the chemical species is very sensitive to the action of radiation. A reference value is the radiochemical yield of Fe^{3+} ions formed in an aqueous, acidic, and aerated 10^{-3} M solution of Fe^{2+} ions. The radiochemical yield $G(Fe^{3+})$ is 15.6 for low LET 1-MeV photons and 5.1 for the highly ionizing 5.3-MeV α particles of ^{210}Po.

If the G value for the transformation of a compound and the amount of energy absorbed is known, the effect of autoradiolysis can in principle be evaluated. In practice, the result is only very approximate for several reasons: radioactive matter generally comprises a mixture of several compounds or a solution with various solutes, the radiations are of mixed nature and energies, the energy deposited can be grossly estimated, and chemical damage may be amplified by secondary reactions. Obviously, the damage increases with the total activity and with the specific activity of radioactive matter.

From the wealth of data accumulated in radiation chemistry, several consequences of self-irradiation for various types of radioactive matter can be anticipated.

a. *Gases*

Radiochemists are mainly concerned with the effect on air during the storage of radioelements. Because of their low density, the primary species are not confined to the particle track, but diffuse rapidly throughout the gas. Accordingly, LET effects are reduced.

The main chemical compounds that appear in irradiated air are ozone, easily detected by its odor even at very low concentrations, and nitrous oxides.

The radiation stability of many compounds in aqueous solutions depends to a large extent on the presence of dissolved oxygen, owing to the formation of peroxides. In such cases, solutions of radioactive matter must be stored under air-free conditions. In any event, the free space above these solutions is rapidly occupied by H_2, O_2, CO_2, and other gaseous radiolytic degradation products.

Radioactive matter containing the radionuclides ^{226}Ra, ^{224}Ra, and ^{223}Ra continuously generate short-lived isotopes of the gaseous element radon. The *emanation power* of the medium is the fraction of the formed radon atoms that is released into the surroundings. It may reach 100% for solid organic compounds or in heated solutions of the precursors.

The longest-lived radon isotope, 3.8-d ^{222}Rn, has been extensively used for investigation of the chemical effects of α particles in gaseous systems. It decays successively to ephemeral isotopes of the solid elements Po, Bi, and Pb (see Table 5.3), which are adsorbed on aerosols and dust particles or deposited on the walls of the containers. When these species are released into the air, they constitute a radiological hazard because they can be inhaled and retained by the broncho-respiratory tract (see Chapter 10). Eventually, the long-lived daughters 22.3 years ^{210}Pb and 140 d ^{210}Po grow in and produce a durable radioactive contamination of the air in addition to the effects of radiolysis.

b. *Aqueous Solutions*

In radiochemistry, the most common forms of radioactive matter are probably those in aqueous solutions. The radiation chemistry of water is well understood. In dilute solutions below 10^{-3} *M*, and *a fortiori* at tracer concentrations, the effect of radiation is essentially indirect. The energy of the radiations is absorbed in the solvent. Along the path of the ionizing particles, water molecules are either ionized,

$$H_2O \rightarrow H_2O^+ + e^-$$

or excited,

$$H_2O \rightarrow H_2O^*$$

The ionized water molecule rapidly (in less than 10^{-15} s) undergoes the ion-molecule reaction

$$H_2O^+ + H_2O \rightarrow H_3O^+ + {}^\circ OH$$

with simultaneous formation of the free radical °OH. Here H_3O^+ stands for the solvated proton and the dot on °OH and on other symbols represents the single electron of a free radical. The electron released in the ionization of a water molecule generally has enough kinetic energy to escape from the attraction sphere of the positive molecular ion and diffuses in the medium. It becomes thermalized in about 1 ps and eventually forms an aquated species, the *hydrated electron*:

$$e^-(\text{fast}) \rightarrow e^-(\text{thermal}) \rightarrow e^-_{aq}$$

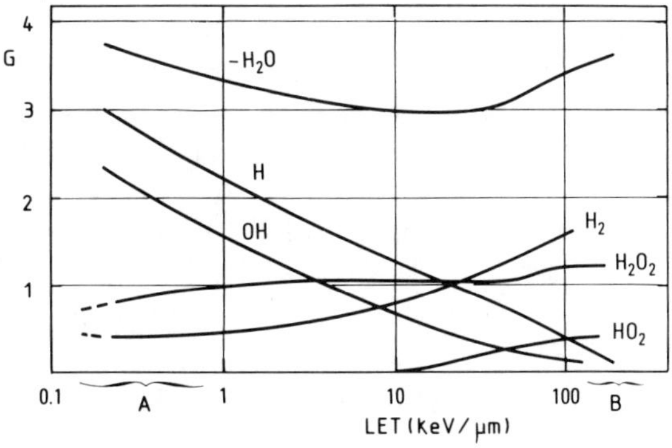

FIGURE 4.5. Radiochemical yields of the radicalar and molecular products formed in the radiolysis of aerated neutral water, as a function of the LET of the incoming radiation. (A) Electrons up to 3 MeV and 1 MeV γ rays; (B) 5-MeV α particles.

The excited water molecule decomposes into two free radicals, as was already suggested by André Debierne in the early period of radioactivity:

$$H_2O^* \rightarrow {}^\circ H + {}^\circ OH$$

The three species °OH, H° and e_{aq}^- are initially localized in the tracks. Their further fate is governed by competition between diffusion out of the tracks and very fast mutual reactions which lead to the formation of *molecular* products:

$${}^\circ H + {}^\circ H \rightarrow H_2$$

$${}^\circ OH + {}^\circ OH \rightarrow H_2O_2$$

$${}^\circ H + e_{aq}^- \rightarrow H_2 + OH^-$$

$$e_{aq}^- + e_{aq}^- \rightarrow H_2 + 2\ OH^-$$

including the back reaction

$${}^\circ H + {}^\circ OH \rightarrow H_2O$$

For particles with high LET, such as α particles and fission fragments, the density of primary species is very high and mutual reactions are favored. For fast electrons on the other hand, more free radicals diffuse into the bulk of the solutions.

The dependence of radiochemical yields of free radicals and molecular products in neutral water as a function of the LET is shown in Figure 4.5. For low LET radiation, yields amount to 2.7 for the aqueous electron, 0.6 for °H radicals, and 2.7 for °OH radicals; the radiochemical yields are 0.45 for H_2 and 0.75 for H_2O_2. The relative yields of the reducing species °H and hydrated electron are pH-dependent because of the reaction

$$e^-_{aq} + H_3O^+ \rightarrow {}^\circ H + H_2O$$

In the presence of air, °H is converted to the hydroperoxyde radical HO_2° and e^-_{aq} to O_2^-; the overall yield of reducing agents decreases.

At concentrations below approximately 10^{-4} *M*, and even more at tracer concentrations, the probability that a solute species will be encountered in the particle track is negligible. The radiolytic transformation of the solute results from reactions with the diffused species. °OH is a powerful oxidizing radical:

$$^\circ OH + e^- \rightarrow OH^- \quad E^\circ = 2.8\ V$$

and e^-_{aq} is a strong reducing agent:

$$e^-_{aq} + H^+ \rightarrow \frac{1}{2} H_2 \quad E^\circ = 2.77\ V$$

Radiochemical yields of the radiolytic species are known for a large variety of LET values.

The redox reactions of multivalent radionuclides in water result from the action of these various species and depend on the redox potentials and on pH. Even at very low radiation doses and at low molar concentrations, the oxidation state of radionuclides may be altered.

This effect should be important for α emitters such as actinides. For instance, an activity of 100 Bq/mL of a nuclide which decays by emission of a 5 MeV α particle, conveys to the solution 5×10^8 eV/s. If the radiochemical yield G is 10, the number of species transformed per second is 5×10^7. The 100 Bq activity per milliliter corresponds to 10^{14} ^{239}Pu atoms, i.e., a 10^{-7} *M* solution. It is estimated that after one year all the Pu atoms will have undergone a radiation-induced change in chemical state.

For a given number of moles of a radioactive species, autoradiolysis increases with the emission rate, i.e., with decreasing half-life. The effects are much less pronounced for the long-lived 24,110 years ^{239}Pu and 7370 years ^{243}Am than for 87.7 years ^{238}Pu and 432 years ^{241}Am.

Radiolytic decomposition products of an acid, such as nitrous acid and oxides of nitrogen in nitric acid media, and Cl_2 in hydrochloric acid solutions add to those of the solvent and all react with Pu. An illustrative example of the complexity of such reactions is provided by the progressive change in oxidation state for Pu(VI) in nitric acid solution. The first step is reduction to Pu(V) by the action of e^-_{aq} followed by reduction to Pu(IV) and Pu(III). When Pu(V) accumulates to a sufficient extent, the following disproportionation occurs:

$$3\ Pu(V) \rightleftarrows Pu(III) + 2\ Pu(VI)$$

In the presence of sufficient Pu(III) the reaction

$$\mathrm{Pu(VI) + Pu(III) \rightleftarrows Pu(V) + Pu(IV)}$$

and the reproportionation reaction

$$\mathrm{Pu(V) + Pu(III) \rightleftarrows 2\ Pu(IV)}$$

take place. This latter reaction increases the autoreduction rate, i.e., the decomposition of Pu(VI) is enhanced by autocatalysis. The steady-state concentration of all Pu species depends on the concentration of both nitric acid and Pu, which in turn determines the dose rate.

The autoreduction of Am(VI) has been ascribed to the action of radiolytic H_2O_2 and that of Am(V) to the radicals HO_2°. Am(V) is reoxidized to Am(VI) by the $^{\circ}OH$ radicals. The autoreduction rate depends on the nature and concentration of the acid.

Reagents and materials used in chromatographic columns may also undergo radiation-induced modifications; sulfonic acid groups of ion exchangers produce sulfate anions which may interfere in the separation process of radioelements.

c. *Labeled Molecules*

Self-radiolysis is important in organic compounds labeled with radionuclides. For instance, 40% of ^{14}C-cholesterol with a specific activity of 100 MBq/mmol is decomposed after one year. For autoirradiation of ^{14}C- and tritium-labeled compounds, typical values of the radiochemical decomposition yield G(− number of molecules) lie in the range from 1 to 20, but sometimes chain reactions are initiated with G values of several hundred or thousand. The extent of decomposition depends on the radiation sensitivity of the compound, on the specific activity, the nature and energy of emitted radiations, and on storage conditions.

d. *Solid Compounds*

Radiation effects of solid compounds were first observed in the coloration of glassware in contact with radioelements, and in pleochroic halos in mica irradiated by the inclusion of radioactive substances. Another consequence of radiation damage in radioactive minerals is metamictization, i.e., the conversion of a crystalline state into an amorphous one. A concentration of U or Th as low as 1% is sufficient to cause complete metamictization of some minerals in the course of geological time.

Self-radiation effects in solids are mainly due to the displacement of atoms by α particles, recoil nuclei, and neutrons. These effects are very important when one is dealing with macroscopic amounts of short-lived radionuclides. In a ^{210}Po compound, each atom is displaced several times per day under the combined effects of α particles and recoiling ^{206}Pb atoms. The chemical effects of self-irradiation depend on the type of compounds. $^{239}PuF_6$ is continuously decomposed by α autoradiolysis to F_2 and lower plutonium fluorides.

Autoradiolysis is expected to be more pronounced in hydrated materials because of the intervention of reactive species from radiolysis of water. On the other hand, the heat released in these materials may anneal part of the damage.

2. Transmutation Effects

The chemical effects of transmutation are much more relevant to radiochemistry than those of the emitted radiations. With the exception of isomeric transition, daughter atoms formed in radioactive decays have a chemical nature different from that of the parent. They must adapt to an environment that may be unusual or incompatible with their normal chemistry. Peculiar situations arise when Ra decays to Rn, Xe to I, C to N, tritium to He, etc. The nucleogenic atom will end up in one of its allowed oxidation states, whence it will necessarily form its own chemical bonds, or occupy a lattice site of an appropriate size. Secondary effects of the transmutation, such as recoil, shake-off, and electronic excitation play a prominent role in stabilization of the daughter.

A great wealth of experimental data has been accumulated since the discovery of the chemical effects of nuclear transformations in 1934. The fate of nucleogenic atoms depends on the nuclear parameters of the decay and on the chemical environment. No generally valid laws can be stated and each mode of decay should be considered separately.

a. *α Disintegration*

The daughter atom formed in α decay acquires a very high recoil energy of the order of 100 keV and all its chemical bonds are likely to be severed; at the same time, the α particle and the recoil nucleus dissipate their energy into the surrounding medium. Depending on the system, bonds may be re-formed; in a crystal lattice, the daughter may end up in an interstitial position or undergo a replacement collision with an atom at a regular site.

The formal reduction which accompanies α emission, such as

$$M_1^{n+} \rightarrow M_2^{(n-2)+} + {}_2^4He^{2+}$$

is in most cases not observed, even in the gas phase. Neutral or positively charged polonium atoms are formed in the disintegration of radon. Reduction also does not occur in the bulk solid phase.

In the filiation ^{253}Es (α, 20.4 d) ^{249}Bk ($β^-$, 220 d) ^{249}Cf (α, 351 years), the first and second descendants are trivalent when the parent Es is present as a trivalent halide. Starting with divalent Es, it is found that the oxidation state 2 and the crystal structure are preserved for the granddaughter, and presumably also for Bk. This system is unique in that after three years the chemical transformations of the actinides are observable by X-ray diffraction and optical spectroscopy rather than by radiochemical methods. In a similar way, the oxidation state is maintained in the $CmBr_3(\alpha)PuBr_3$ decay.

As a rule, the nucleogenic atom has a marked propensity for ending up in the oxidation state of the parent provided this state is shared by the daughter. Thus it is expected that UO_2 should transmute to ThO_2, despite the fact that U^{4+} formally changes to Th^{2+} as a consequence of α decay. The electronic adjustments are very fast, as revealed by Mössbauer emission spectroscopy. NpO_2 is stabilized within less than 1 ns in the decay of ^{241}Am in AmO_2.

On the other hand, maintenance of the oxidation state is not possible when the parent state is incompatible with the chemical properties of the daughter. A typical case is U(VI)(α)Th(IV). Here reduction is straightforward because it satisfies neatly the charge balance. Not surprisingly, the oxydation UO_2^{2+} forms Th^{4+}, the most stable oxidation state of thorium. Similar ''normal'' changes occur in Pa(V)(α)Ac(III) or Th(IV)(α)Ra(II) transitions.

In nature, transmutation effects lie at the origin of radioactive disequilibrium in the natural radioactive families. The first daughter of ^{238}U is ^{234}Th. With its 80-keV recoil energy, it may be expelled from the parent site and decay to ^{234}U at some distance away. As a result, the activity ratio of the two uranium isotopes may differ markedly from unity in many uraniferous materials or in natural waters.

b. *β^- Disintegration*

The molecular consequences of a pure β^- decay are relatively mild. The daughter atom simply replaces the parent in its original site and molecular rearrangements ensure chemical stability. For example, $^{14}C\,H_4$ transforms to $^{14}NH_4^+$, $(C_6H_5)_3{}^{210}Bi$ to $(C_6H_5)_3{}^{210}Po^+$ and CH_3T to the cation CH_3^+. In this latter case, the C–^{3}He bond is broken for obvious chemical reasons:

$$CH_3T \rightarrow [CH_3\text{–}^3He]^+ \rightarrow CH_3^+ + {}^3He$$

The role of the transmutation of ^{14}C to ^{14}N in the biochemical evolution of life has been suggested. Laboratory experiments have demonstrated the formation of methylamine from ^{14}C-labeled ethane and of aniline following decay of ^{14}C bound in the methyl group of toluene. Succinic acid doubly labeled with ^{14}C in the carboxyl groups rearranges to an amino acid

$$^{14}COOH\text{–}CH_2\text{–}CH_2\text{–}^{14}COOH \rightarrow {}^{14}COOH\text{–}CH_2\text{–}CH_2\text{–}^{14}NH_2$$

following decay of one of the ^{14}C atoms. These results may argue in favor of the radiosyntheses of fundamental bricks of life, such as amino acids, resulting from decay of radiocarbon formed by the action of cosmic rays.

Formally, β^- decay is an oxidative process since the parent atom loses an electron:

$$M_1^{n+} \rightarrow M_2^{(n+1)+} + \beta^- + \bar{\nu}$$

The Br(VII) oxidation state was first observed in the decay

$$^{83}SeO_4^{2-} \rightarrow {}^{83}BrO_4^- + \beta^- + \bar{\nu}$$

and was followed by a successful *chemical* synthesis of the perbromate anion. After numerous unsuccessful attempts to oxidize Cm (III) or Cm(IV) to Cm(VI), hexavalent Cm was obtained in the transmutation

$$^{242}AmO_2^+ \rightarrow {}^{242}CmO_2^{2+} + \beta^- + \bar{\nu}$$

In a similar way, ^{249}Bk(IV) forms the unusual ^{249}Cf(V) state.

However, β^- decay does not always lead to oxidation. The path from Th(IV) to Pa(V) and from Pa(V) to U(VI) is ''normal'', but odd cases such as Bk(III) → Cf(III) or Th(IV) → U(IV) are also encountered. Mössbauer data point to U(IV)(β^-)Np(IV) and radiochemical analysis to U(VI)(β^-)Np(IV) rather than to the formally expected oxidized form Np(VII).

Redox reactions arising from transmutation in a solid matrix can be visualized with the help of the following model. A closed system containing the compound M_1X_3 is considered. The nuclide M_1 decays by β^- emission to the daughter M_2 and it is supposed that M_1 and M_2 display reasonably stable oxidation states of 3 and 4. The number of (M_1 + M_2) atoms and the number of X atoms remain constant during decay. Assuming the system is not totally isolated, i.e., electrons can flow in or out, it will carry no net charge. Hence, if M_1^{3+} is converted to M_2^{4+}, an additional monocharged anionic species must appear in the system. The only way this can happen is by trapping an electron in the lattice. Most solids will tolerate the creation of such defects to a limited extent. Thus, if the amount of M_2 is small, it may be found as M_2^{4+}. But there is usually a limit to the introduction of such defects and eventually only M_2^{3+} will be found. If substantially all M_1 nuclides decay, only M_2X_3 will be formed.

c. *Electron Capture*

The daughter atom resulting from EC carries a high positive charge. A typical example is provided by ^{125}I, which decays by EC to the 1.6 ns/0.038 MeV excited level of ^{125}Te; the latter, in turn, decays to the ground state via a highly converted transition (Figure 4.6). Hence, two Auger ionization cascades occur successively in the daughter atom.

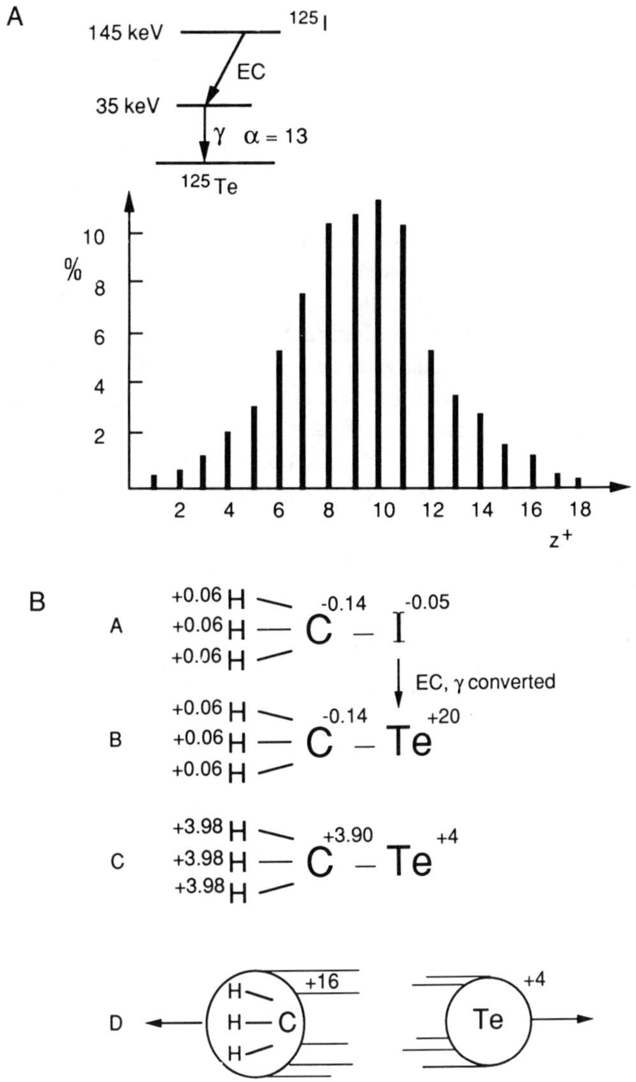

FIGURE 4.6. Molecular fragmentation following EC. A. Charge states of $^{125}Te^{z+}$ ions formed in EC decay of ^{125}I in methyl iodide. B. Molecular explosion following Auger vacancy cascades. Charge distributions are calculated according to the model of electronegativity equalization. (A) Charge distribution in the molecule of methyl iodide; (B) vacancy cascade triggered by the EC decay of ^{125}I (10^{-16}-10^{-15} s); (C) charge redistribution (10^{-15} s); (D) molecular rupture resulting from Coulombic repulsion.

The fate of a molecule in which the iodine atom ^{125}I is bound covalently illustrates the dramatic consequences of the Auger effect. Multiple ionization is achieved in about 10^{-15} s, which is a shorter time than that of the molecular vibration period (10^{-14} to 10^{-12} s). The multiple positive charges initially remain located on the Te atom without affecting the rest of the molecule.

In the gas phase, the molecules may be considered as isolated. The Te^{n+} ion attracts electrons from various parts of the molecule. The positive charges are driven to the extremities of the molecule, which is likely to be destroyed by coulombic repulsion. This model of *Auger explosion* is based on mass spectrometric studies, which show that whenever an Auger effect occurs, the parent molecule is fragmented (see Figure 4.6).

In condensed systems, the charge is transferred to neighboring molecules. The Te^{n+} ion does not survive long enough for an intramolecular redistribution. However, the neutralization of such a high charge is very damaging, even in the solid state, since it may deposit up to 100 eV of excitation energy in the molecule, and this is much higher than the average bond dissociation energy. The resulting chemical effects depend on the time taken in the neutralization processes and how effectively the excited species produced in neutralization can dissipate their energy.

Auger electrons have short range and high LET and dissipate their energy in the vicinity of the decay site. This produces an intense localized autoradiolysis, which adds to the other fragmentation process. Furthermore, radioactive matter may be highly damaged in EC decays and from IC.

B. Isotope Exchange Reactions

Exchange reactions are very important in radiochemistry; first, because of their importance in chemical reactivity and kinetic studies, and second, because redistribution of radionuclides between various species alters the initial chemical form. Isotope exchange reactions can occur in solution, in the solid phase and between two phases. They have been extensively investigated in a wide variety of systems. When dealing with radioactive matter, the possibility of exchange should always be kept in mind.

The thermodynamic justification of an isotope exchange reaction is increase in entropy at equilibrium, when all isotopes are uniformly distributed among the reagents.

The kinetics of a simple exchange reaction of isotopic atoms X and X* between two chemical species AX and BX in a homogeneous system

$$AX + BX^* \rightleftarrows AX^* + BX$$

is expressed by the *Mc Kay* law

$$\ln(1 - F) = -Rt\left(\frac{1}{C_A} - \frac{1}{C_B}\right)$$

where F is the fraction of atoms which have exchanged at time t, R is the rate of exchange in units of concentration per time and C_A and C_B are the total concentrations of AX and BX (see Chapter 7).

Exchange mechanisms in homogeneous solution have been elucidated in many cases from the variation of R with temperature, reagent concentration, pH, and solvent. In solid systems a similar law applies in certain cases, but here the mechanism of exchange appears to be extremely complex.

The most suitable actinide element for observation of isotopic exchange reactions in solution is neptunium, owing to (1) the availability of two convenient isotopes, the long-lived ($T = 2.14 \times 10^6$ years) α emitter ^{237}Np and the short-lived ($T = 2.35$ d) β^--decaying ^{239}Np that can be used for labeling two oxidation states, and (2) the existence of many stable oxidation states. Exchange between the unoxygenated species Np^{3+} and Np^{4+} and between the oxygenated forms NpO_2^+ and NpO_2^{2+} may be considered as an electron transfer and is relatively simple with respect to the mechanism and kinetics of the exchange between Np(V) and Np(IV).

GENERAL REFERENCES

Adloff, J.-P., Gaspar, P. P., Imamura, M., Maddock, A. G., Matsuura, T., Sano, H., and Yoshihara, K., Eds., *Handbook of Hot Atom Chemistry,* Kodansha, Tokyo, 1992.

Choppin, G. R. and Rydberg, J., *Nuclear Chemistry,* Pergamon Press, Oxford, 1980.

Draganic, I. G. and Draganic, Z. D., *The Radiation Chemistry of Water,* Academic Press, New York, 1971.

Evans, R. D., *The Atomic Nucleus,* McGraw-Hill, New York, 1955.

Evans, E. A. and Muramatsu, M., Eds., *Radiotracer Techniques and Applications,* Marcel Dekker, New York, 1977.

Farhataziz and Rodgers, M. A. J., Eds., *Radiation Chemistry, Principles and Applications,* VCH Publishers, New York, 1987.

Friedlander, G., Kennedy, G. W., Macias, E. S., and Miller, J. M., *Nuclear and Radiochemistry,* 3rd ed., John Wiley & Sons, New York, 1981.

Haissinsky, M., *Nuclear Chemistry and Its Applications,* Addison-Wesley, Reading, 1964.

Kroh, J., Ed., *Early Developments in Radiation Chemistry,* Royal Society of Chemistry, London, 1989.

Matsuura, T., Ed., *Hot Atom Chemistry,* Kodansha, Tokyo, 1984.

McKay, H. A. C., *Principles of Radiochemistry,* Butterworths, London, 1971.

Spinks, J. W. T. and Woods, R. J., *An Introduction to Radiation Chemistry,* 2nd ed., John Wiley & Sons, New York, 1976.

Tabata, Y., Ito, Y., and Tagawa, S., Eds., *Handbook of Radiation Chemistry,* CRC Press, Boca Raton, FL, 1991.

Wahl, A. C. and Bonner, N. A., Eds., *Radioactivity Applied to Chemistry,* John Wiley & Sons, New York, 1951.

CHAPTER 5

BRICKS AND TOOLS FOR RADIOCHEMISTRY

I. THE PERIODIC TABLE VIEWED BY THE RADIOCHEMIST

Natural radioactive matter is omnipresent. Every part of our environment contains radionuclides, although the activity level may be very low. However, with the present sensitivity of radiation detection and modern instrumentation, even very weak activities can be investigated.

Artificial radionuclides are available for most natural elements and anthropogenic radioactive matter is even more abundant than the natural species. These radionuclides can be used as models for natural radioactive systems under more convenient conditions of measurement.

In this chapter, an inventory of the main environmental radionuclides, together with mention of their occurrence, will be given. Sources of other radionuclides currently used in radiochemistry will be indicated. Finally, the preparation of radionuclides for measurement and the significance of their recorded activity will be discussed. Description of detectors and their uses, as well as nuclear instrumentation, are outside the scope of this treatise. A list of comprehensive books covering these topics is given in the reference section.

Here, the term *radioelement* ought to be clarified. In the early history of radioactivity, the term was used to designate the newly discovered natural radioelements such as Po, Ra, Rn, and Pa. For instance, when RaC and RaE were recognized as radioactive isotopes of Bi, and RaD and ThB as radioactive isotopes of Pb, these nuclides were not considered as radioelements. Confusion arose after the discovery of artificial radioactivity, when by abusive use of terminology *radioactive isotopes of stable elements*, such as ^{30}P, were called "radioelements" or "artificial radioelements". The misusage was further exacerbated when *true* artificial radioelements, such as Tc and transur-

Table 5.1
Radioelements

Z	Element	Longest-lived isotope	Half-life	Mass of 1 MBq (27 μCi)
43	Technetium	^{98}Tc	4.2 × 10^6 years	31.1 mg
61	Promethium	^{145}Pm	17.7 years	0.19 μg
84	Polonium	^{209}Po	102 years	1.62 μg
85	Astatine	^{210}At	8.1 h	14.7 pg
86	Radon	^{222}Rn	3.8235 d	176 pg
87	Francium	^{223}Fr	22 min	0.706 pg
88	Radium	^{226}Ra	1600 years	27.3 μg
89	Actinium	^{227}Ac	21.773 years	0.373 μg
90	Thorium	^{232}Th	1.40 × 10^{10} years	245 g
91	Protactinium	^{231}Pa	3.28 × 10^4 years	0.572 mg
92	Uranium	^{238}U	4.468 × 10^9 years	80.4 g
93	Neptunium	^{237}Np	2.14 × 10^6 years	38.3 mg
94	Plutonium	^{244}Pu	8.00 × 10^7 years	1.51 g
95	Americium	^{243}Am	7.37 × 10^3 years	135.4 μg
96	Curium	^{247}Cm	1.60 × 10^7 years	295 mg
97	Berkelium	^{247}Bk	1380 years	26.7 μg
98	Californium	^{251}Cf	898 years	17.0 μg
99	Einsteinium	^{252}Es	1.291 years	24.6 ng
100	Fermium	^{257}Fm	100.5 d	5.36 ng
101	Mendelevium	^{258}Md	55 d	2.94 ng
102	Nobelium	^{259}No	58 min	2.16 pg
103	Lawrencium	^{262}Lr	216 min	8.15 pg
104	Rutherfordium	^{261}Rf	1.08 min	40.6 fg
105	Hahnium	^{262}Ha	32 s	20.1 fg
106	Unnilhexium	^{263}Unh	0.9 s	0.51 fg
107	Unnilseptium	^{262}Uns	102 ms	64.1 ag
108	Unniloctium	^{265}Uno	1.8 ms	1.15 ag
109	Unnilennium	^{266}Une	3.4 ms	2.17 ag

anides, were synthesized. In the present context, *radioelements* are those elements, either natural or artificial, *which have no stable isotopes* (Table 5.1).

Since Mendeleev's first attempt of classification on the basis of atomic weight and chemical similarities of the elements and their main compounds, many types of periodic tables have been proposed. The latest to date, adopted by the IUPAC, comprises 18 columns and 7 periods (Figure 5.1). In this presentation, the *lanthanide* and *actinide* elements are separated from the body of the table and constitute two horizontal series starting with La and Ac, respectively. It was suggested that the columns be numbered from 1 (alkaline elements) to 18 (noble gases). The association of the groups with the number of electrons in the atomic subshell being filled is left open. In

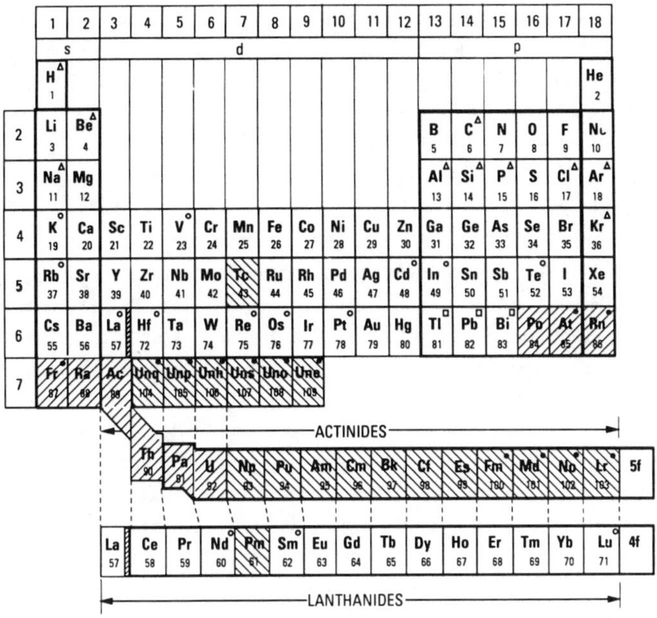

FIGURE 5.1. A radiochemist's periodic table. Names for elements beyond ${}_{103}$Lr are those recommended by IUPAC. The shifted position of the elements Th, Pa, and U with respect to the other actinides reflects the similarities in chemical properties with those of Hf, Ta, and W. The shift of the light actinides up to Cm with respect to lanthanides accounts for differences in chemical properties.

Natural radioelement   Artificial radioelement

where ○, stable element of which at least one isotope is a primordial radionuclide; △, stable element for which at least one isotope is a radionuclide produced by cosmic radiation; ●, natural or artificial radioelement not available in ponderable amount; □, stable element of which isotopes pertain to radioactive families.

the periodic table, it is generally the mass number of the longest-lived isotope of a radioelement that is quoted.

In order to obviate disputes about nomenclature following controversial claims of the discoveries of elements beyond Lr, the International Union of Pure and Applied Chemistry (IUPAC) has recommended a more systematic appellation.[1] The three figures in the atomic number Z are replaced by the syllables: *nil* for 0, *un* for 1, *hex* for 6, etc., which are assembled and terminated by the suffix *-ium*. Element 104 becomes *unnilquadium*, Unq; element 105 *unnilpentium*, Unp, etc. This nomenclature will generally be used in the present context. Periodic tables often still bear the historical names *rutherfordium*, Rf, or *kurchatovium*, Ku, for element 104, and *hahnium*, Ha, or *nielsbohrium*, Ns, for element 105.[1]

Radiochemists are currently confronted with many problems related to the actinides and they consider that the arrangement of the terminal part of the periodic chart is important. The position of elements Ac to Lr and the denomination of the series have given rise to much discussion. The term *actinides* is now generally accepted owing to the convenience of historical designation, but it must be kept in mind that this series is not as homogeneous as that of the lanthanides.[2,3]

Characterization of the oxidation states of transuranium elements was formerly based mainly on radiochemical experiments (see Chapter 10), and initially led to recognition of the concept of the actinide series by analogy with the lanthanides. Later, more subtle distinctions between the actinides and lanthanides, and between the actinides and the earlier 5d elements (Hf, Ta, W, Re) became apparent. Rationalization of the properties of elements of the seventh period is more complicated than for the lighter elements, and reconciliation of spectroscopic with chemical properties is not straightforward.

It is presently accepted that, for an accurate description of properties, the traditional concept of a 5f shell should be replaced by the distinction of a $5f_{5/2}$ relativistic subshell with 6 electrons, and a $5f_{7/2}$ relativistic subshell with 8 electrons. The first 5f electron appears only in Pa, the third member of the actinide series, in contrast to the first 4f electron associated with Ce, the second of the lanthanides. Accordingly, it is inaccurate to designate the actinide series as the "5f" element series. The 6d shell is progressively filled starting with Lr, and the presently known transactinides are considered as "6d" elements.

The energies of the 5f and 6d orbitals, in contrast to those of the 4f and 5d orbitals, are relatively close for the early elements in the actinide series. This is reflected by delocalization of 5f electrons for the metallic elements. As a result, the 6d orbital contains one electron in Ac, is doubly occupied in Th, and then reverts to single occupation from Pa to Np. Only the next members of the series correspond electronically to the lanthanides.

It follows that the chemical properties of actinides do not match those of the lanthanides. The oxidation states of the actinides are more numerous, especially as regards the earlier members Ac to Am in comparison with La to Eu (Figure 5.2). The chemical properties of homologous elements are really similar only for the terminal groups Cm to Lr and Gd to Lu. The elements Th, Pa, and U are chemically closer to the 5d elements Hf, Ta, and W than to the lanthanides.

Radiochemists further distinguish the elements that possess only radioactive isotopes with short half-lives (say, less than a few days) from those available in weighable amounts (see Figure 5.1).

The last element for which radiochemical experiments have been performed is 105, but the table continues up to 109. For the time being, it seems premature to provide space for higher-Z elements. Since 1968, all efforts to synthesize

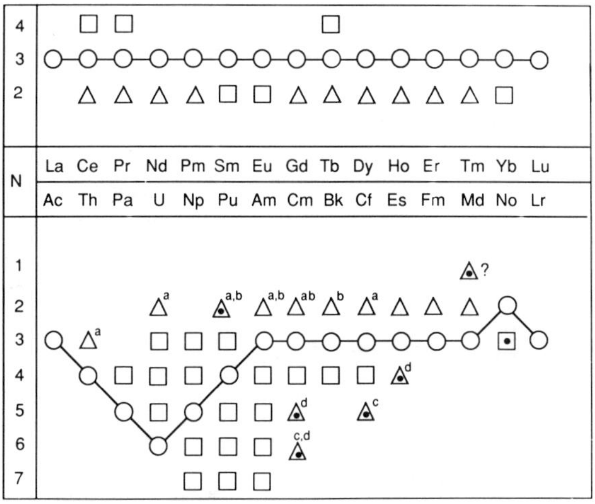

FIGURE 5.2. Comparison of oxidation states in lanthanides and actinides. These states are essentially those that can be obtained in aqueous solution, in solid compounds (except metals), in fused salts, and by chemical means, classical electrochemical procedures or in pulse radiolysis. ○, Most stable oxidation state; □, oxidation state of variable stability which depends on experimental conditions and particularly on the concentration of the element; △, oxidation state observed under special, specified conditions: (a) pulse radiolysis, (b) only in fused salts, (c) state produced in β^- decay of a parent nuclide, (d) only in the gaseous fluoride; ●, oxidation state observed only with nonponderable amounts.

these elements or provide evidence for their existence in nature have been fruitless. From these negative results, it must be concluded that cross sections of the synthetic nuclear reactions must be below 10^{-30} cm^2 and that the amount in ores must be less than 10^{-18} g/g.

These attempts were stimulated by theoretical predictions of the possible existence of an island of nuclear stability located in the Z, N plane beyond the limit of the known nuclides. With increasing Z, the half-lives for α decay and spontaneous fission diminish as a result of increasing repulsive potential between the protons; this repulsion is no longer counterbalanced by the opposing nuclear forces. Accordingly, a nucleon assembly with very high Z, N can no longer be bound, and at best exists only as a very ephemeral combination.

It is known that nuclear stability increases markedly for magic numbers of protons and neutrons. Hence, theory predicts that nuclides close to the double magic configuration Z = 114 and N = 184 should be relatively stable with respect to the nuclides with Z > 103. Ununquadium, $_{114}$Uuq, is the homologue of $_{82}$Pb, which possesses isotopes with a very high nuclear stability. If Uuq were produced during the last step of nucleosynthesis (and has an adequate half-life, see below), it might not be entirely extinct and could remain concealed in rocks and ores. Its chemical properties should be similar to those of the 6p elements Tl, Pb, and Bi, unless relativistic effects favor unexpected oxidation states.

Various sophisticated experiments have been performed in the search for such *superheavy* elements using the most sensitive detection techniques available: recording of fission tracks, multiple neutron emission, Pixe, and activation analysis. Bombardments of ^{248}Cm with accelerated $^{238}U^{n+}$ ions have not shown the formation of these elements although extremely sensitive radiochemical and physical analyses of the reaction products were carried out.[4,5]

II. NATURAL RADIOELEMENTS

Natural radioelements are those elements present in nature that have no stable isotopes. They contribute to the environmental radioactivity and their distribution within the earth has many implications in geology and geochemistry.[6-8] These radioelements include Po, At, Rn, Fr, Ac, Ra, Th, Pa, and U and have atomic numbers ranging from 84 to 92. Astatine is added since it is formed in rare branchings of natural Po isotopes, but this radioelement is generally classed with the synthetic ones; its total amount in the earth's crust is less than 30 g. The list deliberately does not include Tc and Pm, although these are produced in minute amounts in the spontaneous fission of uranium, nor Np and Pu found in uranium ores as the result of natural nuclear reactions; in the present context they are considered as synthetic. Heavy natural radionuclides include isotopes of the stable elements Tl, Pb, and Bi.

These elements would not exist on earth without the occurrence of an island of stability formed by the three radionuclides ^{238}U ($T = 4.468 \times 10^9$ years), ^{235}U ($T = 0.704 \times 10^9$ years), and ^{232}Th ($T = 14.05 \times 10^9$ years). These half-lives are sufficiently long to allow survival since the formation of the earth, 4.5×10^9 years ago, and correspond to the definition of *primordial* radionuclides.

The lowest limit of half-life which would prevent the complete extinction of a primordial radionuclide is roughly estimated in the following way. The mass of the earth is 6×10^{27} g. If it were assumed that the earth at the time of its formation was composed of only one type of radionuclide of mass 250, the initial number of atoms would have been 1.5×10^{49}. If the half-life were 0.027×10^9 years, only about one or several atoms would survive today. This value may be compared to the half-lives of the longest-lived known heavy elements other than the three previous mentioned, i.e., ^{247}Cm ($T = 0.016 \times 10^9$ years), ^{236}U ($T = 0.02 \times 10^9$ years), and ^{244}Pu ($T = 0.083 \times 10^9$ years). Only the latter might have had a slight chance of survival.

This nuclide was first mentioned as an extinct one in order to explain the excess of ^{134}Xe and ^{136}Xe in some meteorites.[9] To get a more realistic figure for the amount of natural ^{244}Pu which could be found on earth, it is assumed that, at the formation of the solar system (at a time very close to the age of the earth), the ratio $^{244}Pu/^{238}U$ was 0.5 and that a total amount of 40.8×10^{12} t of U was located in the earth's crust. This gives a ratio $^{244}Pu/^{238}U = 0.9 \times 10^{-16}$, which leads to 4 kg of surviving ^{244}Pu. If this Pu is homogeneously distributed in the upper layer of the crust, the concentration would be about 10 atoms per cubic centimeter. The reputed discovery of 10^{-18} g of ^{244}Pu/g in the mineral bastnaesite would point to a 10^6 to 10^8-fold enrichment.[10]

It will be noted that the half-life of ^{235}U does not substantially exceed the limit of 0.03×10^9 years. The present isotopic abundance of ^{235}U is about 0.71%; at the time of formation of the earth it was 28% and 2×10^9 years ago it was about 10%, a proportion high enough to trigger a fission chain reaction in a natural nuclear reactor, the *Oklo phenomenon*.[9]

A. The Natural Radioactive Families

The three natural radioactive families are represented in Figure 5.3. The successive members are related by α or β decays; accordingly, values of the mass numbers A differ by multiples of 4.

1. ^{238}U or 4n + 2 Family

The ^{238}U family ends with the stable ^{206}Pb after 8 α and 6 β^- disintegrations (see Figure 5.3A). The longest-lived members are ^{234}U (T = 0.245×10^6 years), which is the third natural uranium isotope (abundance 0.0055% or 55 ppm), and ^{230}Th (T = 0.08×10^6 years), formerly called *ionium* and the parent of the Curies' radium. The weights of nuclides contained at present in 1 ton of natural uranium (992.8 kg of ^{238}U) are given in Table 5.2. They are inversely proportional to the half-lives. Apart from U and Th, only Ra and ^{210}Pb are present in milligram amounts. In this family are found the longest-lived known isotopes of the elements U, Ra, Rn, and the longest-lived natural isotope of Po.

The values in Table 5.2 are calculated for the state of radioactive equilibrium. In rocks and minerals equilibria are frequently perturbed by recoil effects and selective lixiviation. The precursors of Ra are soluble in groundwaters with a high content in SO_4^{2-} ions and are insoluble in media rich in Cl^-, whereas Ra and its daughters have the opposite behavior. Even the activity ratio of the uranium isotopes ^{234}U and ^{238}U may differ markedly from the theoretical value of unity in minerals and waters. The total activity per unit mass of a uraniferous deposit, with a given U content, can vary greatly and members of the series can migrate to long distance from the parent.[11,12]

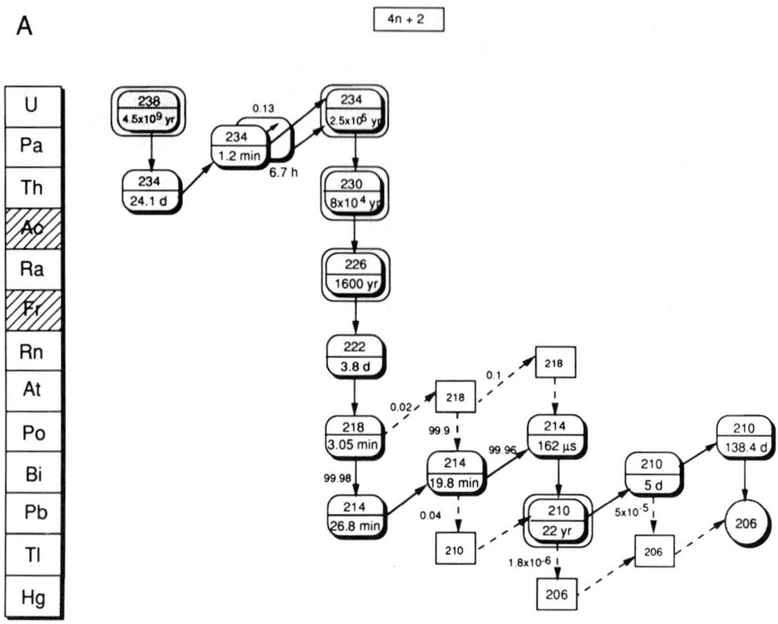

FIGURE 5.3. The three natural radioactive families:

- More than one mg of the member is in equilibrium with one ton of uranium
- Main chain
- Branching decays with percentage
- Branch member
- Stable terminal element

(A) ^{238}U or 4n + 2 family. The branch members are 0.03 s ^{218}Rn, 1.3 s ^{218}At, 1.32 s ^{210}Tl, 4.23 min ^{206}Tl, and 8.6 min ^{206}Hg. (B) ^{235}U or 4n + 3 family. The branch members are 22 min ^{223}Fr, 50 s ^{219}At, 10^{-4} s ^{215}At, 0.5 s ^{211}Po, and 7.4 min ^{215}Bi. (C) ^{232}Th or 4n family.

2. ^{235}U or 4n + 3 Family

The last member of the ^{235}U family, ^{207}Pb, results from 7 α and 4 β^- decays (see Figure 5.3B). This family is peculiar in that it includes the three longest-lived isotopes of Pa, Ac, and Fr, and the longest-lived natural radioactive Tl isotope (^{207}Tl, T = 4.77 min). Fr exists solely in this family. The only member present in ponderable amount in 1 ton of U is ^{231}Pa (see Table 5.3).

U is widely dispersed in the environment, where it is found in the Earth's crust in the form of U(IV) and U(VI) at an average concentration of 3 ppm (estimated limits are 1 to 10 ppm in igneous rocks), and in the oceans in the

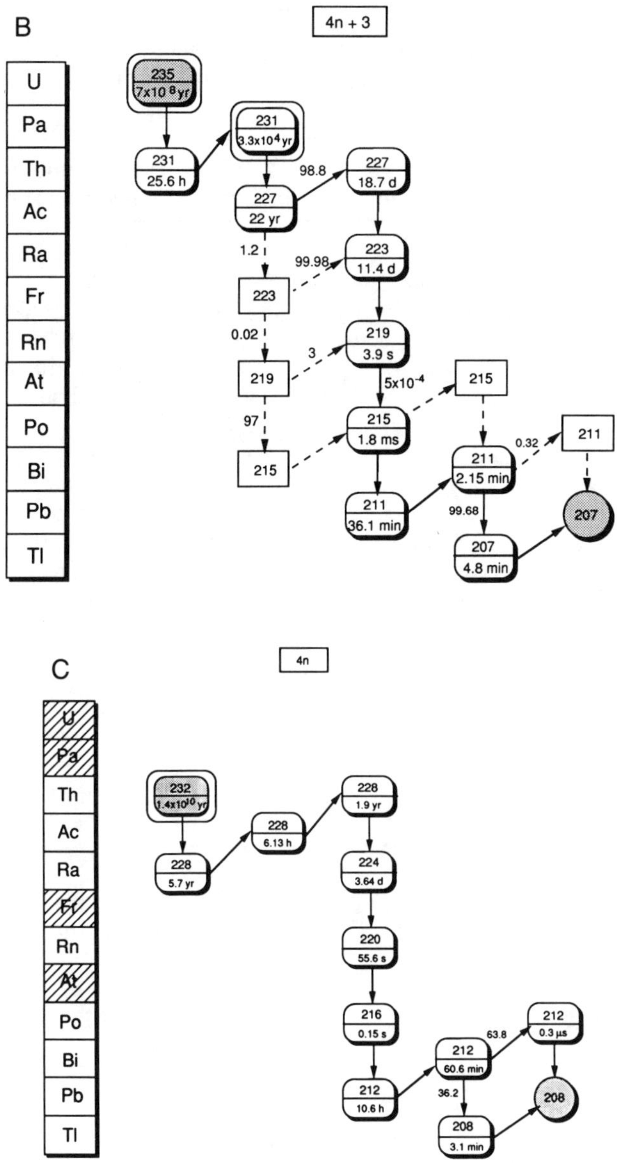

FIGURE 5.3 (continued).

oxidized form U(VI) to the extent of 10^{-3} ppm (limits 10^{-5} to 10^{-2} ppm, depending on the salinity). The U content of the earth's crust to a depth of 20 km is estimated at 10^{14} tons and the content of the oceans at 10^{10} tons. Today, only ores containing at least 300 g/ton are of economic value; locally, the amount of U may reach concentrations up to 700 kg/ton.

Table 5.2
Amounts of Nuclides in Equilibrium with ^{238}U in 1 Ton of Natural Uranium and Estimated Amounts in the Earth's Crust

Nuclide	Half-life	Amount in 1 ton U	Amount in earth's crust
^{234}Th	24.1 d	14 μg	1400 tons
^{234m}Pa	1.17 min	0.42 ng	42 kg
^{234}U	2.45×10^5 years	53.5 g	5×10^9 tons
^{230}Th	8×10^4 years	17.25 g	2×10^9 tons
^{226}Ra	1600 years	337 mg	3.4×10^7 tons
^{222}Rn	3.823 d	2.17 μg	210 tons
^{218}Po	3.05 min	1.2 ng	120 kg
^{214}Pb	26.8 min	10.1 ng	1 ton
^{214}Bi	19.7 min	7.5 ng	750 kg
^{214}Po	1.64×10^{-4} s	1 fg	100 mg
^{210}Pb	22.3 years	4.37 mg	4.3×10^5 tons
^{210}Bi	5 d	2.7 μg	270 tons
^{210}Po	138.4 d	74 μg	7400 tons

Table 5.3
Amounts of Nuclides in Equilibrium with ^{235}U in 1 Ton of Natural Uranium and Estimated Amounts in the Earth's Crust

Nuclide	Half-life	Amount in 1 ton U	Amount in earth's crust
^{231}Th	1.06 d	29.9 ng	3 tons
^{231}Pa	3.25×10^4 years	327 mg	3.2×10^7 tons
^{227}Ac	21.7 years	211 μg	2.1×10^4 tons
^{227}Th	18.7 d	506 ng	50 tons
^{223}Fr	22 min	5.5 pg	550 g
^{223}Ra	11.4 d	303 ng	30 tons
^{219}Rn	3.96 s	1.2 pg	120 g
^{215}Po	1.8×10^{-3} s	0.5 fg	50 mg
^{211}Pb	36.1 min	630 pg	63 kg
^{211}Bi	2.14 min	37.4 pg	3.7 kg
^{207}Tl	4.77 min	79 pg	7.9 kg

The normal abundance of ^{235}U in natural uranium is 0.7202 ± 0.0010 atom-percent. This figure was considered as a constant until the discovery in 1972 of the natural reactor at Oklo in the Republic of Gabon. The first clue toward this remarkable finding came from the measurement of an abundance of 0.7171 atom% in a sample of natural uranium originating from this site.[13]

Table 5.4
Amounts of Nuclides in Equilibrium with 1 Ton of Natural ^{232}Th and Estimated Amounts in the Earth's Crust

Nuclide	Half-life	Amount in 1 ton U	Amount in earth's crust
^{228}Ra	5.76 years	402 μg	10^5 tons
^{228}Ac	6.13 h	48.6 ng	12 tons
^{228}Th	1.913 years	133 μg	3.3×10^4 tons
^{224}Ra	3.66 d	690 ng	172 tons
^{220}Rn	55 s	117 pg	29 kg
^{216}Po	0.15 s	620 fg	155 g
^{212}Pb	10.64 h	79 ng	20 tons
^{212}Bi	60.6 min	7.5 ng	1.9 tons
^{212}Po	3×10^{-7} s	4×10^{-19} g	100 μg
^{208}Tl	3.05 min	133 pg	33 g

Later, uranium with an isotopic abundance as low as 0.440% was discovered. As yet, no other event similar to that at Oklo has been established with certainty, although many natural reactors may have been in operation about 2×10^9 years ago.

3. ^{232}Th or 4n Family

^{232}Th leads to ^{208}Pb after 6 α and 4 β^- decays (Figure 5.3C). The longest-lived member in the family is ^{228}Ra (5.76 years, formerly called *mesothorium I*), but at equilibrium less than 1 mg exists in 1 ton of thorium. (Table 5.4). If for some reason this nuclide is removed, radioactive equilibrium is again attained within about 50 years. On geological time scales thorium is always in equilibrium with its descendants and the activity is constant for a unit weight of the parent. In the other two families, equilibrium would be reestablished only after millions of years following the separation of the longest-lived member.

The average concentration of Th(IV) is 12 ppm in the earth's crust with limits between 5 and 20 ppm and 10^{-7} ppm in ocean waters. On the whole, the earth's crust may contain 2.5×10^{14} tons of the element.

The average α and β^- activity of 1 kg of terrestrial crust is about 800 Bq. Thus, the natural radioelements are very common on earth, and some of their members are more abundant than familiar stable elements. The corresponding radioactivity is not limited to the soil; all waters contain dissolved natural radionuclides in amounts which vary by many orders of magnitude depending on local conditions.

B. Atmospheric Radioactivity

Natural radioactivity in the atmosphere is due mainly to the presence of one Rn isotope in each family: ^{222}Rn (the *true* radon, T = 3.8d), ^{220}Rn (thoron, T = 55 s) and ^{219}Rn (actinon, T = 4 s). Only the first two have lifetimes that are compatible with their release from soils and waters. The most important is ^{222}Rn because its relatively long half-life ensures a higher steady concentration in the environment (see Chapter 10). The rate of release from the surface of the ground is about one atom cm^{-2} s^{-1}.

The concentration of Rn in air varies within several orders of magnitude, depending on the location. The total activity of the atmosphere is 10^{18} Bq (0.8 mol) and its average concentration near the surface is several Bq m^{-3}, corresponding to a few atoms per cubic centimeter. Values between 15 and 70 Bq m^{-3} have been reported for air in dwellings. In certain caves the radon concentration may reach 7500 Bq m^{-3}.

Rn isotopes are in turn parents of their own filiation. In successive decays short-lived isotopes of Po, Pb, and Bi, nuclides of the *fast active deposit*, are produced. Actinon and thoron terminate these rapid filiations with the stable Pb nuclide. In the case of radon, fast filiation leads to 21 years ^{210}Pb, the parent of the triad of the Pb, Bi, and Po isobars which constitutes the *slow* active deposit.

The descendants of radon isotopes are solids, which either condense on aerosols that are always present in the atmosphere, or react with atmospheric gases. Eventually, they fall to the surface of the earth and are thus also taken up by ocean waters. The behavior of radon isotopes is used for the study of atmospheric circulations.

III. ARTIFICIAL RADIOELEMENTS

Artificial (also called synthetic or anthropogenic) radioelements are those which have no stable isotopes and do not exist in nature, or at the most only in minute amounts and often as short-lived ephemeral species. They include $_{43}$Tc, $_{61}$Pm, $_{85}$At (with the previously mentioned restriction), and presently all elements from $_{93}$Np to $_{109}$Une. They are synthesized in nuclear reactions induced by neutrons for elements up to $_{99}$Es and $_{100}$Fm, and by charged particles for the heavier ones. In many cases these reactions lead to a mixture of isotopes of a given element for which the composition varies widely with the target, the nature and energy of the projectile, and the duration of irradiation.

Particular attention should be paid to the atomic weight of artificial radioelements. This is illustrated with Tc: cyclotron-produced ^{97}Tc has an atomic weight of 96.906; the atomic weight of the fission isotope ^{99}Tc is 98.906, and that of the longest-lived isotope of the element, ^{98}Tc, is 97.907. Another striking example is provided by the varying atomic weights of U and Pu produced at large scale in various kinds of nuclear reactors.

All artificial radioelements with $Z < 97$ (except At, whose longest-lived isotope is 8.3 h ^{210}At), i.e., Tc, Pm, Np, Pu, Am, and Cm, are found in large or appreciable amounts in irradiated U or Pu fuels used in nuclear reactors. In a sense, these fuels may be considered as ores of these elements if they were systematically exploited. Like any ore, they contain impurities in addition to the desired elements; in this case, the impurities are the fission products other than Pm and Tc together with vast amounts of uninteresting and unwanted material, such as the uranium isotopes. However, this kind of artificial ore has a very high activity (Tables 5.5 and 5.6).

Among the artificial radioelements, only Pu is regularly extracted from irradiated fuels, either to serve as nuclear fuel or for military purposes. Other radioelements (^{237}Np) and fission products (mainly, ^{137}Cs and ^{90}Sr) are (or have been) occasionally separated. Rather ambitious projects are now being elaborated in countries operating reprocessing plants to recover the actinides from U to Am as well as most of the elements formed in the fission process. Separation of the minor actinides Np, Am, and Cm could markedly improve the management of nuclear wastes.

The stable end products of fission chains include potentially useful and naturally scarce elements such as Pd and Xe and most of the lanthanides. However, some of these elements contain long-lived radioisotopes such as 6.5×10^6 years ^{107}Pd, which may preclude their use in consumer products, although the nuclide mentioned emits only soft β^- rays. The isotopic composition and thus the atomic weight of the stable fission elements differ from those of the natural ones, sometimes by more than 1%. The atomic weight of natural Cs is 132.905; that of fission Cs is 134.92.

The radionuclides formed in a nuclear reactor and indicated later in Figure 5.6A result from many competing reactions induced by thermal and fast neutrons. Spent fuels are a convenient source for the most commonly used light actinide elements (Table 5.7).

^{237}Np, with a half-life of 2.14×10^6 years, is produced in large amounts in nuclear reactors. It is the parent nuclide of the $4n + 1$ radioactive family which is now missing in nature (Figure 5.4). This series has the peculiarity of ending with stable ^{209}Bi via 7 α and 3 β^- decays, and of including in its main branch Fr and At. All isotopes of transuranium elements, with the exception of those decaying solely by spontaneous fission, lead to chains which eventually rejoin the four radioactive families. Ultimately, all these radionuclides end up as one of the four stable nuclides ^{208}Pb, ^{207}Pb, ^{206}Pb, or ^{209}Bi (Figure 5.5).

Isotopes of high Z elements, from $_{97}$Bk to $_{100}$Fm, are obtained by successive slow neutron capture in ^{239}Pu or ^{242}Pu targets. They were also formed in the explosion of a thermonuclear weapon loaded with ^{235}U and ^{239}Pu, but in this case result from fast multiple neutron capture, followed by a cascade of β^- decays.

Table 5.5

Fission Products in Spent Fuel. W_1 is the Total Initial Amount of the Element and W_2 the Initial Amount of the Specified Radionuclide in 1.134 Ton of Spent Fuel (1 Ton of Initial U)

		Nuclide		β Activity		
Z	W_1(g)	A	T	Ci	TBq	W_2(g)
1 H		3	12.3 years	600	22	0.06
34 Se	54.64	79	6.5×10^4 years	0.327	0.012	4.709
35 Br*	20.32					
36 Kr	358.5	85	10.7 years	7900	292	20.1
37 Rb*	350.6					
38 Sr	842.8	90	28.5 years	6.816×10^4	2522	499.6
39 Y	461.6	90^+	64.1 h	6.816×10^4	2522	0.125
40 Zr*	3583	93	1.5×10^6 years	1.79	0.066	713
41 Nb*	5.3×10^{-4}					
42 Mo	3332					
43 Tc	813.8	99	2.1×10^5 years	13.7	0.51	813.8
44 Ru	2164	106	369 d	6.4×10^4	2368	19.3
45 Rh	484.9	106^+	2.2 h	6.4×10^4	2368	<0.001
46 Pd*	1240	107	6.5×10^6 years	0.10	0.004	200.6
47 Ag*	76.87					
48 Cd	76.93	113m	14.6 years	24.5	0.91	0.10
49 In*	1.479					
50 Sn	51.05	123	129.2 d	3.22	0.12	4×10^{-4}
		126	1×10^5 years	0.57	0.02	20.3
51 Sb*	11.13	125	2.8 years	3478	129	3.36
		126	12.4 years	0.58	0.02	7×10^{-6}
52 Te*	471.4	125m	58 d	804	29.7	0.044
		127m	109 d	10	0.37	0.001
		127	9.35 h	9.8	0.35	4×10^{-6}
53 I*	207.8	129	1.6×10^7 years	0.03	0.001	169.5
54 Xe*	4301					
55 Cs	3621	134	2.06 years	5×10^4	1850	38.68
		135	3×10^6 years	1.5	0.055	1307
		137	30.17 years	9.78×10^4	3619	1130
56 Ba*	1562	137m	2.55 min	9.26×10^4	3426	0.0016
57 La	1204					
58 Ce	2350	144	284.3 d	7.5×10^4	2700	23.52
59 Pr	1108	144^+	17.3 min	7.5×10^4	2700	0.008
60 Nd*	3997					

Table 5.5 (continued).
Fission Products in Spent Fuel. W_1 is the Total Initial Amount of the Element and W_2 the Initial Amount of the Specified Radionuclide in 1.134 Ton of Spent Fuel (1 Ton of Initial U)

		Nuclide		β Activity		
Z	W_1(g)	A	T	Ci	TBq	W_2(g)
61 Pm	85.87	147	2.62 years	7.96×10^4	2945	85.87
62 Sm**	776	151	93 years	420	15.5	15.97
63 Eu**	133.4	152	12 years (EC 72%)	0.05	0.002	2.6×10^{-4}
		154	8.5 years	5169	191	19.59
		155	4.9 years	5855	216.6	12.5
64 Gd	73.12					
65 Tb	1.86					
Total	33,820			7.615×10^5	2.82×10^4	3780

Note: Data refer to a standard PWR fuel containing initially 3.50 mass percent of ^{235}U, 3 years after removal from the reactor. This fuel had produced an energy of 33,000 MWday and per ton, which corresponds for a 1 GWe reactor to fission of 33.820 kg of U and Pu per initial ton of U and to a 0.003% decrease in mass. The data computed and checked experimentally are reproduced by permission from the CEA-ANDRA "Catalogue des déchets conditionnés", reference IRDI/DERCA/DIP/84/1076/, 1984. Elements or radionuclides produced in amounts less than 1 g (1 ppm) have been omitted in the table, unless they are daughters of recorded parents. They include the elements Ge, As, Dy, and Ho; the radionuclides ^{89}Sr (0.211 Ci, 0.008 TBq), ^{91}Y (2.19 Ci, 0.081 TBq), ^{95}Zr (9.38 Ci, 0.35 TBq), ^{95}Nb (20.82 Ci, 0.77 TBq); the filiations $^{103}Ru \rightarrow {}^{103}Rh$ (0.0056 Ci, 2×10^{-4} TBq) and $^{110m}Ag \rightarrow {}^{110}Ag$ (3507 Ci, 129.8 TBq), ^{124}Sb (0.0018 Ci, 6×10^{-5} TBq), ^{123m}Te (0.0155 Ci, 5.7×10^{-4} TBq), and ^{160}Tb (0.014 Ci, 5×10^{-4} TBq).

* Includes stable daughter of Z − 1 parent.
** Includes stable daughter of Z + 1 parent.
+ Radioactive daughter of Z − 1 isobar.

The isotopic composition of *artificial* U and of Pu of various origins in the industrial nuclear fuel cycle is shown in Tables 5.8 and 5.9. Several different varieties of uranium are now available. Large quantities of U enriched in ^{235}U between 1.8 and 3.7% are produced as fuel for pressurized water reactors (PWR); high flux reactors for research and material testing use U enriched from 20 to 94%. Uranium recovered in fuel reprocessing plants contains the new isotopes ^{232}U (posing safety problems owing to the presence of the 2.6-MeV γ emitter ^{208}Tl) and ^{236}U (<2%), and increased amounts of ^{234}U (80 ppm) and ^{235}U (≈ 0.83%) with respect to the natural abundances. ^{232}U is presently at the parts per billion level, but with higher burn-up this figure may increase to the parts per million range.

Table 5.6
Major Elements and Residual Radionuclides Contained in 1.13 Ton of a "Spent Fuel Ore" 300 Years after Removal from the Reactor

Element	Total (g)	Radionuclide	Amount in g	Activity Ci	Activity TBq
Se	55	^{79}Se	4.7	0.32	0.012
Kr	340				
Rb	370				
Sr	340	^{90}Sr	0.5	70	2.6
Y	460				
Zr	3580	^{93}Zr	715	1.8	0.067
Mo	3330				
Tc		^{99}Tc	813	13.7	0.51
Ru	2160				
Rh	445				
Pd	1460	^{107}Pd	200	0.1	0.004
Ag	76				
Cd	77				
Sn	50	^{126}Sn	20	0.57	0.02
Te	470				
I	207	^{129}I	170	0.03	0.001
Xe	4300				
Cs	2450	^{135}Cs	1300	1.5	0.055
		^{137}Cs	1.13	98	3.6
Ba	2800				
La	1200				
Ce	2325				
Pr	1100				
Nd	4000				
Sm	870	^{151}Sm	5	132	4.9
Eu	150				
Gd	100				

Residual uranium, depleted in ^{235}U, from isotopic enrichment plants, may serve in the metallic state as efficacious material for protection against γ rays because of its high density, i.e., 19 g cm^{-3} as compared to that of the commonly used Pb (11.3 g cm^{-3}).

With increasing burn-up of UO_2 in nuclear reactors and longer storage times of spent fuels before reprocessing, present-day Pu becomes richer in ^{238}Pu and in ^{241}Am. The composition of this Pu is 1.2% ^{238}Pu, 61.1% ^{239}Pu, 2.3% ^{240}Pu, 8.8% ^{241}Pu, and 4.2% ^{242}Pu together with 1.2% ^{241}Am. In this material, there is enhancement in the emission of neutrons from (α,n) reaction on PuO_2 and of 60-keV γ photons from ^{241}Am; the release of heat is also increased.

It is possible that, on the basis of new political agreements, military Pu may become available for civil uses.

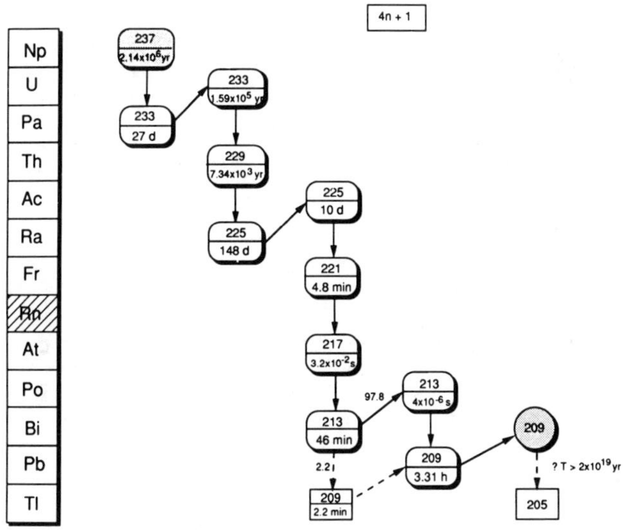

FIGURE 5.4. Artificial ^{237}Np or 4n + 1 family. Isotopes of Rn are missing. The establishment of radioactive equilibrium in the series will require about one million years.

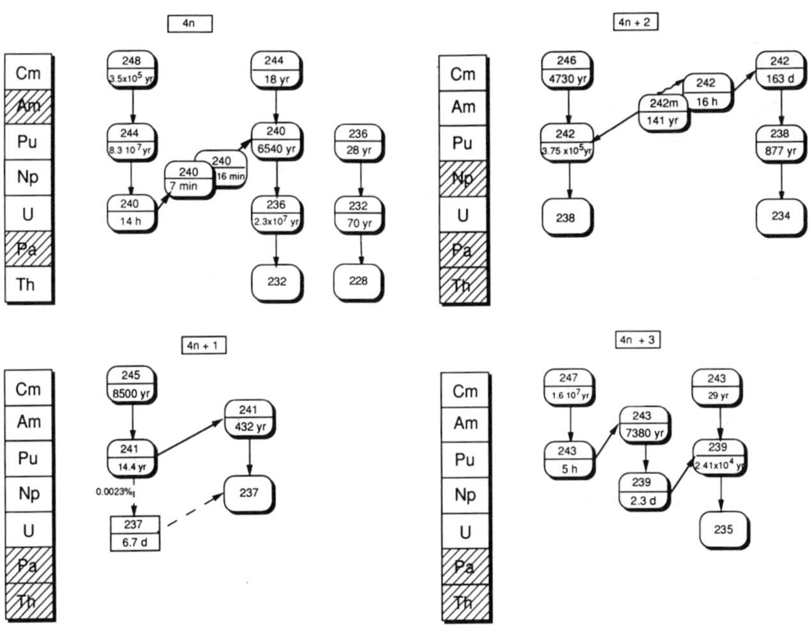

FIGURE 5.5. Connection of actinide nuclides with the four radioactive families 4n, 4n + 1, 4n + 2 and 4n + 3.

Table 5.7
Actinide Isotopes in 1.134 Ton of Spent Fuel

Radionuclide	Amount (g)	T_α (unless specified)	Activity Ci	Activity TBq	Neutrons from SF and α,n (s^{-1})
U 232	975×10^{-6}	72 years	0.02	7.4×10^{-4}	20
234	165.7	2.45×10^5 years	1.02	0.037	640
235	10,270	7.04×10^8 years	0.022	8.1×10^{-4}	15
236	4383	2.34×10^7 years	0.276	0.012	150
237	35.8×10^{-6}	β^- 6.75 d	2.91	0.11	
238	940,600	4.47×10^9 years	0.314	0.016	10^4
Np 237	433.2	2.14×10^6 years	0.30	0.011	194
239	88×10^{-6}	β^- 2.35 d	20.6	0.76	
Pu 236	0.55×10^{-3}	2.87 years	0.29	0.010	400
238	175.9	87.7 years	3011	111.4	3.63×10^6
239	5673	24,100 years	347.9	12.87	3×10^5
240	2214	6560 years	504.5	18.67	2.44×10^6
241	490	α,β^- 14.4 years	α 2.874 β^- 1.23×10^5	0.11 4551	2000
242	490	3.75×10^5 years	1.87	0.069	8.24×10^5
Am 241	222.5	432 years	762.5	28.2	8×10^5
242m	0.731	β^- 141 years	0.0337	0.001	140
242	8.6×10^{-6}	β^- (84%) EC 16 h	0.0337	0.001	
243	101.3	7370 years	20.1	0.74	1.86×10^4

Cm 242	0.134	162.8 d	444.4	16.4	3.42×10^{6}
243	0.321	29.1 years	14.78	0.55	1.8×10^{4}
244	24.03	18.1 years	1944	71.9	2.8×10^{8}
245	1.215	8500 years	0.208	0.008	200
246	0.18	4760 years	0.0558	0.002	1.23×10^{7}
248	0.19×10^{-3}	3.5×10^{5} years	8×10^{-7}	3×10^{-8}	7600
Cf 250		13.08 years	4×10^{-5}	1.4×10^{-6}	4000
		3×10^{-7}			
252		2.654 years	2×10^{-5}	0.7×10^{-6}	9×10^{4}
Total	966,000		α 7057	261.1	3.06×10^{8}
			β^{-} 1.23×10^{5}	4551	

Note: The spent fuel has the characteristics described in the note of Table 5.5. The theoretical mass balance is the following: 1 ton natural U [35 kg ^{235}U + 965 kg ^{238}U] = 966.18 kg U [940.6 kg ^{238}U + 10.27 kg ^{235}U + 4.38 kg ^{236}U + other isotopes] + 33.8 kg fission products [19.5 kg from ^{235}U, 1.55 kg from ^{238}U, 11.8 kg from ^{239}Pu, and 1.25 kg from ^{241}Pu] + 9.74 kg Pu + 433 g Np + 324.5 g Am + 26 g Cm. The theoretical activity balance is: 0.3 Ci (0.01 TBq) → 784,500 Ci (29,030 TBq) β^{-} activity + 7060 Ci (150.2 TBq) α activity.

Table 5.8
Isotopic Composition of Natural and Artificial (Reprocessed) Uranium

	Natural uranium		**Reprocessed uranium**[a]	
Isotope	**Isotopic composition (%)**	**α activity Bq per g of U**	**Isotopic composition (%)**	**α activity**[b] **Bq per g of U**
232[c]	0		1.25×10^{-7}	—
234	0.055	12,300	0.0163	35,560
235	0.72	570	0.83	660
236	—	—	0.36	8,810
237[d]	—	—	—	—
238	99.275	12,300	98.81	12,210
Total	100	25,170	100	58,240

Note: Additional α activity in Bq per g of U: 1760 from isotopes of Pu to Cm; 1750 from ^{228}Th and its daughter products formed in decay of ^{232}U; 350 from fission products.

[a] The composition depends on burn-up. Values from La Hague reprocessing plant.
[b] Activity of uranyl nitrate. Fluorination yields UF_6 without transuranium nuclides and the ^{232}U daughter activity is reduced by a factor of 20.
[c] The amount depends on the intermediate storage time of the spent fuel owing to the decay ^{236}Pu (2.8 years) $\alpha \rightarrow$ ^{232}U(70 years).
[d] Traces of ^{237}U (6.75 d) decay to ^{237}Np.

Table 5.9
Isotopic Composition (in %) of Plutonium as a Function of Burn-Up

Burn-up $MWdt^{-1}$	4,000	7,500	33,000	80,000
^{238}Pu	0.1	0.1	1.5	2.4
^{239}Pu	76.6	68.4	55.7	56.1
^{240}Pu	20.1	25.6	24.5	28
^{241}Pu	2.8	4.6	13.4	6
^{242}Pu	0.4	1.4	4.9	7.5

The main part of man-made radioelements produced in large amounts is confined in and around nuclear reactors, in facilities for fuel fabrication and in reprocessing plants. Sealed sources of radioelements are widely distributed because of their numerous applications in research, industry, and in hospitals, and their use is subject to stringent regulation.

IV. RADIONUCLIDES FOUND IN NATURE

Radioactive isotopes of stable elements and isotopes of radioelements are found in the environment. They result from natural nuclear reactions or from contamination due to anthropogenic activities.

A. *Natural Origin*

1. Primary Radionuclides

Long-lived radionuclides formed in the process of nucleogenesis are called primary or *primordial* radionuclides. A list of them is given in Table 5.10. The most important is ^{40}K, half-life 1.25×10^9 years. Its present isotopic abundance is 0.0117% as compared to 0.145% in the early earth. Potassium is highly abundant in nature; there are 10 kg of K in 1 ton of sandstone and 35 kg in 1 ton of granite. The average activity is 630 Bq/kg. The 1.232 MeV β^- radiation of ^{40}K is the main contributor to heat generated by radioactive substances in the earth's crust.[7]

About 20 primary radionuclides have been recorded; they possess very long β^- half-lives ($T > 10^{10}$ years, except for ^{40}K) and α half-lives ($T > 10^{11}$ years). Most of them are very difficult to detect because (1) the isotopic abundance is sometimes below 1%, (2) the half-lives are associated with low energies of the emitted α and β particles, and (3) the γ photons are highly converted. For these reasons the primary radionuclides are of little concern in radiochemistry. They are even considered as stable nuclides in the isotopic composition of the corresponding elements. The possible occurrence of primary ^{244}Pu has already been mentioned.

Table 5.10
Primordial Natural Radionuclides

Nuclide	% in element	T (years)	Decay mode	Energy (MeV)	Daughter nuclide
^{40}K	0.0117	1.25×10^9	β^-	1.312 (89%)	^{40}Ca
			β^+,EC	γ 1.451 (10.7%)	^{40}Ar
^{50}V	0.250	$>3.9 \times 10^{17}$	EC	γ 1.554	^{50}Ti
			β^-	γ 0.793	^{50}Cr
^{87}Rb	27.835	4.89×10^{10}	β^-	0.273	^{87}Sr
^{113}Cd	12.22	9×10^{15}	β^-		^{113}In
^{115}In	95.7	4.4×10^{14}	β^-	0.496	^{115}Sn
^{123}Te	0.908	1.3×10^{13}	EC		^{123}Sb
^{138}La	0.09	1.06×10^{11}	EC	Ba K X (13%)	^{138}Ba
			β^-	γ 0.789 (34%)	^{138}Ce
				γ 1.436 (66%)	
^{144}Nd	23.80	2.1×10^{15}	α	1.83	^{140}Ce
^{147}Sm	15.0	1.08×10^{11}	α	2.23	^{143}Nd
^{148}Sm	11.3	7×10^{15}	α	1.96	^{144}Nd
^{174}Hf	0.162	2.0×10^{15}	α	2.50	^{170}Yb
^{176}Lu	2.59	3.7×10^{10}	β^-	0.57	^{176}Hf
			γ	0.308(93%)	
^{187}Re	62.60	4.5×10^{10}	β^-	0.0026	^{187}Os
^{186}Os	1.58	2×10^{15}	α	2.75	^{182}W
^{190}Pt	0.01	7×10^{11}	α	3.18	^{186}Os
^{232}Th	100	1.40×10^{10}	α	4.01 ...	$\rightarrow$ ... ^{208}Pb
^{235}U	0.7200	7.04×10^8	α	4.401 ...	$\rightarrow$ ... ^{207}Pb
^{238}U	99.274	4.47×10^9	α	4.196 ...	$\rightarrow$ ... ^{206}Pb

Note: Further nuclides for which only the lower limits of half-life have been estimated are ^{128}Te, ^{130}Te, ^{142}Ce, ^{150}Nd, ^{149}Sm, ^{156}Dy, and ^{209}Bi.

2. Isotopic Composition and Atomic Weights of Primary Radionuclides and their Daughters

A consequence of existing long-lived primary radionuclides is progressive nucleogenesis of stable elements, which started at the time of the Big Bang event about 10^{12} years ago, with formation of hydrogen, the "oldest" element in the universe. Most elements were formed about 5×10^9 years ago. The continuing decay of parent nuclides since nucleogenesis progressively changes the isotopic composition and hence the atomic weight of the parent and daughter elements. Strictly speaking, the constancy of isotopic composition and atomic weight over time for samples of a given origin holds only for the mononuclidic stable elements. All other elements are (or may have been) of varying composition and atomic weight. This phenomenon was first observed for the composition of lead, for which isotopic proportions in rocks and minerals vary widely depending on the age of the material and on the content of the three precursors of the radioactive families.

In 11% of the decays, ^{40}K forms ^{40}Ar. In the course of geological time, the latter nuclide has accumulated to the extent that it now constitutes the heaviest and most abundant isotope of the element, i.e., 99.600%, as compared to the abundances of ^{36}Ar, 0.337%, and ^{38}Ar, 0.063%. Hence, the element $_{18}Ar$ has the peculiar property of having an atomic weight of 39.948, which is heavier than that of its next neighbor $_{19}K$ (39.0983) in the periodic chart. Similarly, the isotopic composition of $_{20}Ca$ is determined by β^- branching of ^{40}K and the atomic weight of the former is 40.078, i.e., closer to that of $_{19}K$ rather than of $_{21}Sc$ (44.956). Owing to the low actual abundance and very long half-life of ^{40}K, the atomic weight of the element can be safely assumed to be constant. Two other inversions of nuclear origin in the sequence of atomic weights occur for the pairs Co-Ni and Te-I.

The isotopic abundance of the primary radionuclides ^{115}In (95.7%) and ^{187}Re (62.60%) suggest that the respective daughters ^{115}Sn and ^{187}Os should have reasonable abundances. In fact, the values of 0.36% for ^{115}Sn and 1.6% for ^{187}Os reflect the long half-lives of the parents, which exceed the age of the universe by several orders of magnitude.

Under stellar conditions, in which matter is highly ionized and excited, the half-life of nuclides decaying by EC or by emission of low energy β^- rays may not be constant, but on earth, half-life values remain essentially constant and because the kinetics of radioactive growth and decay are easily computed, the parent-daughter relationships of primordial radioelements can be conveniently used for dating geological materials and events.

3. Radionuclides from Reactions of Cosmic Radiation

A dozen radionuclides are produced in the interaction of cosmic radiations with ^{14}N, ^{16}O, and ^{40}Ar atoms in the upper levels of the atmosphere (Table 5.11). The stationary amount is the highest for the long-lived cosmogenic nuclides ^{10}Be, ^{14}C, ^{26}Al, and ^{36}Cl, but the average volumic activity is very low because of the enormous dilution factor.

The nuclides are formed with an excess of kinetic energy and react like hot atoms, mainly with O_2, to form a variety of labeled molecules; some of these are fixed on aerosols. After a variable residence time in the atmosphere, the nuclides reach the surface of the earth in rain and snow. Owing to isotope exchange and biological reactions, ^3H and ^{14}C are found in many hydrogen- and carbon-containing compounds. Activities of the cosmonuclides are very low and difficult to measure under ordinary conditions.

Table 5.11
Major Radionuclides Produced by Cosmic Radiation

Nuclide	Half-life	Nuclear-reaction	Production rate (atoms m^{-2} y^{-1})	Decay mode and energy (MeV)	Daughter nuclide
^{3}H	12.3 years	$^{14}N(n,^{3}H)^{12}C$	1.3×10^{11}	β^- 0.018	^{3}He
^{7}Be	63 d	Spallation	2.5×10^{10}	EC γ 0.477	^{7}Li
^{10}Be	1.6×10^{6} years	Spallation	1.3×10^{10}	β^- 0.555	^{10}B
^{14}C	5730 years	$^{14}N(n,p)^{14}C$	7×10^{11}	β^- 0.155	^{14}N
^{22}Na	2.605 years	Spallation	2.2×10^{7}	β^+ 0.545 γ 1.27	^{22}Ne
^{26}Al	7.2×10^{5} years	Spallation	4.8×10^{7}	β^+ 1.16 γ 1.83	^{26}Mg
^{32}Si	100 years	Spallation	5×10^{7}	β^- 0.1	^{32}P
^{32}P	14.3 d	Spallation	2.5×10^{8}	β^- 1.71	^{32}S
^{33}P	25.3 d	Spallation	1×10^{9}	β^- 0.249	^{33}S
^{36}Cl	3×10^{5} years	Spallation	5×10^{8}	β^- 0.714	^{36}Ar
^{39}Ar	269 years	$^{38}Ar(n,\gamma)^{39}Ar$	4.2×10^{11}	β^- 0.57	^{39}K
^{81}Kr	2.1×10^{5} years	$^{80}Kr(n,\gamma)^{81}Kr$	3×10^{5}	EC	^{81}Br

Note: Production rates are approximate.

4. Neutron-Induced Nuclear Reactions

In minerals containing α-emitting radioelements, neutrons are released in (α,n) reactions with light elements such as O, N, and F. In turn, these neutrons, together with neutrons emitted in the spontaneous fission of ^{238}U, produce reactions similar to those occurring in nuclear reactors, e.g., ^{235}U(n,f), ^{238}U(n,γ), and ^{238}U(n,2n). They account for traces of ^{237}Np and ^{239}Pu detected in uraniferous rocks with ratios ^{239}Pu/^{238}U about 3×10^{-12}, and ^{237}Np/^{238}U of the order of 1.8×10^{-12}. Fission products from spontaneous and neutron-induced fission are also found in very small amounts in uranium minerals.

On the other hand, all radioactive products formed in the Oklo reactors two billion years ago are now extinct. The only witnesses of this remarkable event are the stable end products of the fission chains and the ^{235}U-depleted uranium. The location of these species close to the reactor site is of relevance to the storage of radioactive wastes.[14]

B. Anthropogenic Sources of Radionuclides in Nature

Human activities increase the spread of natural radioelements in the environment.[6] Technologically enhanced natural radiation is primarily due to mining of uranium and thorium. These operations release large amounts of radon, which otherwise would have remained buried. Once the elements of interest U and Th have been separated from the ores, the residues are stockpiled or abandoned without much supervision. The latter contain radium isotopes which continue to emanate radon, together with other long-lived natural radionuclides which may be dissolved in groundwaters or lixiviated by rainfall. The release rate of ^{222}Rn is about 1 GBq/ton of ore containing 1% U_3O_8. The world extraction of uranium ore is presently 40,000 tons/year, equivalent to a release of 20 TBq of radon per gigawatt-year of thermal reactor operation.

Many oil fields have formed over organic shales which contain uraniferous phosphates with 1000 ppm or more of U. The underlying sandstones are porous and contain brine which dissolves Ra and its daughters. These nuclides are transported to natural oil and gas reservoirs and constitute the major radioactive contaminants of petroleum. Initially, oil and gas removed from wells are dry, but soon they become mixed with water, forming an emulsion or aerosol. This water contains many ions which eventually deposit a scale that carries Ra, with a global yearly activity estimated at 3.6 to 7.2 TBq. Nuclides not incorporated in the scale are discharged in solution into seas and rivers, with an additional contamination of 2.1 to 21 TBq in radium isotopes.

Coals always contain radionuclides of the U and Th series that are closer to equilibrium than in oil and gas fields. Combustion in thermal power plants distributes the radionuclides according to their volatility and leaves uraniferous ashes. ^{210}Pb and ^{210}Po are preferentially fixed in the finer fly ash, which is the most difficult to retain. The global Rn emission can be directly determined from the amount of coal consumed. It is estimated at 75 TBq, which is about 1000th of the natural release from soils.

The phosphate industry is another source of uranium-bearing wastes, and the widespread use of potassium-rich fertilizers has substantially increased the radioactivity of arable soils owing to the contribution from ^{40}K.

Contamination with artificial nuclides may be accidental or deliberate. In 1957, the chemical explosion of a concrete tank containing 160 tons of high-level wastes from nuclear fuel reprocessing at Kyshtym in the Southern Urals led to the release of 74 PBq of activity into the atmosphere, mainly from ^{144}Ce (T = 284 d) and ^{90}Sr (T = 28 years) which is a major threat to health. Curiously, no ^{137}Cs was released. Less than a month later, a reactor accident in Windscale (U.K.) resulted in the release of 22 TBq of ^{137}Cs, 740 TBq of ^{131}I, and smaller amounts of ^{210}Po, ^{239}Pu, ^{106}Ru, etc. Very little radioactivity was released in the first nuclear power plant accident in 1979 at Three Mile Island (U.S.), as opposed to the Chernobyl event in 1986. After 10 days following the latter accident, some 10^3 PBq activity of nonvolatile fission products and transuranium nuclides were released into the environment, representing 3 to 4% of the content in the reactor core. All the fission rare gases and 20% of the iodine available in the core (670 PBq) were dispersed in the atmosphere.

On two occasions, reentry of space satellites contributed to the contamination of the upper atmosphere. In 1964, an isotopic power unit loaded with about 1 kg of ^{238}Pu (629 TBq) exploded over the Indian Ocean. A Soviet satellite powered by a nuclear reactor, Cosmos 954, broke apart over northern Canada in 1978, spreading radioactive debris over 1000 km^2. This reactor contained 3.1 TBq of ^{90}Sr, 181 TBq of ^{131}I, and 3.18 TBq of ^{137}Cs. However, contamination on the earth's surface was not noticeable.

Between 1958 and 1981, atmospheric tests of nuclear fission and fusion weapons have disseminated into the atmosphere, troposphere, and stratosphere large amounts of radionuclides, which include many fission products, transuranium elements, activation products, and ^{3}H with local or global fallout. The amount of Pu released in the tests is estimated at 4.2 tons, of which 2.8 were dispersed in the atmosphere and 1.4 tons deposited locally. About 90% of the total amount of Pu has now returned to the surface of the earth. Isotopic ratios of fallout Pu are 3×10^{-4} for ^{238}Pu/^{239}Pu and 0.18 for ^{240}Pu/^{239}Pu.

Underground nuclear explosions have released — and still are releasing — the same radionuclides together with ^{45}Ca and ^{59}Fe from neutron activation in localized areas of the geosphere. The total amount of Pu involved probably exceeds 1.5 tons.

Large amounts of radioelements and concomitant fission products are confined in nuclear reactors or in reprocessing plants. The present world inventory of Pu is several hundred tons (probably over 600 tons) for civil uses and as much for military purposes. Most of the civil Pu is still contained in irradiated fuels together with other actinides. An increase in reprocessing is foreseen by the turn of the century.

Disposal of wastes from the nuclear fuel cycle is a major problem. The practice of immersion of solid materials in ocean waters was stopped in 1982 after the disposal of 170 to 200 kg of Pu which will eventually be disseminated. Today, the annual release from reprocessing plants is less than 100 g.

Every day several milligrams of the long-lived ^{99}Tc ($T = 2.13 \times 10^5$ years) appear as radwaste from the worldwide use of ^{99m}Tc in nuclear medicine, which contributes to the ubiquitous occurrence of this radioelement. Liquid and gaseous effluents from nuclear industry are strictly controlled, but some dispersion, although limited, is unavoidable. The dilution of most radionuclides in the environment is extremely high, although reconcentration may occur locally. The corresponding activities can still be easily detected.

V. HOW TO OBTAIN RADIONUCLIDES

The ideal radionuclide with the most convenient half-life and type of radiation, well adapted for a specific purpose, is not always at hand. It may happen that this nuclide does not exist, or cannot be produced with available irradiation facilities, or is unobtainable in a desired state of chemical or radiochemical purity.

Among the various requirements, there is generally one criteron which is the most exacting. The lower practicable limit of time for achieving a chemical experiment is about one minute; hence the half-life, in many cases, determines the choice of the nuclide. Efficient detectors are available for counting all types of radioactivity, although the measurement of γ photons is frequently the most convenient procedure. In any event, a radiochemically pure nuclide with high specific activity is a particular asset.

There are many ways of preparing a radionuclide, but in nearly all cases one or several steps of chemical and radiochemical separations are necessary. Quite often, chemistry gives rise to more difficult problems than radioactivity itself. Radiochemical methods, which will be considered in Chapter 8, are extensively used, but they must be adapted to the half-life in question.

It is a common practice to use carriers when a radionuclide in very small amount must be separated from a liquid phase. The carrier may be a stable or long-lived isotope of the radionuclide, or a chemically homologous element in the periodic chart, for instance Ba for Ra, Cs for Fr, and Te for Po. When a separation is performed by coprecipitation, the carrier must, of course, be added in weighable amounts. The use of an isotopic carrier decreases the final specific activity. Occasionally, a radionuclide may be separated without use of a carrier by collection and transfer on a single bead of an ion exchanger.

At extremely low concentrations, when the number of radioactive species is less than a few hundred, a carrier must be added to ensure that the radioactive matter will follow the expected thermodynamic and kinetic laws.

Table 5.12
Convenient Sources of Isotopes of Natural Radioelements Suitable for Radiochemical Investigations

4n + x x	Radionuclide	Decay mode	Half-life	Source
2	^{210}Pb	β^-	22 years	Aged solution of ^{226}Ra*, old ampules of ^{222}Rn
2	^{210}Bi	β^-	5 d	^{210}Pb solution
2	^{210}Po	α	138 d	^{210}Pb solution, $^{209}Bi(n,\gamma)^{210}Bi \ \beta^- \rightarrow$
2	^{222}Rn	α	3.823 d	Solid compounds or solution of ^{226}Ra
3	^{223}Fr	β^-	22 min	Solution of ^{227}Ac or actino-ferous rare earths
3	^{223}Ra	α	11.4 d	^{227}Ac solution
3	^{227}Ac	β^-,α	22 years	^{231}Pa** solution, $^{226}Ra(n,\gamma)^{227}Ra\beta^-$*** $\rightarrow$
2	^{234}Th	β^-	24 d	Solution of natural U free from ^{230}Th
0	^{228}Th	α	1.9 years	In effluents of reprocessed fuels; daughter of ^{232}U
3	^{227}Th	α	18.7 d	^{227}Ac solution
1	^{233}Pa	β^-	27 d	$^{232}Th(n,\gamma)^{233}Th\beta^- \rightarrow$
	^{230}U	α	20.8 d	$^{232}Th(p,3n)$
	^{232}U	α	70 years	$^{232}Th(p,n)$
	^{237}U	β^-	6.75 d	In effluents of Pu recycling; α-daughter of ^{241}Pu

* $^{226}Ra(\alpha$, 1,600 years) is presently a by-product of the industrial production of natural U.

** $^{231}Pa(\alpha$, 34,100 years) concentrates in the extraction residues of natural U; the world stock is $\approx$120 g Pa_2O_5. It is also produced by $^{230}Th(n,\gamma)\ \beta^- \rightarrow$ on ^{230}Th separated from aged solutions of natural U.

*** About 1 kg of ^{227}Ac has been produced from Ra.

A. Natural Sources of Radionuclides

With the availability of a very large choice of artificial nuclides, natural ones have been somewhat overlooked. The reservoir of natural nuclides includes several members with very convenient half-lives and in carrier-free states, e.g., ^{234}Th, ^{228}Ac, ^{212}Pb, ^{208}Tl, and ^{223}Fr (Table 5.12). Separation procedures of these nuclides are known since the early period of radiochemistry and have been improved with the aid of more recent techiques such as solvent extraction and high performance chromatography. The very long-lived natural radionuclides are more difficult to handle.

Several natural radionuclides are more conveniently obtained in nuclear reactions, but at the expense of lower specific activity.

B. Nuclear Reactions in a Neutron Flux

Neutron irradiation is the easiest and most common method for obtaining radionuclides. Activation reactions frequently involve high cross sections. Reactors used for the production of radionuclides have fluxes in the range of 10^{12} to 10^{14}, in exceptional cases, 10^{15} neutrons cm^{-2} s^{-1} (High-Flux Isotope Reactor at Oak Ridge). The available flux is a very important parameter since the yield of a nuclide produced in n successive neutron captures increases with the nth power of the flux.

Neutrons penetrate easily into bulky materials; the irradiation is fairly homogeneous and the target does not require special preparation. The use of high-purity materials is recommended in order to avoid activation or impurities which would reduce the radiochemical purity of the desired radionuclide. If a compound rather than the pure element is to be irradiated, care must be taken to use heteroatoms which are not activable, or which produce only short-lived activities.

The most common reaction with thermal neutrons is the (n,γ) radiative capture, which leads to a β^- emitter, an isotope of the target element. Accordingly, the radionuclides are highly diluted in inactive isotopic matter and the specific activity is reduced. It can be increased by (1) increasing the irradiation time in the case of moderately long or very long half-lives, (2) irradiation in a higher flux (although activation of impurities increases in the same proportion), (3) using a target enriched in the activable isotope (which is always expensive, and of course impossible with mononuclidic elements), or (4) exploiting the Szilard-Chalmers effect.[15]

This last effect is based on the recoil energy of the nascent radionuclide and subsequent hot chemical reactions of the nucleogenic atom inside the target. Eventually, the radionuclide ends up in an oxidation state different from that of the isotopic target atoms, from which it can be separated. A typical example is the formation of $^{51}Cr^{3+}$ in irradiated solid potassium chromate. Following irradiation, the target is dissolved and the reduced species separated from the bulk by ion exchange. The yield of isotope separation and the specific activity depend on many factors, which together with the associated mechanisms have been extensively investigated. Procedures of this kind have been applied for radiohalogens, ^{32}P, ^{76}As, and a few other radionuclides but are now becoming obsolete with the availability of reactors with high fluxes. More favorable cases than the (n,γ) reactions exist (1) when the β^- emitter produced decays to another radionuclide which has an atomic number Z + 1 with respect to the target atom Z and can be obtained in a carrier-free state with maximum specific activity and (2) in Z-changing reactions such as (n,p) or (n,α).

An example for the first case is the preparation of the very important ^{99m}Tc, extensively used in nuclear medicine:

$$^{98}\text{Mo}(n,\gamma)\ ^{99}\text{Mo}\ \beta^-,\ 66\ \text{h} \rightarrow\ ^{99m}\text{Tc IT},\ 6\ \text{h} \rightarrow\ ^{99}\text{Tc}$$

Thermal neutron-induced reactions other than radiative capture generally have low cross sections, with the exception of ^{6}Li(n,α)^{3}H, σ = 940 b, used on a large scale for the production of tritium for (future) fusion reactors. Nuclides with relatively high specific activity are obtained in the ^{14}N(n,p)^{14}C, ^{35}Cl(n,α)^{32}P, and ^{35}Cl(n,p)^{35}S processes, but for these reactions thermal neutron activation cross sections are of the order of 1 b or less.

Reactions of this type induced by fast particles produced with neutron generators in the very exoenergetic ^{3}H(d,α)n reaction lead to much higher yields. The average energy of these neutrons is 14 MeV. Radioisotopes of the lighter elements, which are not activable in a nuclear reactor or are produced with very low activities, are conveniently obtained with fast neutrons, e.g., ^{14}N(n,2n)^{13}N and ^{16}O(n,p)^{16}N. These nuclides are short-lived and in many cases β^+ emitters. Neutron generators are not suitable for long-term irradiations, i.e., for the preparation of high activities of long-lived nuclides.

Neutron sources that incorporate radioelements are termed *isotopic* or radioactive neutron sources. They are based on (α,n) or (γ,n) reactions, or on spontaneous fission decay. Formerly, they were prepared from a mixture of Be and a natural α emitter, in particular ^{226}Ra; today, the most frequently used source is 433 years ^{241}Am mixed with Be. The neutrons are produced in the ^{9}Be(α,n)^{12}C reaction with a yield of about 80 neutrons per MBq. The 4π-emission may be as high as 10^7 to 10^8 n s^{-1} depending on the Am content. The neutrons have a broad energy spectrum and are thermalized in water or paraffin.

Photoneutron (γ,n) sources utilize a mixture of Be and 60.3 d ^{124}Sb (γ rays up to 2.090 MeV). They release nearly monoenergetic 24-keV neutrons with total yields up to 10^8 n s^{-1}.

The most important spontaneous fission source is ^{252}Cf, of half-life 2.65 years. Fission occurs in 3.1% of the decays, releasing an average of 3.76 n. The output of a 1 g source is 2.3×10^{12} n s^{-1}, about that of a small nuclear reactor. This nuclide is still scarce and very expensive; commercial sources are available with microgram rather than milligram loads.

Isotopic neutron sources are compact, portable, easy to handle, and reliable, but owing to the low neutron yield have little use in radiochemistry. They have applications in industry for well logging and process control, particularly when simplicity, rapidity, and low cost prevail over precision. On the other hand, they offer excellent opportunities for experimental training in institutes where reactors are not available.

The preparation of radionuclides of heavy elements in reactors merits special attention. First, radioisotopes of natural radioelements that may be difficult to extricate from their radioactive families, or that are required in large amounts, can be more conveniently synthesized. The starting material may be a stable element, as in the preparation of polonium,

$$^{209}Bi(n,\gamma)\ ^{210}Bi\ \beta^-, 5\ d \rightarrow\ ^{210}Po\ (T = 138.4\ d)$$

or a long-lived radionuclide, e.g., protactinium and actinium, prepared in the sequences

$$8 \times 10^4\ \text{years}\ ^{230}Th(n,\gamma)\ ^{231}Th\ \beta^-, 25.6\ h \rightarrow\ ^{231}Pa\,(T = 3.25 \times 10^4\ \text{years})$$

$$1600\ \text{years}\ ^{226}Ra(n,\gamma)\ ^{227}Ra\ \beta^-, 41\ \text{min} \rightarrow\ ^{227}Ac\,(T = 21.7\ \text{years})$$

The preparation of a highly radioactive target for reactor irradiation is not a simple task, nor is the separation of the desired nuclide.

Weighable amounts of radioelements with $Z > 92$ are obtained in high flux reactors by successive neutron capture reactions and β^- decays, starting with U or ^{239}Pu (Figure 5.6A).[16] In a first stage, the ^{239}Pu target is irradiated for about 1 year, in order to convert it into ^{242}Pu by a third-order neutron capture reaction. After this time, ^{242}Pu is present to maximum extent, but is highly contaminated with the products of the fission of ^{239}Pu and ^{241}Pu, and contains ^{243}Am and ^{244}Cm. The ^{242}Pu (and/or ^{243}Am and ^{244}Cm) is subsequently purified from these contaminants and used in a second stage for the synthesis of heavier actinides. The dual operation is necessary because of the high fission cross section of ^{239}Pu. Even in a flux as high as 10^{15} n cm^{-2} s^{-1} the build up of ^{252}Cf from ^{239}Pu does not exceed 0.3%. ^{242}Pu, which is not fissionable, is a more convenient material for the synthesis of heavy actinides in a reactor.

The sequence of reactions (Figure 5.6B) terminates with $_{100}Fm$. Resolution of the impressive mixture of several actinides with many isotopes and large amounts of fission products is a challenge for radiochemists. The yields decrease with increasing Z because of the high order of the (n,γ) reactions and blocking due to high fission cross sections and/or short half-lives of intermediate nuclides. The production rate is slow for the same reasons (Table 5.13).

Specific actinide nuclides can be prepared independently of the general sequence, e.g., ^{238}Pu from neutron-irradiated ^{237}Np via $^{237}Np\ (n,\gamma)^{238}Np\ \beta^-$, 2.1d → ^{238}Pu, ^{249}Cf from β^- decay of extracted ^{249}Bk and ^{248}Cm from α decay of ^{252}Cf (Table 5.14).

C. Nuclear Reactions with Charged Particles

The preparation of radionuclides in nuclear reactions induced by accelerated charged particles differs in many ways from reactor irradiation.

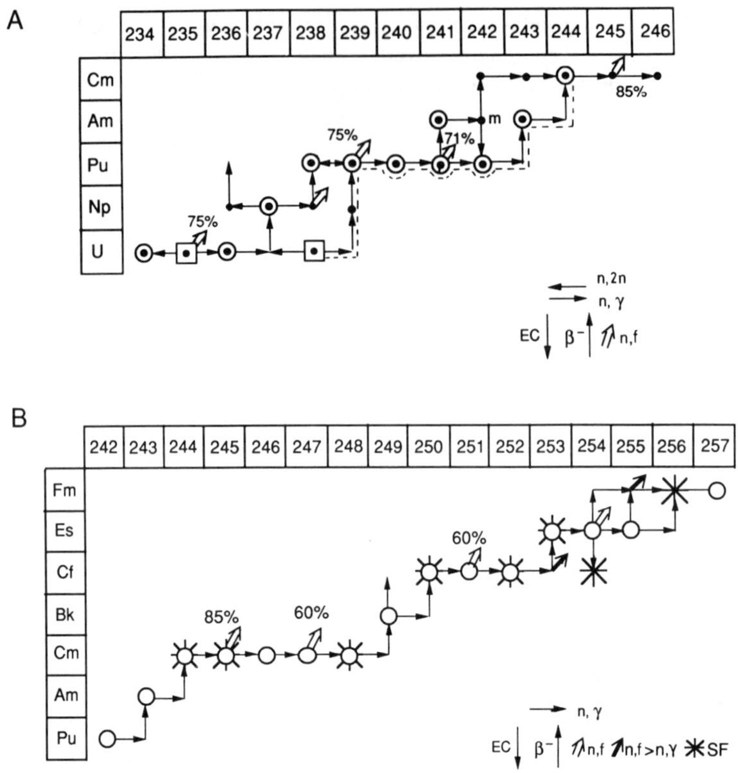

FIGURE 5.6. Production of transuranium elements in a neutron flux. (A) Formation of isotopes of actinide elements up to Cm in neutron-irradiated uranium. Half-lives are indicated in Table 5.7. The reaction paths indicated are those prevailing during a 3-year irradiation period. The nuclides reported are those found in U in amounts greater than 1 mg after 3 years of cooling. ⊙, Radionuclides formed in ponderable amounts (> 1 mg); ⊡, Radionuclides present initially in the target; --->, Main pathway. α-Decays are not shown. Those intervening within a period of 3 years in the accumulation of actinide isotopes are $^{236}Pu \rightarrow ^{232}U$, $^{242}Cm \rightarrow ^{238}Pu$ and $^{244}Cm \rightarrow ^{240}Pu$. (B) Build-up of isotopes of transplutonium actinides in neutron-irradiated ^{242}Pu. ○, Radionuclides produced in amounts greater than 1 pg. α-Decays contributing to the accumulation of actinide nuclides during irradiation are $^{253}Es \rightarrow ^{249}Bk$, $^{254}Fm \rightarrow ^{250}Cf$ and $^{255}Fm \rightarrow ^{251}Cf$.

Owing to the limited range of the particles in solid materials, the target must generally be mounted in the form of a thin foil or as a deposit on a holder, or, in some cases, as a gas. Technical aspects of the preparation of targets will not be considered here.

The cross section of the reaction depends on the particle energy, which in turn is a function of the penetration depth. These parameters must be reconciled in order to obtain a maximum yield of the desired nuclide. Since many reactions are endoenergetic and have energy thresholds, reaction types may not be the same in the superficial as in the deeper target layers. If the

Table 5.13
Actinides Produced in High Flux Reactors

Nuclide	Decay mode	Half-life	Amount produced	Comments
$^{249}Bk^{a}$	β^-	320 d	<0.5 g	Parent of ^{249}Cf (α, 351 years)
^{252}Cf	α,SF	2.64 years	<0.5 g	Parent of ^{248}Cm (α, 3.48×10^5 years); contains traces of ^{249}Cf and ^{250}Cf
$^{253}Es^{b}$	α	20.5 h	mg	Parent of ^{249}Bk
^{254}Es	α	275 d	μg	
^{254m}Es	β^-	1.6 d	μg	
^{255}Es	β^-	40 d	μg	
^{255}Fm	α	20 h		Parent of ^{251}Cf(α, 898 years)
^{257}Fm	α,SF	100 d	pg	

[a] Approximate amounts obtained in 2-years irradiation of a few hundred grams of ^{242}Pu, ^{243}Am, and ^{244}Cm separated from kilogram amounts of ^{239}Pu irradiated in a flux of 3×10^{14}n cm^{-2} s^{-1}

[b] Approximate amounts obtained in 1-month irradiation of about 1 g of ^{252}Cf separated from the preceding irradiation products.

Table 5.14
Convenient Sources of Isotopes of Transuranium Elements and Applications[a]

^{238}Pu

α, 87.7 years; specific power, 0.54 W g^{-1}; annual amount, gram to kilogram
Source: neutron irradiation of $^{237}NpO_2$; ^{237}Np available in large amounts from reprocessing of spent fuels
Uses: electric power, SNAP, ionization

^{241}Am

α, 432 years; specific power, 0.11 W g^{-1}; annual amount, gram to kilogram
Source: aged $^{241}PuO_2$, recycling of effluents from preparation of Pu metal
Uses: Am-Be neutron source for start of nuclear reactors and activation analysis; source for X-ray fluorescence spectroscopy, thickness gauge, calibration of detectors

^{243}Am

α, 7951 years; annual amount, gram to kilogram
Source: effluents from spent fuel reprocessing, slightly contaminated with ^{241}Am; neutron irradiation of $^{239}PuAl$ or $^{239}PuO_2$[b]
Use: chemistry of Am

Table 5.14 (continued).
Convenient Sources of Isotopes of Transuranium Elements and Applications[a]

^{242}Cm

α, 163 years; SF, 7.2 × 10^6 years; specific power, 120 W g^{-1}
Source: neutron irradiation of ^{241}Am and β^- decay of ^{242}Am
Uses: electric power, source of high-grade ^{238}Pu

^{244}Cm

α, 17.6 years; SF, 1.3 × 10^7 years; specific power, 2.8 W g^{-1}
Source: high-flux neutron irradiation of Am/Cm targets
Uses: Cm-Be neutron source, electric power

^{248}Cm

α, 4.7 × 10^5 years; SF, 4.6 × 10^6 years
Source: high flux irradiation of Am/Cm target
Uses: research on chemistry of Cm, target for heavy element production

^{249}Bk

β^-, 314 years; annual amount, milligrams to grams
Source: high-flux irradiation of Am/Cm target
Uses: research on chemistry of Bk, source of high-grade ^{249}Cf

^{249}Cf

α, 360 d; SF, 1.5 × 10^9 years; annual amount, ≈g
Source: high-flux irradiation of Am/Cm target
Use: research on chemistry of Cf

^{252}Cf

α, 2.65 years; SF, 85 years; specific power, 39 W g^{-1}, annual amount, ≈g
Source: high-flux irradiation of Am/Cm target
Uses: intense neutron sources, *in situ* activation analysis, neutron therapy, analysis of ^{235}U by induced fission

[a] Production depends on country and needs; values indicated are annual productions.
[b] Other actinide nuclides formed include ^{242m}Am (141 years), ^{243}Cm (28.5 years), ^{245}Cm (8500 years), ^{246}Cm (4700 years), ^{247}Cm (1.60 × 10^7 years), ^{250}Cm (≈10^4 years), ^{247}Bk (1380 years), ^{250}Cf (13.08 years), ^{251}Cf (898 years), ^{253}Es (20.4 d), ^{254}Es (275.7 d), and ^{257}Fm (100.5 d).

cross section is compatible with the required reaction over the full energy range, a thick target can be used in which the particle beam is fully stopped. But more often a mixture of nuclides is formed that will require chemical separation and radiochemical purification.

In contrast to neutron irradiation, for which the target temperature does not exceed a few ten degrees above ambient, a particle beam releases a high amount of energy in the target. The latter must be cooled if melting or destruction is to be avoided (or volatilization loss of the product or of the target material). Quite often, the beam intensity must be reduced for the sole purpose of counteracting thermal effects.

Common machines used for preparation of radionuclides are cyclotrons, which deliver light particles such as protons, deuterons, tritons, ^{3}He, and ^{4}He with energies up to several ten MeV and beam intensities up to several milliamperes. Differences in conditions of preparation and properties of these radionuclides compared to those formed in reactors are

1. There is flexibility in choice of type and energy of the projectile.
2. Most nuclides result from Z-changing reactions and are obtained with high specific activities.
3. The nuclides are neutron deficient and decay by EC or β^+ emission.
4. Radioisotopes of the lighter elements (C, O, N, F) for nuclear medicine, in particular for positron emission tomography, can only be obtained with cyclotron irradiation.
5. The absolute yield is limited by the irradiation time, which cannot be maintained over periods of days or weeks as in reactors (the production of weighable amounts is generally out of the question).
6. The nuclides are considerably more expensive than reactor-produced ones.

Charged particle reactions have been extensively used in transuranium research. Most of the actinides, and all transactinides were discovered with such reactions. Above $_{100}$Fm, the syntheses require heavy ions (of elements $_5$B, $_7$N, $_8$O, etc., up to $_{26}$Fe) which permit increase of Z in the target element by several units in a single step (Table 5.15). High-intensity beams of heavy ions are available in a few laboratories in the U.S., Russia, and Germany. The reactions which form transactinides have very low yields for several reasons: (1) scarcity of the target material, (2) extremely small cross sections (in the microbarn range), and (3) preferential de-excitation of the compound nucleus by spontaneous fission rather than by neutron evaporation. Furthermore, half-lives become extremely short with increasing Z (Table 5.16).

D. Radionuclide Generators

A major advance in the investigation and use of short-lived radionuclides in the last decades has been possible through the use of radionuclide gener-

Table 5.15
Accelerator-Produced Heavy Actinides

Nuclide	Decay mode	Half-life	Amount produced	Comments
^{256}Md[a]	EC,α	1.6 h	10^6 atoms	^{253}Es(α,n); detected via SF of ^{256}Fm daughter
^{255}No	α	3.1 min	10^3 atoms	^{249}Cf(^{12}C,α2n)
^{259}No	α	28 s	10^3 atoms	^{248}Cm(^{18}O,α5n) or ^{254}Es(^{22}Ne, 3α p 4n) (considered)
^{256}Lr	α	28 s	10 atoms	^{249}Cf(^{11}B, 4n)
^{260}Lr	α	3 min	10^2 atoms	^{254}Es(^{22}Ne, 3α 4n) (considered)

[a] Approximate amounts for targets of a few micrograms irradiated during one to three half-lives of the produced radionuclide.

Table 5.16
Production of Isotopes of Transactinide Elements Suitable for Fast Radiochemical Experiments

Nuclide	Decay mode	Half-life	Nuclear reaction	Yield
^{261}Unq (^{261}104)	α SF, EC	1.1 min	^{248}Cm(^{18}O, 5n)	10 atoms detected in 180 experiments
^{259}Unq (^{259}104)	SF	4.5 s	^{242}Pu(^{22}Ne, 5n)	16 atoms detected in 80 irradiation hours
^{260}Unp (^{260}105)	$\alpha \approx 90\%$ SF $< 10\%$ EC $< 2\%$	1.5 s	^{249}Cf(^{14}N, n) ^{243}Am(^{22}Ne, 5n)	
^{261}Unp (^{261}105)	$\alpha \approx 75\%$ SF$\approx 25\%$	1.8 s	^{243}Am(^{22}Ne, 4n) ^{249}Bk(^{16}O, 4n)	18 atoms detected in 180 irradiation hours
^{262}Unp (^{262}105)	SF$\approx 75\%$ $\alpha \approx 22\%$ EC $< 5\%$	32 s	^{249}Bk(^{18}O, 5n)	

ators. Short-lived carrier-free daughters are periodically milked from long-lived parents dispatched to points far removed from the production site. In many cases, the generator consists of a column of inorganic exchangers, since these are more resistant to radiation. The column is loaded with the parent, and the daughter is separated simply by passage of a suitable eluant. The radiochemical purity of the short-lived nuclide is very high if the parent itself

Table 5.17
Reactor-Produced Parent Nuclides for Preparation of Radioactive Generators

Parent	Half-life	Daughter	Half-life
^{28}Mg	20.9 h	^{28}Al	2.24 min
^{42}Ar	32.9 years	^{42}K	12.36 h
^{47}Ca	4,54 d	^{47}Sc	3.35 d
^{90}Sr	28.8 years	^{90}Y	64.0 h
^{99}Mo	66.02 h	^{99m}Tc	6.007 h
^{113}Sn	115.1 d	^{113m}In	99.5 min
^{115}Cd	53.5 h	^{115m}In	4.5 h
^{125}Sb	2.77 years	^{125m}Te	58 d
^{132}Te	78.2 h	^{132}I	2.3 h
^{137}Cs	30.1 years	^{137m}Ba	2.55 min
^{140}Ba	12.7 d	^{140}La	40.3 h
^{144}Ce	284.9 d	^{144}Pr	17.3 min
^{188}W	69.4 d	^{188}Re	17 h

Adapted from Mani, R. S., Reactor production of radionuclides for generators, *Radiochim. Acta,* 41, 103, 1987.

is free from contaminants, and breakthrough can be avoided in spite of repeated milkings.

Radionuclide generators mainly use artificial parents, but a few convenient systems exist for natural radioisotopes, such as $^{238}U \rightarrow {}^{234}Th$, $^{227}Ac \rightarrow {}^{223}Fr$, and $^{210}Pb \rightarrow {}^{210}Bi$. A particular type of generator is provided by the parents of the radon isotopes, so-called *emanation* sources (see later Figure 5.8). The short-lived Po nuclides formed by α decay of radon are received on negatively charged metallic surfaces, where they further disintegrate to the active deposit nuclides, which are isotopes of Pb, Bi, and Tl. In the 4n family, ^{220}Rn generates 10.6h ^{212}Pb, a convenient radioactive tracer of lead. Very useful ^{220}Rn emanation sources are made from 5.6-year ^{228}Ra.

The most important generator-produced radionuclide is ^{99m}Tc because of its many uses in nuclear medicine.[17] The parent nuclide is ^{99}Mo, produced in reactors either via the (n,γ) reaction on Mo or by fission of ^{235}U. The daily worldwide demand amounts to tens of TBq.

More than 20 radionuclide generators utilize parent nuclides produced in reactors, and over one dozen are based on cyclotron-produced radionuclides (Tables 5.17 and 5.18).

E. Purity of Radionuclides

Radiochemical purity is more easily checked than chemical purity. In normal practice it is sufficient to run a γ or α spectrum of the sample. The nature

Table 5.18
Cyclotron-Produced Nuclides for Preparation of Generators

Parent	Half-life	Daughter	Half-life
^{44}Ti	48.2 years	^{44}Sc	3.9 h
^{68}Ge	271 d	^{68}Ga	68.3 min
^{72}Se	8.4 d	^{72}As	26.0 h
^{82}Sr	27 d	^{82}Rb	1.3 min
^{87}Y	80.3 h	^{87m}Sr	2.8 h
^{113}Sn	115.1 d	^{113m}In	1.7 h
^{118}Te	6.0 d	^{118}Sb	3.5 min
^{122}Xe	20.1 h	^{122}I	3.6 min
^{178}W	21.5 d	^{178}Ta	9.3 min

Note: Most of these nuclides decay by positron emission or by electron capture; the daughters have many applications in nuclear medicine.

Adapted from Qaim, S. M., Cyclotron production of generator radionuclides, *Radiochim. Acta*, 41, 111, 1987.

of radioactive impurities is inferred from the energies and half-lives of the various components of the spectra.

The definition of chemical and radiochemical purity is not straightforward. A simple case is that of a mixture of two elements, two compounds or two species, each labeled with one radionuclide. If p is the subscript of the main compound (or species) and nuclide, and s that of the second, unwanted, compound or radionuclide, one defines:

The *chemical* purity,

$$P_c = \frac{\text{Moles of compound of interest}}{\text{Sum of all moles in sample}} = \frac{n_p}{n_p + n_s}$$

which is generally close to unity;

The *radiochemical* purity, P_{rc}, expresses the relative number N_p of atoms of the desired radionuclide

$$P_{rc} = \frac{N_p}{N_p + N_s}$$

The *radioactive* purity, P_r, is the relative activity of the desired radionuclide

$$P_r = \frac{A_p}{A_p + A_s}$$

To illustrate these terms, the neutron irradiation of Ta containing 10 mol% Nb is considered. The activity of ^{182}Ta (T = 0.3 year) is, say, 10^7 Bq and the associated ^{94}Nb (T = 2 × 10^4 years) activity is 2 × 10^3 Bq. The consumption of Ta and Nb is quite negligible. The target is dissolved in hydrofluoric acid and the purity parameters of the solution are

$$P_c = \frac{n_{Ta}}{n_{Ta} + n_{Nb}} = \frac{1}{1 + 0.1} = 0.9$$

$$P_{rc} = \frac{N(^{182}Ta)}{N(^{182}Ta) + N(^{94}Nb)} = \frac{10^7 \times 0.3}{10^7 \times 0.3 + 2 \times 10^3 \times 2 \times 10^4} = 0.07$$

$$P_r = \frac{A(^{182}Ta)}{A(^{182}Ta) + A(^{94}Nb)} = \frac{10^7}{10^7 + 2 \times 10^3} \approx 1$$

In the frequently encountered case of irradiation of an element containing impurities at the parts per million level, all purity terms are practically equal to unity.

A more complicated case is that of a sample containing two compounds or species E_1 and E_2 of an element M and two radioactive isotopes X_p and X_s of element M. Two extreme situations may be envisaged in which (1) the isotopes X_s and X_p are randomly distributed among E_1 and E_2 and (2) all the X_s atoms are in E_1 and all X_s in E_2.

In both cases the chemical and radioactive purities are equal, but the radiochemical purity differs.

Consider an acidic solution containing 10^{-6} *M* Pu(IV) and 2 × 10^{-8} *M* Pu (VI). Let the activity of ^{238}Pu (T = 87.7 years) be 150 MBq and that of ^{236}Pu (T = 2.85 years), 80 MBq. The chemical purity of Pu is equal to one. That of Pu(IV) is

$$P_c = \frac{n_{Pu(IV)}}{n_{Pu(IV)} + n_{Pu(VI)}} = \frac{10^{-6}}{10^{-6} + 2 \times 10^{-8}} = 0.98$$

For a random distribution of the Pu isotopes, the radiochemical purity of Pu(IV) is

$$P_{rc} = \frac{N(^{238}Pu(IV))}{N(^{238}Pu(IV)) + N(^{236}Pu(IV))} = \frac{150 \times 87.7}{150 \times 87.7 + 80 \times 2.85} = 0.98$$

The radioactive purity of ^{238}Pu is

$$P_r = \frac{A(^{238}Pu)}{A(^{238}Pu) + A(^{236}Pu)} = \frac{150}{150 + 80} = 0.65$$

and a similar calculation gives P_r = 0.35 for ^{236}Pu.

When the solution is prepared by mixing $^{238}Pu(IV)$ and $^{236}Pu(VI)$, the radiochemical purity of $^{238}Pu(IV)$ is obviously equal to 1 and the radioactive purity is still 0.65.

Depending on the type of experiment, one is interested in the chemical or radiochemical purity, and possibly in both. In practice, these parameters can never be strictly equal to unity. Rather complicated cases may be encountered.

VI. PREPARATION OF RADIOACTIVE MATTER FOR MEASUREMENT

Any experiment in radiochemistry involves at least one step in which the relative or absolute intensity of the emitted radiations must be determined. For this purpose aliquot homogeneous parts are separated from the investigated system. Two operative modes are open to choice: (1) incorporation of the sample *inside* the detector, such as a gas in a proportional counter or ionization chamber, or a solution in a liquid scintillator; or (2) deposit of the sample on a holder or on a backing material positioned *outside* the detector. This latter practice is the more common, but it requires special attention owing to interaction of the radiations with the source holder or with the dry material obtained from an evaporated solution. Details of the preparation and handling of radioactive sources are found in many radioisotope laboratory manuals. Only more elaborated methods than direct evaporation and painting will be briefly described. It should be noted that clean and well-adapted sources are crucial for the measurement of activities.

A. Electrodeposition

The deposit of radionuclides on a suitable metallic substrate from an aqueous solution that permits reduction to the metallic state at potentials above -1.8 V is easily monitored. The deposit depends on the reduction potential of the M(N)/M redox couple and on the intensity and duration of electrolysis. The yield may not be quantitative. This procedure is applicable to at least 30 of the 70 known metallic elements (Figure 5.7A).

Deposit of heavy radioelements and actinides which require a potential below -1.8 V is more difficult to control because of the decomposition of water and reactions of anions and cations at the electrodes. However, even in this case experimental conditions can be found for which the deposit is reproducible. The metal deposited at the cathode reacts with water or OH^- anions. It is essential to avoid hydrolysis of the element, otherwise electrodeposition is precluded. In practice, complexing agents are added and the solution is buffered at a pH between 1.5 and 5. Each element requires optimization of its own procedure.[18–22] The final deposit is calcined in order to form an adhesive layer of oxide.

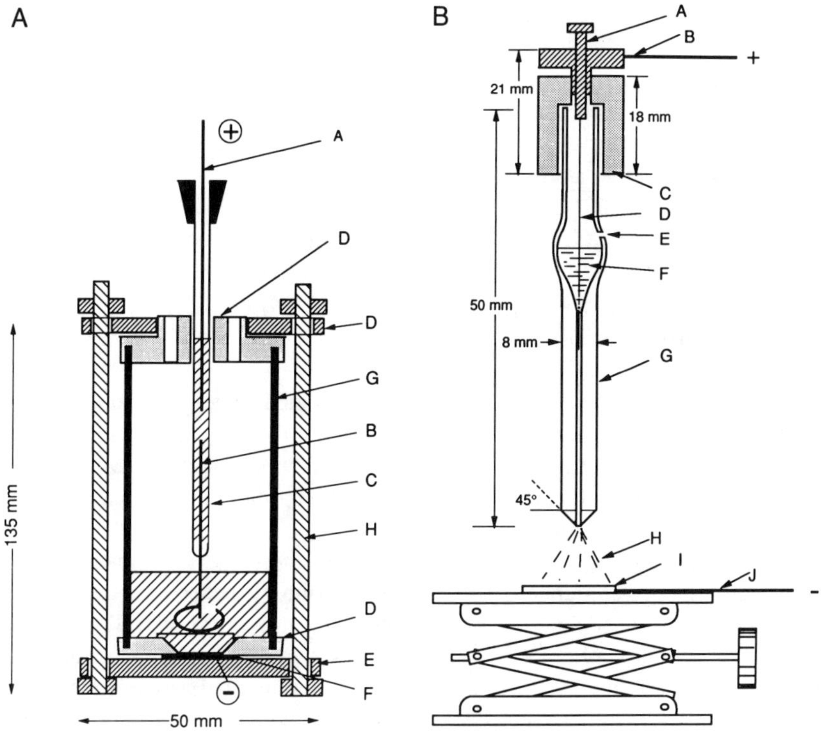

FIGURE 5.7. Deposit of radionuclides for activity measurement. A. Preparation of sources by electrodeposition: (A) copper wire connected to the positive pole; (B) Pt wire (anode); (C) mercury; (D) Teflon cover with three holes; (E) copper rings (diameter 50 mm × 3 mm thick); (F) Pt foil (support for deposit); (G) glass tube (H) copper holders. The inter-electrode distance is 10 mm. B. Preparation of sources by electrospraying: (A) screw for wire adjustment; (B) connection to positive potential (0 to 10 kV); (C) Teflon; (D) Ta or Pt wire; (E) vent (diameter, 3 mm); (F) solvent (200 to 300 μl); (G) glass capillary; (H) spraying cone; (I) carbon film; (J) negative pole (ground).

The elegant method of electrospraying is also widely used for preparing sources.[23–25] It involves a capillary (i.d., 450 μm; length, 1.5 cm) containing a solution of the radioelement in a volatile solvent such as alcohol, acetone, or acetic acid. A high potential of several thousand volts is applied between the tube and the backing material (Figure 5.7B). The solution spreads continuously out of the capillary at a flow of 1 mL/h in micron-sized, charged droplets numbering up to 5×10^{11}/mL. The solvent evaporates instantaneously. The residual solid particles are accelerated towards the backing material. As before, the deposit is calcined.

In both methods the efficiency of deposition is determined with a radionuclide that serves as a tracer. In electrospraying, it may be necessary to add a carrier for the radioelement to be deposited if the latter has a propensity for being adsorbed on the walls of the capillary. The application of electrospraying is limited by the solubility of the elements in the organic solvent.

A radionuclide can be electrosprayed directly onto the gold window of a detector or onto a thin foil, such as a few ten μg cm^{-2} carbon foil, which does not absorb α particles. This type of source is convenient for 4π counting; for this purpose it is sandwiched between the windows of two opposite detectors.[26]

Electrospraying is also used for the preparation of target material that is available only in extremely small amounts, such as the heavy actinides.

B. Nuclear Recoil

Radionuclides are born with an excess of kinetic energy in radioactive decays or in nuclear reactions. Owing to momentum conservation, the nuclides are ejected from the parent or from the target material. This behavior offers a very convenient and fast procedure for obtaining carrier-free, radiochemically pure sources (Figure 5.8).

In radioactive decays, α emission is the most appropriate for this purpose. The recoil energy amounts to 100 keV and recoiling species are received on a catcher foil located in front of the emitter, as for instance ^{208}Tl from an electroplated source of ^{212}Bi. The yield increases with an applied electric field because the recoiling species are rapidly fixed on charged aerosols from the air. Radionuclides of the active deposits are collected via the recoil momentum of α daughters.

Nuclides formed in β^- decays can be separated in the same way, although the recoil range is substantially shorter. It can be increased by placing the emitter and catcher in a vacuum.

In the case of spontaneous fission, the recoil fragments have energies up to 100 MeV. Each decay is accompanied by sputtering of the material and the fission fragments are carried by the ejected microparticles.

The products of nuclear reactions induced with energetic light or heavy ions are also emitted from thin target materials, preferentially in the direction of the incident beam. After thermalization, they adhere to aerosols and can be transported very rapidly and with high yields using a supersonic jet stream. Nuclides with half-lives of less than 1 s can be continuously extracted from the target during irradiation and their nuclear properties determined on line by α or γ spectroscopy.

The element Md was first identified in a recoil experiment. ^{256}Md atoms produced at a rate of one species per hour from a target of 10^9 Es atoms were projected by recoil onto a gold catcher foil.

As a consequence of recoil, radioactive matter may be dispersed from its original location by self-sputtering. This effect results in contamination of the immediate surroundings, including the detectors.

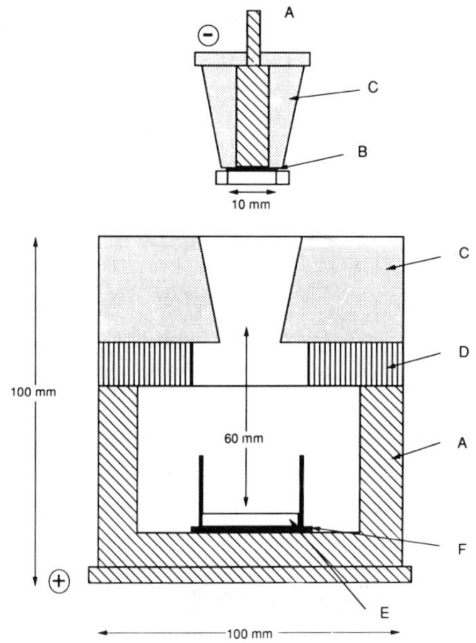

FIGURE 5.8. Preparation of sources by nuclear recoil. (A) Connection to the negative potential; (B) collecting support (Al, Pt); (C) Plexiglas™; (D) Teflon™; (E) emanation source; (F) gold container. This arrangement is used with emanation sources, for instance ^{228}Th (about 10^5 Bq) electrodeposited or coprecipitated as thorium phosphate. The collection yield of ^{220}Rn (54.5 s) and its daughters ^{212}Bi and ^{212}Pb is increased if the electric potential of the support is negative with respect to the source container. Active deposit sources are carrier-free and are the most suitable for calibration of α spectrometers with ^{212}Bi (6.050 MeV, 70.2%; 6090 MeV, 26.8%), and ^{212}Po (8.780 MeV, 100%). Both nuclides are in equilibrium with 10.6 h ^{212}Pb.

C. Absorption and Self-Absorption of the Emitted Particles

The interaction of emitted radiations inside the radioactive material, as well as with the source holder, must be taken into account. For electromagnetic radiations, corrections for self-absorption are required only when the energy of X- and γ-rays is below 100 keV.

Sources for α counting must be very thin and homogeneous. The thickness must be less than a few ten mg cm^{-2}, otherwise the particles will be absorbed in the material. The range of a 5-MeV α particle in a uranium compound is 20 mg cm^{-2}. The energy loss in 100 $\mu g\ cm^{-2}$ is 20 keV and the energy dispersion is 11.5 keV. Under these conditions the peak appears in the spectrum at the appropriate energy and the peak width, expressed as the full width at half maximum (fwhm) is 12 keV. The chemical purity of the source is important since it determines the amount of residual solids.

Positron emitters are readily measured using the 0.511-MeV annihilation γ rays. On the other hand, pure β^- emitters require self-absorption corrections for samples with varying thickness. These corrections are less stringent than for α rays because the range of β^- particles is substantially greater. Calculated corrections using external absorption data are not usually very accurate and an experimentally determined correction curve is to be preferred.

The superficial activity A_s (in Bq cm^{-2}) of a soft β^- emitter in a source of thickness m (in g cm^{-2}) is

$$A_s = \frac{S_o}{m}(1 - e^{-\mu m})$$

S_o is the specific activity (in Bq g^{-1}) and μ (in cm^2 g^{-1}) is the $E_{\beta-}$ dependent mass absorption coefficient of the source material. For a sufficiently high thickness, the emerging radiations reach a limit and A_s becomes independent of the source thickness (the sample is then said to be *infinitely* thick).

VII. ACTIVITY MEASUREMENTS

The result of activity measurement for a sample with a given detector is usually expressed by a number N of pulses recorded per unit time. This number is related to the absolute activity A by an overall constant efficiency coefficient η that depends on the properties and interaction of the radiation and on the geometry. It is also related to the probability k (or detector efficiency) that the detector will deliver one pulse per disintegration (or one pulse per ray hitting the detector)

$$N = \eta k A$$

Radioactive measurements may be absolute or relative. *Absolute* determinations require the detection of all radiations emitted in 4π steradians; absorption must be avoided. Depending on the type of radiation, absolute measurements are more or less difficult to perform. They are based on 4π counting systems and, in favorable cases, on coincidence measurements of radiations emitted in cascade, such as β^- rays followed by γ photons.

In radiochemistry, the normal practice is the *relative* measurement, for which it is very important to keep constant the proportionality factor between registered pulses and activity. This requires constant geometry of the sample in relation to the detector and, as seen, constant thickness and composition in the case of samples that emit soft radiation.

The factor k is assumed to be independent of time; hence the number of pulses N and the activity follow the same time law.

The binomial distribution law of the number of pulses or counts registered during the interval dt at time t is characterized by the terms

$$\overline{N} = k\lambda N \, dt$$

$$\sigma = \sqrt{Nk\lambda \, dt(1 - k\lambda \, dt)}$$

In a first approximation $(k \lambda \, dt)^2$ is neglected and the preceding relation reduces to

$$\sigma = \sqrt{Nk\lambda \, dt} = \sqrt{\overline{N}}$$

The precision of the measurement is independent of k and this parameter can be ignored.

For pulses or counts registered during a time t the binomial distribution law leads to

$$\overline{N} = \sqrt{N_0 \, k(1 - e^{-\lambda t})}$$

$$\sigma = \sqrt{N_0 k(1 - e^{-\lambda t})\,(1 - k + k\,e^{-\lambda t})} = \sqrt{\overline{N}\,(1 - k + k\,e^{-\lambda t})}$$

In this case, σ *depends on k.*

In the absence of a radioactive sample, the detector registers a certain number of pulses per unit time, which defines the background. The latter may have several origins, including cosmic radiation, atmospheric radioactivity (mainly radon), radionuclides (^{40}K, ^{238}U, ^{232}Th, and daughters) present in the building materials of the detector, the sample holder, and the surroundings, or accidental contamination. Determination of the activity of a sample becomes increasingly difficult when it approaches the background level.

The latter is characterized by a Poisson distribution of the number of counted pulses. In the presence of a radioactive sample, and with the exception of rare cases which are considered in Chapter 9, the recorded pulses follow a Gaussian distribution.

A. Approximations of the Binomial Distribution Law

In common radiochemical practice, the number of radioactive atoms is always very large. The binomial distribution law, which carries all information on the time dependence of radioactivity, is not very convenient and is replaced by the more manageable, approximate expressions of the Poisson and Laplace-Gauss distributions.

The mean value $\overline{N}_i$ of the number of atoms which decay at time t during the interval dt given by these expressions is

$$\overline{N}_i = N \, \lambda \, dt$$

and the exponential decay law still applies. Here, it is implicit that dt and the measurement time are very short with respect to the half-life ($\overline{N}_i << N$). If this is not the case, the mean value of measurements made during a given time t should be used (see Chapter 3).

The Poisson law applies for low values of $\overline{N}_i$, i.e., below 20, and N < 50. This implies that the decay probability λ is low and that the N_i values are discrete. The probability that N_i atoms will decay during dt at time t, when the average value is $\overline{N}_i$, is expressed as

$$\mathcal{P}(N_i) = \frac{(\overline{N}_i)^{N_i}}{N_i!} \exp(-\overline{N}_i)$$

In terms of the standard deviation

$$\sigma = \sqrt{\overline{N}_i}$$

the Poisson law reads

$$\mathcal{P}(N_i) = \frac{\sigma^{2N_i}}{N_i!} \exp(-\sigma^2)$$

The same relation holds for the number of pulses with $\sigma = \sqrt{\overline{N}}$.

The Laplace-Gauss law is valid for large $\overline{N}_i$ values. No restriction is placed on λ and the N_i values may be considered continuous. The probability that a number N_i of atoms, lying between N_1 and N_2, will decay during dt at time t when the average value is $\overline{N}_i$, is given by the expression

$$\mathcal{P}(N_1 < N_i < N_2) = \int_{N_1}^{N_2} \sigma \frac{1}{\sqrt{2\pi}} \exp\left(-\frac{(N_i - \overline{N}_i)^2}{2\,\sigma^2}\right) dN_i$$

where

$$\sigma = \sqrt{\overline{N}_i\,(1 - \lambda\,dt)}$$

Again, the same relation holds for pulses. Several ways of establishing the Poisson and Laplace-Gauss expressions starting from the binomial law of radioactive decay have been proposed. An accurate demonstration expliciting the relation between σ and $\overline{N}_i$ is given in Appendix 5.1.

B. Activity Measurements and Precision

In the following, the terms N_i and N can be exchanged. The experimental mean value $\overline{N}_i$ of n repetitive measurements is

$$\overline{N}_i = \frac{1}{n} \sum N_i$$

and the experimental standard deviation is given by

$$\sigma^2 = \frac{1}{n - 1} \sum (N_i - \overline{N}_i)^2$$

Owing to the statistical nature of radioactive decay and the difficulty of determining the distribution law of N_i, it is customary to measure a single value to which an inherent error is attributed.

The latter represents the confidence level which is assigned to the unique experimental determination. It is expressed by the probability $G(\alpha)$ of fulfilling the condition

$$|N_i - \overline{N}_i| \leq \alpha\sigma$$

in which α is a dimensionless number. By convention, the error on N_i is defined by the value $\alpha = \sqrt{2}$. Hence, $G(\sqrt{2})$ is the probability that the unique experimental value N_i is comprised between the limits

$$\overline{N}_i - \sigma\sqrt{2} \quad \text{and} \quad \overline{N}_i + \sigma\sqrt{2}$$

It expresses also the probability that the mean value $\overline{N}_i$, which cannot be measured, but which represents the best value which *could* be measured, is comprised between the limits $N_i \pm \sigma\sqrt{2}$.

1. Poisson Distribution Law

In this case, and if $dt << T$

$$\sigma = \sqrt{\overline{N}_i}$$

and

$$G(\sqrt{2}) = \mathcal{P}(\sigma^2 - \sigma\sqrt{2} < N_i < \sigma^2 + \sigma\sqrt{2}) = \sum_{N_i=\sigma^2-\sigma\sqrt{2}}^{N_i=\sigma^2+\sigma\sqrt{2}} \frac{\sigma^{2N_i}}{N_i} \exp(-\sigma^2)$$

This expression must be calculated step by step for the various values of σ. For instance, for $\overline{N}_i = 4$, $\sigma = 2$ and, $\sigma\sqrt{2} = 2.8$. Hence, the summation limits can be taken as equal to 4 ± 2, from which

$$G(\sqrt{2}) = \sum_{N_i=2}^{N_i=6} \frac{2^{2N_i}}{N_i!} \exp(-4) = 0.79$$

In other words, when the mean value is 4, there is a 79% probability that a measured value is comprised between 2 and 6. With summation limits 4 ± 3, $G(\sqrt{2}) = 0.92$.

2. Laplace-Gauss Distribution Law

Here N_i and $\overline{N}_i$ are continuous variables. Setting

$$z = \frac{|N_i - \overline{N}_i|}{\sigma}$$

it follows that

$$dz = \frac{|- dN_i|}{\sigma} = \frac{dN_i}{\sigma}$$

The integration limits become

$$- \sqrt{2} < z < \sqrt{2}$$

Owing to the symmetry of the Laplace-Gauss distribution with respect to N_i, the G-term becomes

$$G(\sqrt{2}) = \mathcal{P}(z < \sqrt{2}) = 2 \int_0^{\sqrt{2}} \frac{1}{\sqrt{2\pi}} \exp(- \frac{z^2}{2} dz)$$

This is the value of the error function erf(x) for $x = 1$, i.e., 0.842. It follows that the probability for a measured N_i to lie between the limits

$$\overline{N}_i - \sigma\sqrt{2} \quad \text{and} \quad \overline{N}_i + \sigma\sqrt{2}$$

is 85%. Assuming $\sigma \approx \sqrt{\overline{N}_i} \approx \sqrt{N_i}$, represents also the probability that the mean value that *could* be measured is within the interval $N_i \pm \sqrt{2N_i}$. This is the common way to express the result of activity measurements.

These calculations can be performed for other values of α. In particular for $\alpha = 0.6745$, the confidence level is 50%; this defines the *probable error*. The value $\alpha = 1$ corresponds to the confidence level 68.3%, which represents the standard error. Finally, the *double precision* for $\alpha = 2$ leads to a confidence level of 95%.

When the results are to be represented by a curve, it is customary to quote the *standard error*.

C. Errors in Combined Measurements

Whatever the decay rate, the uncertainty in a measurement is usually expressed by $\sqrt{2N_i}$, where N_i is the number of disintegrations recorded in a single experiment. In principle, this error can be decreased at will by increasing the value of N_i, provided there is no limit in the time available for measurement.

The theory of error propagation states that the standard deviation σ_x for a function $x = f(y_i)$ of random variables y_i, as for the number of pulses or the activity A, is

$$\sigma_x^2 = \sum_{i=0}^{i} \left(\frac{df}{dy_i}\right)^2 \sigma_{y_i}^2$$

Assuming that this expression applies for the errors, i.e.,

$$\sigma_{y_i} = \Delta y_i = \sqrt{2N_i}$$

the standard deviation $\sigma_x = \Delta x$ can be calculated for all possible combinations of the random variables y_i. The most frequently encountered functions are $x = y_1 - y_2$ (which expresses the actual activity of a sample in the presence of a background y_2), the ratios $x = y_1/y_2$ and the products $x = y_1y_2$ of activities.

Once the error Δx has been estimated, the optimal experimental conditions for minimizing error can be derived, taking into account the time available for the measurement. It is also possible to calculate the amount of radionuclide which must be used in order to obtain a given error Δx.

For a function $z = f(x_i,X_i)$ which depends on the combined random variables x_i and other nonrandom variables X_i,

$$\Delta z = \sum_{i=0}^{i} \left(\frac{\delta f}{\delta_{xi}}\right) \Delta x_i + \sum_{i=0}^{i} \left(\frac{\delta f}{\delta X_i}\right) \Delta X_i$$

The values $\Delta x_i = \sigma_{xi}$ are determined as before. All error calculations can be performed with these expressions.

D. Detection Limits

The number of species of a radionuclide which can be detected at time t during the time interval dt is

$$N = \frac{1}{\lambda}\frac{dN}{dt} = \frac{1}{\lambda} A$$

The smallest number N is obtained from the smallest meaningful value of the activity A, which itself depends on the background A_2, since

$$A = A_1 - A_2$$

where A_1 is the measured activity. Since the number of pulses is proportional to the activities, it is necessary for each specific experiment to compare the relative values of N_1 and N_2. To be significant, the measurement must satisfy the condition

$$\frac{N_1}{t_1} > x \frac{\sqrt{2N_2}}{t_2}$$

where x is equal to 2 or 3 and the difference between the measurement times t_1 and t_2 reasonably small. The detection limit is generally imposed by the background. The latter results from the natural contamination of the detector material, from Rn and its decay products if measurements are performed in the presence of air, and from cosmic radiation either directly or via (n,α) and (n,p) reactions.

The easiest procedure for measuring radionuclides is based on the detection of γ photons. However, the γ-background is often high, and the sensitivity is limited. In the measurement of α emission or spontaneous fission, on the other hand, the corresponding backgrounds are very weak but the preparation of sources is more demanding. Experimental conditions can be worked out for which the background remains virtually nil over periods of days or weeks.

Measurement of *low* activities is a common practice in radiochemistry and under certain conditions, even *extremely low* α activities can be measured. An example is provided by ^{256}Cf, for which several α particles per week have been recorded.[26] Similar levels of activities are expected to be found in the double β^- decay of ^{238}U, which is estimated to produce 10 Bq per year of ^{239}Pu (9000 atoms) per hundred kilograms of uranium.

Elaborate procedures have been set up for the detection of other rare types of disintegration, such as the emission of ^{14}C particles from Ra isotopes.[27-29] The branching ratio of this event is 10^{-10} to 10^{-11}.

Radiochemists have greatly benefited from the improvement of nuclear instrumentation during the last decade and from progress in data treatment. In addition, highly sensitive non-nuclear methods have become available for the measurement of long-lived nuclides. These include time-resolved laser-induced fluorescence spectroscopy and laser resonance ionization mass spectrometry.

The ultimate conceivable limit of detection would be that of a unique radioactive atom. This is conceptually not a very edifying situation, since this unique nuclide will have disappeared at the moment of its detection. Moreover, the decay of a single atom would be extremely difficult to observe since the instant of the event is intrinsically unpredictable.

REFERENCES

1. **Hyde, E. K., Hoffman, D. C., and Keller, O. R., Jr.,** A history and analysis of the discovery of elements 104 and 105, *Radiochim. Acta,* 42, 57, 1987.
2. **Katz, J. J., Seaborg, G. T., and Morss, L. R., Eds.,** *The Chemistry of the Actinide Elements,* Vol. 1 and 2, 2nd ed., Chapman and Hall, New York, 1986.
3. **Seaborg, G. T. and Loveland, W. D.,** *The Elements beyond Uranium,* John Wiley & Sons, New York, 1990.
4. **Kratz, J. V.,** The search for superheavy elements, *Radiochim. Acta,* 32, 25, 1983.
5. **Oganessian, Y. T., Lobanov, Y. V., Hussonnois, M., Kharitonov, Y. P., Gorski, O., Contantinescu, O., Popeko, A. G., Bruchertseifer, H., Sagaidak, R. N., Tretyakova, S. P., Buklanov, G. N., Rykhlyuk, A. V., Gulekyan, G. G., Pleve, A. A., Ivanov, G. N., and Plotko, V. M.,** The experiments aimed at synthesizing element 110, *Trends Nucl. Phys., Bologna,* 258, 1989.
6. **Kathren, R. L.,** *Radioactivity in the Environment: Sources, Distribution and Surveillance,* Harwood Academic Publishers, Chur, 1984.
7. **Durrance, E. M.,** *Radioactivity in Geology, Principles and Applications,* Ellis Horwood, Chichester, 1986.
8. **Eisenbund, M.,** *Environmental Radioactivity,* Academic Press, New York, 1973.
9. **Kuroda, P. K.,** *The Origin of the Chemical Elements and the Oklo Phenomenon,* Springer-Verlag, Berlin, 1982.
10. **Hoffman, D. C., Lawrence, F. U., Mewherter, J. L., and Rourke, F. M.,** Detection of plutonium 244 in nature, *Nature,* 234, 132, 1971.
11. **Fritz, P. and Fontes, J. Ch., Eds.,** *Handbook of Environmental Isotope Chemistry, Volume 1, The Terrestrial Environment, A,* Elsevier, Amsterdam, 1980.
12. **Ivanovich, M. and Harmon, R. S., Eds.,** *Uranium Series Disequilibrium. Application to Environmental Problems,* Clarendon Press, Oxford, 1982.
13. **Bouzigues, H., Boyer, R. J. M., Seyve, C., and Teulier, P.,** Contribution à la solution d'une énigme scientifique, in *The Oklo Phenomenon,* IAEA, Vienna, 1975.
14. **Hagemann, R. and Roth, E.,** Relevance of the studies of Oklo natural reactors to the storage of radioactive wastes, *Radiochim. Acta,* 25, 241, 1978.
15. **Ebihara, H. and Collins, K. E.,** Production of radionuclides using hot atom chemistry, in *Handbook of Hot Atom Chemistry,* Adloff, J. P., et al., Eds., Kodansha, Tokyo, 1992, 378.
16. **Baybarz, R. D.,** Recovery and application of the transuranium elements Np to Cf, *At. Energy Rev.,* 8/2, 327, 1970.
17. **Boyd, R. E.,** Technetium generators: status and prospects, *Radiochim. Acta,* 41, 59, 1987.
18. **Pupal, K. W. and Olsen, D. R.,** Electrodeposition of alpha-emitting nuclides from a mixed oxalate-chloride electrolyte, *Anal. Chem.,* 44, 284, 1972.
19. **Talvitie, N. A.,** Electrodeposition of actinides for alpha spectrometric determination, *Anal. Chem.,* 44, 290, 1972.
20. **Mitchell, R. F.,** Electrodeposition of actinide elements at tracer concentrations, *Anal. Chem.,* 32, 326, 1960.
21. **Aunnann, D. C. and Mullen, G.,** Quantitative electrodeposition of actinides from dimethylsulfoxyde, *Nucl. Instrum. Methods,* 75, 1151, 1974.
22. **Beceril, V. A., Rojas, H. A., and Meas, V. Y.,** Application of the species existence — predominance diagram to samarium electrodeposition, *Radiat. Phys. Chem.,* 38, 250, 1991.
23. **Lauer, K. F. and Verdingh, V.,** Preparation by electrospraying of thin uranium, plutonium and boron samples for neutron cross section measurements in 4π geometry, *Nucl. Instrum. Methods,* 21, 161, 1963.

24. **Lauer, K. F. and Verdingh, V.,** The preparation of uranium layers by high voltage electrophoresis, *Nucl. Instrum. Methods,* 60, 125, 1967.
25. **Lauer, K. F. and Verdingh, V.,** Equipment for electrospraying, *Nucl. Instrum. Methods,* 49, 179, 1967.
26. **Hussonnois, M., Brillard, L., and Constantinescu, O.,** Preparation of thin α sources by electrospraying (in French), *Nucl. Instrum. Methods Phys. Res.,* 224, 355, 1984.
27. **Houranni, E., Hussonnois, M., Stab, L., Brillard, L., Gales, S., and Schapira, J. P.,** Evidence for the radioactive decay of ^{226}Ra by ^{14}C emission, *Phys. Lett. B,* 160, 375, 1985.
28. **Hussonnois, M., Le Du, J. F., Brillard, L., Dalmasso, J., and Ardisson, G.,** Search for a fine structure in the ^{14}C decay of ^{222}Ra, *Phys. Rev. C.,* 43, 2599, 1991.
29. **Ardisson, G.,** New radioactivities: exotic radioactivity or magic radioactivity?, *Appl. Rad. Isot.,* 34, 1, 1992.

DETECTORS AND RADIOCHEMICAL TECHNIQUES

Birks, J. B., *The Theory and Practice of Scintillation Counting,* Pergamon, New York, 1964.

Bransome, E. D., Ed., *Liquid Scintillation Counting,* Grune & Stratton, New York, 1970.

Braun, T., *Radiochemical Separation Methods,* Elsevier, Amsterdam, 1975.

Debertin, K. and Helmer, R. G., *Gamma- and X-Ray Spectrometry with Semiconductor Detectors,* North-Holland, Amsterdam, 1988.

Dyer, A., Ed., *Liquid Scintillation Counting,* Vol. 1, Heyden and Sons, London, 1971.

Dyer, A., *Liquid Scintillation Counting Practice,* Heyden and Sons, London, 1980.

Faires, R. A. and Boswell, G. G. J., *Radioisotope Laboratory Techniques,* 4th ed., Butterworths, London, 1981.

Geary, W., *Radiochemical Methods,* John Wiley & Sons, New York, 1986.

Horrocks, D. L., *Applications of Liquid Scintillation Counting,* Academic Press, New York, 1974.

Kowalski, E., *Nuclear Electronics,* Springer-Verlag, Berlin, 1970.

Krüger, J., Ed., *Instrumentation in Applied Nuclear Chemistry,* Plenum Press, New York, 1973.

Mann, W. B., Ayres, R. L., and Garfinkel, S. B., *Radioactivity and its Measurement,* 2nd ed., Pergamon Press, New York, 1980.

Meinke, W. W., Choppin, G. R., and Ballou, N. E., Eds., Nuclear Science Series, Radiochemical Techniques, Subcommittee on Radiochemistry, National Academy of Sciences, National Research Council, NAS-NS, 3110 ff, Office of Technical Services, Department of Commerce, Washington, D.C.

Meinke, W. W., Starting Ed., Nuclear Science Series, Radiochemistry of the Elements, Subcommittee on Radiochemistry, National Academy of Sciences, National Research Council, NAS-NS, 3001 ff, Office of Technical Services, Department of Commerce, Washington, D.C.

Ouseph, P. J., *Introduction to Nuclear Radiation Detectors,* Plenum Press, New York, 1975.

Overmann, R. T. and Clark, H. M., *Radioisotopes Techniques,* McGraw-Hill, New York, 1960.

Siegbahn, K., *Alpha-, Beta- and Gamma-Ray Spectroscopy,* (Vol. 2), North-Holland, Amsterdam, 1966.

Tait, W. H., *Radiation Detection,* Butterworths, London, 1980.

APPENDIX 5.1
DERIVATION OF THE POISSON AND LAPLACE-GAUSS DISTRIBUTIONS

The binomial distribution law reads

$$P(N_i) = \frac{N!}{N_i!(N - N_i)!} p^{N_i}(1 - p)^{N-N_i} = \frac{p^{N_i}}{N_i!} AB$$

with $\overline{N}_i = p\,N$ and $\sigma^2 = \overline{N}_i\,(1 - p)$.

For $p << 1$ and $N_i << N$,

$$A = \frac{N!}{(N - N_i)!} \approx N^{N_i} \qquad B \approx e^{-Np}$$

The Poisson law follows immediately:

$$P(N_i) = \frac{\overline{N}_i^{N_i}}{\overline{N}_i!} e^{-\overline{N}_i}$$

with $\sigma \approx \sqrt{\overline{N}_i}$

In the case where $0 < p < 1$, $N_i << N$ and both N and N_i are large enough to be considered as continuous variables, $P(N_i)$ attains a maximum value for $N_i = N_i^{max}$ for which $\ln P(N_i)$ can be expanded. Setting $N_i = N_i^{max} + \eta$, it follows

$$\ln P(N_i) = \ln P(N_i^{max}) + \left[\frac{d \ln P(N_i)}{d N_i}\right]_{N_i^{max}} \eta + \frac{1}{2}\left[\frac{d^2 \ln P(N_i)}{d N_i^2}\right]_{N_i^{max}} \eta^2 + \dots$$

or

$$\ln P(N_i) = \ln P(N_i^{max}) + A + B$$

For $N_i = N_i^{max}$, $A = 0$; B is always negative since N_i tends toward a maximum. Setting β the absolute value of the second derivative of $\ln P(N_i)$ one obtains

$$P(N_i) = P(N_i^{max} + \eta) = P(N_i^{max})\, e^{-(1/2)\beta\eta^2}$$

$P(N_i^{max})$ is a constant which is explicited from the normalization condition

$$\sum_{i=0}^{i=N} P(N_i) = \int_0^{\infty} P(N_i)\, dN_i = \int_{-\infty}^{+\infty} P(N_i^{max} + \eta)\, d\eta = 1$$

In the term at left, N_i is a discontinuous variable. From

$$P(N_i^{max}) \int_{-\infty}^{+\infty} e^{-(1/2)\beta\eta^2}\, d\eta = P(N_i^{max}) \sqrt{\frac{2\pi}{\beta}}$$

it follows that

$$P(N_i) = \sqrt{\frac{\beta}{2\pi}}\, e^{-(1/2)\beta(N_i - N_i^{max})^2}$$

N_i^{max} and β are obtained through straightforward calculations and appropriate approximations as shown below:

$$\ln P(N_i) = \ln N! - \ln N_i! - \ln (N - N_i)! + N_i \ln p$$
$$+ (N - N_i) \ln (1 - p)$$
$$\ln N! = \ln\sqrt{2\pi}\, N_i + N_i \ln N_i - N_i$$
$$\frac{d \ln N_i!}{d N_i} = \frac{1}{2}\frac{\sqrt{2\pi}}{\sqrt{2\pi}\, N_i}\frac{1}{\sqrt{N_i}} + \frac{N_i}{N_i} - \ln N_i - 1 = \frac{1}{2\, N_i} + \ln N_i \approx \ln N_i$$

Since N is a constant,

$$\frac{d \ln P(N_i)}{d N_i} = - \ln N_i - \ln (N - N_i) + \ln p - \ln(1 - p)$$
$$= \ln \frac{(N - N_i)\, p}{N_i\, (1 - p)}$$
$$\frac{d^2 \ln P(N_i)}{d N_i^2} = - \frac{1}{N_i} - \frac{1}{N - N_i}$$

The expression of N_i^{max} is obtained from

$$\ln \frac{(N - N_i^{max})\, p}{N_i^{max}\, (1 - p)} = 0$$

which gives

$$N_i^{max} = N\, p = \overline{N}_i$$

With the approximation $N_i \approx \overline{N}_i$

$$\beta = \left| - \frac{1}{\overline{N}_i (1 - p)} \right| = \frac{1}{\sigma^2}$$

and finally

$$P(N_i) = \frac{1}{\sigma\sqrt{2\pi}} \exp - \frac{(N_i - \overline{N}_i)^2}{2\,\sigma^2}$$

REFERENCE

Reif, F., *Fundamentals of Statistical and Thermal Physics*, McGraw-Hill, New York, 1965, 7.

CHAPTER 6

THE RADIOCHEMIST'S VIEW OF CHEMICAL REACTIONS

I. FROM MACROCHEMISTRY TO TRACER CHEMISTRY

A specific characteristic of radiochemistry is the vast range of concentrations, in a given phase, which are experimentally accessible.

In the following, M denotes a metallic element and X a nonmetallic element whose concentrations may vary between wide limits. In the case of mixtures, a specific element will be labeled M_j or X_j. The accessible lower limit for detection of amounts or concentrations of *stable* elements M and X is relatively low thanks to the sensitivity of present-day analytical methods.[1-3] It is now a common practice to measure nanograms or picograms of an element. In terms of concentration, and with regard to constraints imposed by sample preparation, these figures correspond to parts per million (μg/g or mg/kg) or parts per billion (ng/g or μg/kg) levels. In extreme cases, the concentration may be of the order of 10^{-8} *M* for an element with atomic mass 100. Even at this dilution, however, the number of species per gram or milliliter is still enormous, and may amount to about 10^{12}.

The dilution of a stock solution to this concentration range, whatever the nature of the phase, requires many precautions in order to prevent introduction of impurities in amounts comparable to those of the elements M or X.

The terminology of the concentration range is rather vague. Quite generally, *traces* cover the levels from 100 ppm to 1 ppm and microtraces from 1 ppm to 1 ppb (Table 6.1). This more or less indicates the frontiers of ultramicrochemistry, which is still "classical" chemistry but reduced to the microgram scale. For example, 1 μg can be handled in 10^{-2} mL with the aid of a microbalance and capillary test tubes to make up a 10^{-4} *M* solution.

Radiochemists now commonly perform experiments with microgram amounts of radioelements using microscopes, micromanipulators, and beads of ion exchange resins in operations with quantitative transfer. Up to 10 μg of the elements from Am to Cf can be fixed on a single cationic resin bead from a

Table 6.1
Concentration Range of a Solute of Atomic Mass 100

M	Entities (per mL)	Weight (per mL)	Weight fraction	Half-life	Range
10^{-1}	6×10^{19}	10 mg	$1{:}10^{2}$	1.3×10^{12} years	Macroscale
10^{-3}	6×10^{17}	100 mg	$1{:}10^{4}$	1.3×10^{10} years	
10^{-4}	6×10^{16}	10 mg	$1{:}10^{5}$	1.3×10^{9} years	Microscale
10^{-6}	6×10^{14}	100 ng	$1{:}10^{7}$	1.3×10^{7} years	
10^{-7}	6×10^{13}	10 ng	$1{:}10^{8}$	1.3×10^{6} years	Submicroscale
10^{-9}	6×10^{11}	100 pg	$1{:}10^{10}$	1.3×10^{4} years	
10^{-10}	6×10^{10}	10 pg	$1{:}10^{11}$	1.3×10^{3} years	Tracer scale
10^{-16}	6×10^{4}	10 ag	$1{:}10^{17}$	11.6 h	
10^{-17}	6×10^{3}	1 ag	$1{:}10^{18}$	1.16 h	Subtracer scale
10^{-20}	6	10^{-3} ag	$1{:}10^{21}$	4.2 s	
10^{-21}	1	10^{-4} ag	$1{:}10^{22}$	0.7 s	Single atom

Note: The half-life is that of the nuclide which would yield an activity of 1 Becquerel/mL; 1 ag = 1 attogram = 10^{-18} g.

volume of several milliliters. The first absorption spectrum of Bk(IV) was measured on a drop containing 0.6 μg of ^{249}Bk, and the first sample of metallic Pu was obtained by the reduction of 35 μg of PuF_4 with Ba (in 1910, M. Curie and A. Debierne had prepared the first sample of metallic Ra by electrolysis of a solution containing 106 mg of $RaCl_2$!). The structure of BkO_2 was established with 0.02 μg and the magnetic susceptibility of ^{254}Es was measured with 0.01 μg loaded on a resin bead. In these determinations, no use was made of radioactivity.

Concentration ranges and amounts of *radioactive* matter detected by the emitted radiations open new vistas. At present, concentrations of solutions well below 10^{-8} mol L^{-1} are common practice and the lower limit is set by the measurements of very weak activities. Determinant factors are the half-life of the radionuclide, the mode of decay, the properties of the emitted radiations and the time available for completing an experiment, rather than the performance of modern detectors. Concentrations down to 10^{-16} mol L^{-1} are frequently handled, but the solution still contains several thousand species per milliliter. Of course, the amounts of elements M and X will be imponderable.

The *tracer scale*, also commonly termed *tracer level*, refers to amounts less than 10^{-8} g, which usually correspond to concentrations below 10^{-10} mol L^{-1}. The origin of the term lies in the practice of labeling a stable element M (or X) by one of its radioactive isotopes M* (or X*) in order to follow its *trace* along a physical, chemical, or biological pathway. The weight ratio of M* to M is of the order of 10^{-10}, e.g., 10^{-10} g of tracer in 1 g of element M.

Since the first *labeling* experiment of Pb with ^{212}Pb, the historical expression *tracer level* has become by extension a general designation for very small amounts of radionuclides. When dealing with radioactive tracers in solution, concentrations of impurities are usually greatly in excess with respect to the investigated elements M or X.

The term *carrier-free tracer* encountered in the literature appears ambiguous. It is intended to designate an element represented by a short-lived radionuclide in the absence of a stable or long-lived isotope. ^{32}P (T = 14.5 d) with an activity of 1 MBq has a mass of 10^{-10} g; the nuclide is both carrier-free *and* at tracer concentration. The mass of carrier-free ^{239}Pu (T = 2.4×10^4 years) with the same activity is 4×10^{-4} g, well above a tracer amount.

Compounds of elements at the tracer level cannot be isolated and only certain characteristics can be established. Determination of data that require a solid phase of pure M or X or one of their solid compounds is precluded. In consequence, the behavior of tracers can only be inferred from investigations in the gaseous phase or in liquid or solid solutions, including interaction of radionuclides with any solid phases that may be present. Sometimes, however, as will be shown, some characteristics of "virtual" compounds which *may* exist at the macroscopic level can be established at tracer concentration.

Upon dilution of a concentrated solution of M down to the tracer scale, acidobasic and redox processes, as well as dissociative reactions of dimers, may change the nature of the species of M (see Chapter 8).

The *sub-tracer level* represents the ultimate conceivable concentration level and terminates with one single atom. Indeed, systems containing a few ten or hundred atoms in reasonably accessible volumes for which α or spontaneous fission decay events can be measured, do occur. The behavior of matter at the sub-tracer level is discussed in Chapter 9.

II. DISTRIBUTION OF ELEMENTS AMONG VARIOUS SPECIES

An element in aqueous or organic solution, or in a solid solution, may exist in the form of a variety of more or less complex species characterized by their overall stoichiometry, constitution in sub-units and structure. In most cases, only the first of these parameters is accessible. The purpose of radiochemical experiments is to identify these species as far as possible, determine their relative concentrations under given conditions, and obtain their rate of formation and disappearance.

In the following discussion, a convenient notation will be used, and for the sake of simplicity it will be introduced for the commonest case of aqueous solutions. Because the majority of elements involved in radiochemistry are metals, the notations refer to such elements M_j, but they may also be extended to nonmetallic elements.

A. Aqueous Solutions

An aqueous solution is prepared by introducing into water various reagents such as acids, bases, and salts. In the absence of complexing agents (e.g., in perchloric acid media which are considered as noncomplexing), the solution contains H^+ (as short form for H_3O^+ or solvated protons) and OH^- ions. If a single complexing agent H_pL (e.g., H_3PO_4) is introduced, either as such, or as one of its salts, the solution contains various entities $E_i(L)$, which are derivatives of the complexing agent. These include H_pL molecules and the anions resulting from successive deprotonations, i.e., $H_{p-1}L^-$, $H_{p-2}L^{2-}$, until the anion L^{p-}. The ligand L may be a more or less complex atomic group, including parts of polyfunctional compounds of high atomic weight such as fulvic and humic acids. *Spectator* elements which are not involved in the chemical reactions considered (Li^+, Na^+, and possibly others) are indicated as M_j', to distinguish them from the chemically active M_j elements investigated.

Consider an element M (e.g., U) in the oxidation state N written M(N) [e.g., U(IV)] in a solution without or containing only one complexant H_pL. The element M is distributed among various species denoted E_i[M(N), L], for instance E_i[U(VI), PO_4]; in cases where the oxidation state is well defined, a shorter notation E_i(M) such as E(U) will be used. These species include the aqua ion, hydrolyzed forms, monomeric complexes and possibly polymeric ones.

All monomeric entities will be represented by the symbol C_l^z in which l designates the order of the complex and z designates the charge (>0, <0, or nil) of the species. In other words, l is the number of ligands, excluding water and the ligands OH and O derived from water. For noncomplexed aqua ions or hydrolyzed forms, $l = 0$. In the following, depending on the case, the charge z will be used with its algebraical or absolute value.

At this point, the notation is still somewhat ambiguous because it does not specify the nature of the ligands that constitute the entities, e.g., oxo, hydroxo, and ligands which derive from the complexant. This point will be discussed in a later section.

An arbitrary reference solution is considered for illustration. It is prepared by dissolution of the following compounds in water:

NaH_2PO_4, H_3PO_4, $HClO_4$, NaOH, UCl_4, $UO_2(NO_3)_2$, $FeCl_3$, and $FeCl_2$.

The notation of all species in solution is the following:

$E_i(PO_4) = H_3PO_4, H_2PO_4^-, HPO_4^{2-}, PO_4^{3-}, H_6P_2O_8, H_5P_2O_8^-$
$E(Cl) = Cl^-$, $E(ClO_4) = ClO_4^-$, X = Cl, O, P, N; L = Cl, ClO_4, PO_4, P_2O_8, NO_3
$M' = Na$; M = U, Fe

Using the notation C_l^z various species are characterized as follows:

$E_i[U(IV)]$, e.g., $U^{4+} = C_0^{4+}$, $UOH^{3+} = C_0^{3+}$
$E_i[U(IV), PO_4]$, e.g., $U(HPO_4)_2(H_2PO_4)_2^{2-}$ and $U(HPO_4)_3\ (H_3PO_4)^{2-} = C_4^{2-}$
$E_i[U(VI)]$, e.g., $UO_2^{2+} = C_0^{2+}$
$E_i[U(VI), PO_4]$, e.g., $UO_2(H_2PO_4)_2(H_3PO_4)^+$, $UO_2(H_2PO_4)\ (H_6P_2O_8)^+ = C_3^+$
$E_i[Fe(II)]$ and $E_i[Fe(III)]$ are the aqua and hydrolyzed species of Fe(II) and Fe(III)
$E_i[Fe(II), PO_4]$ and $E_i[Fe(III),PO_4]$ are the phosphoric complexes of Fe(II) and Fe(III)

An advantage of this notation is its possible extension to more complicated solutions containing:

An element in different oxidation states: $E_i[M(N)]$, $E_i[M(N+1)]$
A mixture of several elements $E_i(M_j)$
Mixed complexes $E_i(M_1, M_2)$ or more often $E_i(M, L, Q) = C_{l,q}^z$ formed in competing reactions with the complexants H_pL and $H_{p'}Q$. In the symbol, the subscript q refers to the number of ligands deriving from $H_{p'}Q$.

If H_2SO_4 is added to the reference solution, further entities appear:

$E_i[SO_4] = HSO_4^-$, SO_4^{2-}
$E_i[U(IV), SO_4]$ and $E_i[U(VI), SO_4] = C_q^z$
$E_i[(U(IV), PO_4, SO_4]$ and $E_i[U(VI), PO_4\ SO_4] = C_{l,q}^z$

The solution may also contain the species $E_i(M_j')$. The entities $E_i(L)$ generally possess an element M′ and oxygen atoms or an element X′ alone or associated with oxygen. The solutions described are convenient for the modeling of natural aquatic systems containing multiple components.

The notations C_M, C_H, and C_L are used for the total concentrations of element M, hydrogen atoms and ligand groups L, respectively, and C(x) for the total concentration of reagent x. Square brackets indicate equilibrium concentrations. For the reference solution defined above,

$$C_U = C_{UCl_4} + C_{UO_2(NO_3)_2} \qquad C_H = 3\,C_{H_3PO_4} + 2\,C_{NaH_2PO_4} + C_{HClO_4}$$
$$C_{PO_4} = C_{H_3PO_4} + C_{NaH_2PO_4} \qquad C_{Cl} = 4\,C_{UCl_4} + 3\,C_{FeCl_3} + 2\,C_{FeCl_2}$$

In a "true" aqueous solution, thermodynamic equilibrium is established between all entities $E_i(M_j)$, $E_i(L_j)$ and H^+. This implies that the chemical potential is defined for all entities. Quite generally, the equilibrium concentrations $[E_i]$ of all entities involved in the equilibria depend on C_H, C_L, and C_M, which are themselves a function of the concentration C of the reagents. If the pH is known, the concentrations $[E_i]$ can be expressed uniquely in terms of C_L and C_M. The relations between the various $[E_i]$ are often complicated; they include the equilibrium constants, which are assumed to be known, and the thermodynamic activity coefficients, which are always subject to some degree of uncertainty. The attainment of equilibrium may not be instantaneous.

At the very low concentrations encountered with radionuclides, the possible appearance of new species, called *radiocolloids*, complicates the overall description. This phenomenon is observed under conditions for which elements M at much higher concentrations would be hydrolyzed. Because radiocolloids appear progressively and are not in thermodynamic equilibrium, their chemical potential cannot be defined.

B. Non-Aqueous Solutions

Virtually only neutral species exist as solutes in organic solvents of low dielectric constant. They may include molecules, e.g., $E_i(M_j) = C_l^0$ or $E_i(L_j) = H_pL$, and ion pairs $E_i(M_jQ_j) = C_{l,q}^0$.

For example, a chelate derivative of U(IV) in benzene may be $UO_2A_2 = C_2^0$ (HA is the chelating agent); in a benzene solution of $R_3NH^+Cl^-$ (prepared by salification of the amine R_3N with an aqueous solution of HCl) and U(IV), one of the entities may be $UCl_6^{2-}(R_3NH^+)_2 = C_{6,2}^0$.

Similarly, in solid solutions containing associated groups of opposite signs, only neutral entities prevail.

III. MACROCOMPONENTS AND MICROCOMPONENTS

A. Definitions

In radiochemistry it is important to distinguish species at the tracer level from those present at higher concentrations, since the occurrence of reactions will depend on their relative amounts. Hence, a more quantitative definition of tracer and nontracer concentrations is required. This will be illustrated for an aqueous solution and is readily extended to other systems.

Let us consider a true aqueous solution at equilibrium that contains a complexant H_pL and the various entities $E_i(M_2)$ of an element M_2. An element M_1, at a given concentration, is added to the solution; it will react with the $E_i(M_2)$ and $E_i(L)$ entities present. If, after reaction, changes in the equilibrium concentrations of the latter are not measurable, M_1 is said to be a *microcomponent* with respect to M_2 and with respect to the elements which constitute H_pL. To distinguish the element M_1, it will be designated m_1. Conversely, M_2 and the elements included in H_pL are *macrocomponents* with respect to m_1. These definitions of micro- and macrocomponents are valid if $C_{m_1} << C_{M_2}$ and $C_{m_1} << C_L$.

For example, if the reference solution contains the tracer ^{230}U (T = 18 d) and if the reagent concentrations are $C_{H_3PO_4} = 0.5\ M$ and $C_{FeCl_3} = 10^{-3}\ M$, U is a microcomponent with respect to the reagents, since its addition does not change the existing equilibria, despite the possible oxidation of U(IV) by Fe(III) and the complexation of U(IV) and U(VI) by phosphoric acid.

More complicated situations may arise, e.g., in the case where $C_{m_1} \approx C_{m_2} << C_{M_j}$, with j = 3, 4, . . . and $C_{m_1} \approx C_{m_2} << C_L$, m_1 and m_2 are microcomponents. On the other hand, if $C_{m_1} << C_{M_j}$, j = 2, 3, . . . and $C_{m_1} << C_L$, C_Q there is *one* microcomponent, m_1. The microcomponent in a radiocolloidal solution is easily identified (see Chapter 7).

B. Graphical Representation of a Microcomponent Solution

The description of an aqueous solution is rather complicated and it is frequently expedient to represent graphically the composition of a system that contains an element M alone or in the presence of a complex H_pL.[4] Then that 3-D representation illustrates the regions in which the various entities $E_i(M)$ of the element M can exist as a function of the independent parameters which define the solution: pH, $[H_pL]$ (or any other pH- and C_L-dependent equilibrium concentration of $H_{p-x}L^{-x}$ entities) and C_M. These solutions can be obtained by simple dissolution of appropriate reagents in water.

The representation is quite straightforward for the entities of a microcomponent in the solution because the presence of such a component does not change the equilibrium concentrations of the macrocomponent entities. For this reason, the concentration C_m of the microcomponent vanishes, and the solution can be represented in an x-y plane. Moreover, the equilibrium concentrations $[E_i(L)]$ can be calculated from C_L and the pH. In the following, for reasons which will be apparent later, the graphs will be drawn with the independent parameters $[H^+]$ and $[H_pL]$.

All solutions, whether or not they contain complexing agents, are now represented by points in the pH-log $[H_pL]$ plane. The $E_i(m)$ equilibrium concentrations depend only on these two parameters and their domain of existence in the plane are delimitated by isoconcentration curves.

If only one reagent is introduced into water, for example, an acid, the two parameters pH and $[H_pL]$ are related and cannot vary independently. All acid solutions are represented by a curve in the plane. The intersection of this curve with the domain of existence of the $E_i(m)$ entities indicates the nature of the predominant species at a given H_pL concentration. Figure 6.1 represents aqueous solutions of an element at tracer concentration in presence of sulfuric, citric, or phosphoric acid, as well as in the presence of CO_2 and carbonates. Calculation of equilibrium concentrations in simple cases are given in Appendix 6.1.

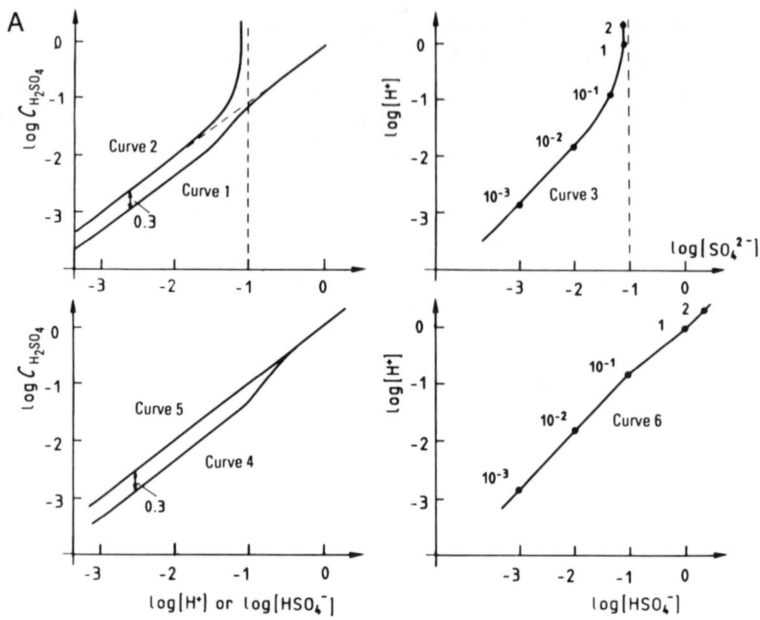

FIGURE 6.1. Graphical representations of solutions of microcomponents. (A) Sulfuric acid media. The variations of log $[H^+]$ and log $[SO_4^{2-}]$ in H_2SO_4 (curve 3) are derived from the computed variations of log $C_{H_2SO_4}$ with log $[H^+]$ and log $[SO_4^{2-}]$, curves 1 and 2. The same procedure is used for the variations of log $[H^+]$ and log $[HSO_4^-]$ (curve 6) in H_2SO_4, derived from curves 4 and 6. The number on the curves denote $C_{H_2SO_4}$. All other points in the x-y plane correspond to independent values of $C_{H_2SO_4}$ and pH. (B) Phosphoric (L = PO_4) or citric (L = Cit) media. Log $[H^+]$ and log $[H_3L]$ are independent, except in phosphoric (curves 1 and 2) or citric (curve 3) acids. The numbers on the curves denote C_{H_3L}. Curve 1: log $[H^+]$, variable ionic strength; curve 2: H_3PO_4, $\mu = 0.2$ (ionic strength for $C_{H_3PO_4} = 4.2$ *M*); curve 3: H_3Cit, $\mu = 1$. To each value which determines the pKa of H_3L corresponds a specific log $[H^+]$/log $[H_3PO_4]$ area. All other points in the x-y plane correspond to independent values of C_{H_3L} and pH. Arbitrary isoconcentration curves are drawn for a C_i^z species of the microcomponent: (a) 80%, (b) 25%, (c) 10%. (C) Carbonate media. In the picture at left, numbers on curve 1 denote P_{CO_2}; in the picture at right, $P_{CO_2} = 10^{-3.5}$ atoms and the numbers on curve 2 denote the concentration C_{CO_3} in solution. The log $[HCO_3^-]$ scale is only for left figure.

IV. THE NATURE OF REACTIONS

A. Reaction Pathways

The acid-base and redox processes which can occur in a given system depend on all possible reactions between the E_i species of the micro- and macrocomponents. Figure 6.2 shows that three pathways are open for these reactions.

Path 1 is the reaction between $E_i(M_j)$ and/or $E_i(L_j)$ entities of macrocomponents. It is the realm of "classical" chemistry at usual concentrations. Path

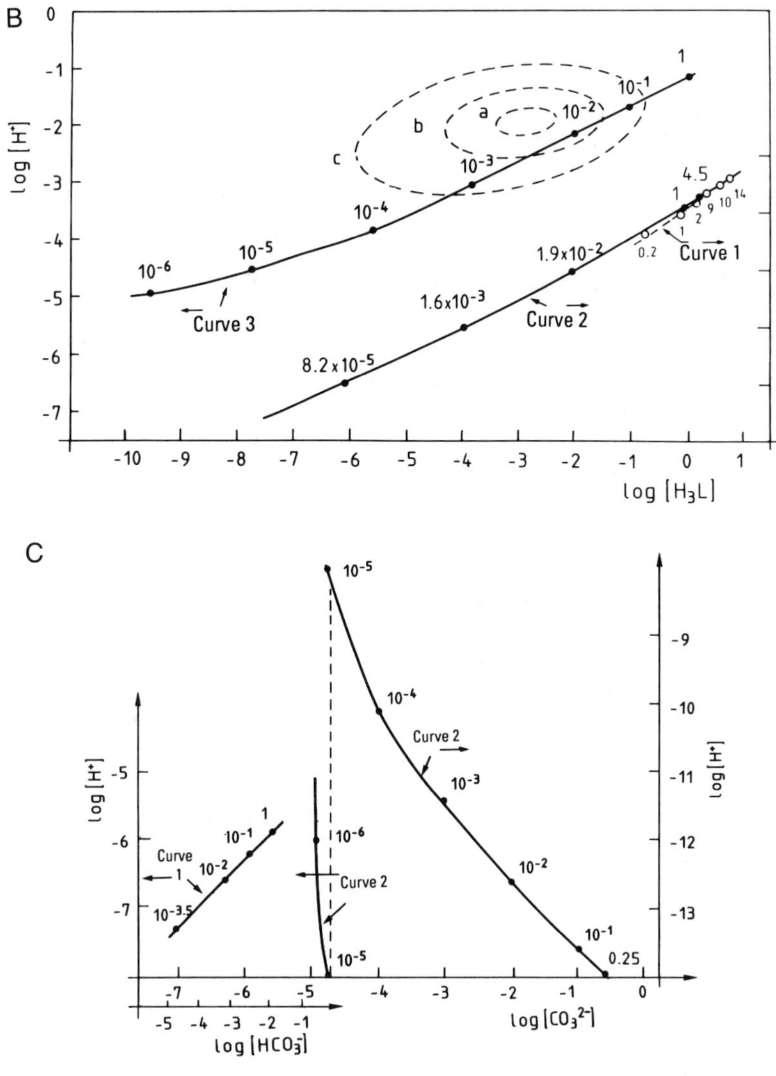

FIGURE 6.1 (continued).

2 between the $E_i(m_j)$ species of the microcomponent and the $E_i(M_j)$ and/or $E_i(L_j)$ species of the macrocomponent is typical for tracer chemistry. The third pathway between $E_i(m_j)$ entities of microcomponents is also pertinent to both tracer and sub-tracer chemistry. However, the reactions are peculiar in that their rates are rather slow. In particular, the probability of encounter between two microcomponent entities, which depends on the prevailing volume, is generally extremely low.

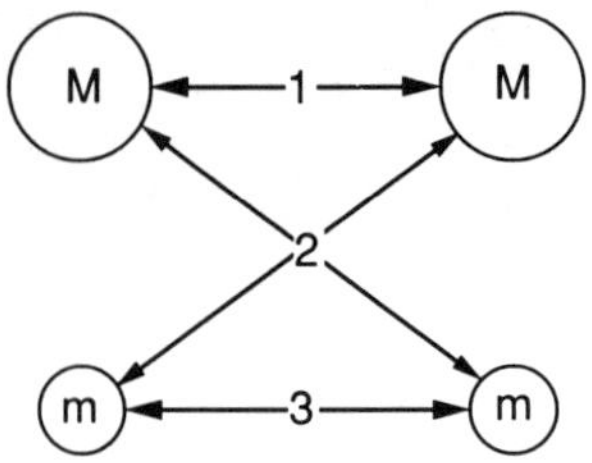

FIGURE 6.2. Reaction pathways between microcomponents m and macrocomponents M. (1) Ordinary chemical reactions; (2) and (3) reactions relevant to radiochemistry.

B. Observable and Observed Reactions

Any discussion on reactions in highly diluted systems is meaningful only if such reactions can really occur and can be observed. A prognosis can be established as a function of various time parameters:

1. The half-life T_j of the radionuclide representing the elements M_j
2. The time t allowed for the chemical species to react
3. The reaction half-time τ for attaining an equilibrium or pseudoequilibrium state
4. The time interval Δt required for collecting significant data on the extent of the reaction

The relative values of these time parameters determine the conditions under which a reaction can be observed. The equilibrium state of a reaction involving elements M_j is only accessible if the various times satisfy the sequence

$$T_j > t > \tau > \Delta t$$

Depending on the value of the time t available for the reaction, two classes of reactions, denoted A and B, are distinguished. Class A includes the usual chemical reactions performed in the laboratory on time scales ranging from minutes to, say, several months. This class also includes natural reactions on earth that develop in the course of several years. In class B, the duration of reactions is very long, extending from several years to millions of years. Such events occur in the geochemical realm or in interstellar space.

With these definitions, it is possible to characterize the reactions according to combinations of the pathways 1, 2, or 3 and the classes A and B, and on the basis of the number of species, or N_{m_j} or N_{M_j}.

Reactions 1A and 1B are not relevant to the context and serve only as a reference. Reactions 3A between radioactive m_i species are generally not observable in the laboratory because of kinetic hindrance, i.e., $t << \tau$ (see Chapter 7). They are even less amenable to observation for stable elements in microamounts. On the other hand, completed 3B reactions can be observed because the reaction time t (and necessarily the half-life T) greatly exceeds τ and kinetics are no longer an obstacle.

For radioactive microcomponents, the observation time Δt may be quite long, and data can be collected on natural systems of limited confinement. Otherwise, as in space chemistry, the observation must be performed on very extended systems. The extent and characteristics of 3A and 3B reactions depend on the number of species N_{m_j} rather than on the half-time values τ.

2A and 2B reactions between micro- and macrocomponents are always kinetically allowed; they may occur and are actually observed. The advancement of the reaction, or the equilibrium state, must be controlled by measurements made on the radioactive microcomponent because the variation in the number of N_{M_i} species in the course of the chemical reaction is not perceptible. The 2A and 2B reactions depend mainly on t, and only to a much lesser extent on N_{m_j}.

Examples of these reactions are very numerous. All anthropogenic and natural reactions which control the behavior of radionuclides, whether artificial or not, belong to types 2A and 2B. They intervene in the atmosphere, geosphere, biosphere, and in all living organisms.

V. THE LIMIT OF KNOWLEDGE OF A CHEMICAL SYSTEM WITH CONVENTIONAL METHODS

Complete and perfect knowledge of a chemical system S would require *in situ* analysis of the oxidation states of the elements and determination of the species in which these elements are incorporated, together with their relative abundances and formation and disappearance rates. If all this information is at hand, the evolution of the system can be predicted.

Here, *in situ* means determination without changing any property of the system. For an aqueous solution of a tracer element M it implies the identification of the preponderant complexes C_l^z of M in its various degrees of oxidation state, and the determination of their formation constants and the rate constants of reactions which lead to equilibria.

A common practice for obtaining chemical information on a system consists in perturbing the electronic environment of the species of interest. The perturbation may be chemical, electrochemical, or electromagnetic, the latter being the most common. The response signal to the perturbation should carry the information desired; however, when dealing with tracers one encounters a concentration limit below which information is no longer retrievable.[5]

With chemical and electrochemical methods, the limit generally lies in the range 10^{-4} to 10^{-5} *M*; a unique exception is the measurement of $[H^+]$, which is performed down to 10^{-13} *M*. The sensitivity of pulsed electrochemical methods for the measurement of a multivalent element in a particular oxidation state attains 10^{-8} *M* or possibly 10^{-11} *M* in the case of pulsed anodic redissolution.[6] However, secondary electrochemical reactions may occur at the electrodes, for example as redox waves or formation of O_2. In the present context, these methods are not reliable owing to the risk of alteration of the actual oxidation state of the elements investigated.

Electromagnetic radiation interacts in many different ways with electronic cores, but only the interaction with outer valence shells is informative. The response signal may consist of an electromagnetic radiation or of electrons if the interaction leads to ionization. Useful spectroscopic methods for the determination of oxidation states include: X-ray absorption and fluorescence (Exafs, Xanes with synchrotron radiation), optical (UV and visible) absorption and fluorescence, IR absorption and Raman emission, electron spectroscopies (Auger, Esca, XPS), and Mössbauer spectroscopy. These methods are not convenient for all elements and in some cases provide the atomic numbers of the species rather than their chemical states.

The minimum amount of a species that can be detected is determined by the number dN of response entities in the exit channel which is required to record an acceptable signal. For an observation time dt,

$$dN = \sigma \, \Phi \, N_{E_i(M)} \, dt$$

in which σ is the cross section (in square centimeter or in barn per atom or per nucleus) of the interaction between the incident beam and the $E_i(M)$ species, Φ the entrance fluence rate ($cm^{-2}\ s^{-1}$) and $N_{E_i(M)}$ is the number of species M in a given oxidation state. The sensitivity increases with the product $\sigma\Phi$ and is highest for the method that permits detection of the smallest response dN/dt.

Cross sections vary in a very broad range from 10^8 b for UV photon-electron interaction to a few barns for photon-nucleus interactions. The values of dN depend on the background and on the means of detection, which may involve measurement of a current or pulses. For a given set of experimental conditions, the sensitivity increases with the particle flux in the entrance channel. With the substitution of conventional sources by synchrotron radiation or lasers, the minimum number of entities that can be detected under the most favorable conditions lies in the range of 10^{12} to 10^{13} per milliliter or corresponds to a concentration of about 10^{-8} *M*, substantially higher than the amount of matter present at the tracer level. Sensitivity can be improved by several orders of magnitude in optical methods by using laser beams.[7]

In the present context, the use of laser-based techniques in actinide chemistry must be emphasized. Resonance ionization spectroscopy is reputed to have a sensitivity down to the detection of one atom.[8-10] With laser resonance ionization mass spectrometry, the present detection limit is 10^7 atoms for ^{239}Pu and ^{99}Tc.[11,12] With laser-induced fluorescence spectrometry, it is possible to detect 2×10^9 molecules per milliliter of the long-lived ^{129}I.[13]

In conventional optical spectroscopy, only the *difference* between the intensities of the incident and transmitted radiation is measured, and this is independent of the initial intensity; under these conditions the sensitivity cannot be increased. On the other hand, very small degrees of absorption on minute amounts of matter can be detected from the direct or indirect measurement of the energy effectively absorbed in the medium at a given wavelength. In this case, sensitivity can be readily improved by simply increasing the power of the excitation source.

A common procedure is the excitation of electronic fluorescence by a pulsed laser beam, followed by measurement of the lifetime of a characteristic level. Time-resolved laser-induced fluorescence spectroscopy is commonly used for the speciation of Am, Cm, Bk, and Es in aqueous solutions[14-17] (the sensitivity for Cm^{3+} is 8.5 ppt at room temperature).[15] In this way, it is possible to identify the oxidation state of a multivalent element and sometimes the species itself and the formation of complexes. The method is also applicable to lanthanides.[18,19] In any event, the element of interest can be distinguished from impurities, especially organic ones, which are always present and frequently fluorescent.

When fluorescence is weak, which means the excited levels decay by nonradiative processes, the absorbed energy is converted to heat that is dissipated into the medium.[20-22]

This phenomenon is utilized in photoacoustic spectroscopy, which measures the pressure wave generated by a tunable laser beam.[23] The pressure is measured with piezoelectric microphones, and for a given wavelength it is proportional to the number of absorbing species. The resulting spectrum is equivalent to a classical absorption spectrum and delivers the same information. Aqua ions and complexes of the actinides U(IV, VI), Np(IV, V, VI), Pu(III, IV, VI), and Am(III) have been detected at concentrations down to 10^{-7} or 10^{-8} *M*.[24-27] The sensitivity of classical spectrophotometry is thus improved by several orders of magnitude.

A second method for probing the energy absorbed by a medium uses the thermal lensing effect, based on the change of refractive index with an increase in temperature.[128] Absorption of the laser beam produces a local heating effect, and the temperature gradient generates a concave lens arising from the transverse gradient of the refractive index. The power of the lens is determined with a second laser beam. The intensity of the thermal lensing effect is proportional to the energy initially absorbed by the species and it provides a sensitive means of detecting very small degrees of absorption. The second laser beam eventually provides a spectrum equivalent to a classical absorption spectrum.

Thermal lensing is even more sensitive than photoacoustic spectroscopy. Fe(II) in chloroform has been measured at a concentration of 3×10^{-11} mol L^{-1}, equivalent to 0.002 ng mL^{-1}. On the other hand, aqueous solutions are much less favorable, as evidenced by the detection limit of 4×10^{-6} mol L^{-1} for U(VI).[29,30]

In conclusion, the concentration limit for which perturbation of the species in solution no longer delivers a perceptible signal is about 10^{-6} *M*. This limit can be extended to 10^{-8} *M* for a few elements under special conditions or by using laser beams. Consequently, classical methods do not provide access to a chemical system at very low concentration. The information obtained by chemistry at higher concentrations can only be extrapolated at the tracer scale, and this may involve risk and uncertainty. Alternatively, other methods of investigation must be sought.

REFERENCES

1. **Rastegar, F., Maier, E. A., Heimburger, R., Christophe, C. L., Ruch, C. L., and Leroy, M. J. F.,** Simultaneous determination of trace elements in serum by energy dispersive X-ray fluorescence spectrometry, *Clin. Chem.*, 30, 1300, 1984.
2. **Ruch, C. L., Rastegar, F., Heimburger, R., Maier, E., and Leroy, M. J. F.,** Energy dispersive X-ray fluorescence spectrometry with direct excitation at picogram levels, *Anal. Chem.*, 57, 1691, 1985.
3. **Amsel, G., Heitz, C. H., and Menu, M.,** MeV ion beam techniques, an outline, *Nucl. Instrum. Methods, B*, 14, 30, 1986.
4. **Rossotti, H.,** *The Study of Ionic Equilibria*, Longman, London, 1978.
5. **Guillaumont, R., Chevallier, P., and Adloff, J. P.,** Identification of oxidation states of ultra-trace elements by radiation detection, *Radiochim. Acta*, 40, 191, 1987.
6. **Benes, P. and Majer, V.,** *Trace Chemistry of Aqueous Solutions*, Elsevier, Amsterdam, 1980, 14.
7. **Kliger, D. S., Ed.,** *Ultrasensitive Laser Spectroscopy*, Academic Press, New York, 1983.
8. **Hurst, G. S., Payne, G., Kramer, S. D., and Young, J. P.,** Resonance ionization spectroscopy and one atom detection, *Rev. Mod. Phys.*, 51, 767, 1979.
9. **Travis, J. C.,** Limits to sensitivity in laser enhanced ionization, *J. Chem. Educ.*, 59, 909, 1982.
10. **Peuber, P., Herrmann, G., Rimke, H., Sattelberger, P., Trautmann, N., Ruster, W., Ames, F., Bonn, J., Kluge, H. J., Kronert, U., and Ohen, E. W.,** Trace detection of Pu by three steps photoionisation with a laser system pumped by a copper vapor laser, *Appl. Phys. B*, 38, 249, 1985.
11. **Trautmann, N., Herrmann, G., Mang, M., Mühlbeck, C., Rimke, H., Sattelberger, P., Ames, F., Kluge, H. J., Otten, E. W., Rehklau, D., and Ruster, W.,** Low level detection of actinides by laser resonance spectroscopy, *Radiochim. Acta*, 44/45, 107, 1988.
12. **Sattelberger, P., Mang, M., Herrmann, G., Riegel, J., Rimke, H., Trautmann, N., Ames, F., and Kluge, H. J.,** Separation and detection of trace amounts of technetium, *Radiochim. Acta*, 48, 165, 1989.
13. **Goles, R. W., Fukuda, R. C., Cole, M. W., and Brauer, F. P.,** Detection of iodine-129 by laser-induced fluorescence spectrometry, *Anal. Chem.*, 53, 776, 1981.
14. **Klenze, R., Kim, J. I., and Wimmer, H.,** Speciation of aquatic actinide ions by pulsed laser spectroscopy, *Radiochim. Acta*, 52/53, 97, 1991.
15. **Beitz, J. V., Bowers, D. L., Doxtader, M. M., Maroni, V. A., and Reed, D. T.,** Detection and speciation of transuranium elements in synthetic groundwater via pulsed-laser excitation, *Radiochim. Acta*, 44/45, 87, 1988.

16. **Moulin, C., Decambox, P., Mauchien, P., Moulin, V., and Theyssier, M.,** On the use of laser-induced time-resolved spectrofluorometry for interaction studies between organic matter and actinides: application to curium, *Radiochim. Acta,* 52/53, 119, 1991.
17. **Kim, J. I., Wimmer, H., and Klenze, R.,** A study of Cm(III) humate complexation by time resolved laser fluorescence spectroscopy, *Radiochim. Acta,* 54, 35, 1991.
18. **Bidoglio, G., Grenthe, I., Qi, P., Robouch, P., and Omenetto, N.,** Complexation of Eu and Tb with fulvic acids studied by time resolved laser induced fluorescence, *Talanta,* 38, 999, 1991.
19. **Sinha, S. P.,** Fluorescence spectra and lifetimes of the lanthanide aquo ions and their complexes, in *Systematics and the Properties of the Lanthanides,* Sinha, S. P., Ed., Reidel, Kluwer Academic Press, Dordrecht, 1983, 451.
20. **McDonald, F. A. and Wetsel, G. C.,** Theory of photothermal and photoacoustic effect in condensed matter, in *Physical Acoustics,* Mason, W. and Thurston, R., Eds., Academic Press, New York, 1988.
21. **Patel, C. K. N. and Tam, A. C.,** Pulsed optoacoustic spectroscopy of condensed matter, *Rev. Mod. Phys.,* 53, 517, 1981.
22. **Sawada, T. and Kitamoni, T.,** Analytical applications of photoacoustic spectroscopy to condensed phase substance, in *Physical Acoustics,* Mason, W. and Thurston, R., Eds., Academic Press, New York, 1988.
23. **Kim, J. I. and Klenze, R.,** Laser-induced photoacoustic spectroscopy for the speciation of transuranic elements in natural aquatic systems, *Top. Curr. Chem.,* 157, 129, 1990.
24. **Klenze, R. and Kim, J. I.,** A direct speciation of transuranium elements in natural aquatic systems by laser-induced photoacoustic spectroscopy, *Radiochim. Acta,* 44/45, 77, 1988.
25. **Pollard, P. M., Liezers, M., McMillan, J. W., Phillips, G., Thomason, H. P., and Ewart, F. T.,** Some actinide speciation using laser induced photoacoustic spectroscopy, *Radiochim. Acta,* 44/45, 95, 1988.
26. **Okajima, S., Reed, D. T., Beitz, J. V., Sabau, C. A., and Bowers, D. L.,** Speciation of Pu(VI) in near-neutral solutions via laser photoacoustic spectroscopy, *Radiochim. Acta,* 52/53, 111, 1991.
27. **Stumpe, R., Kim, J. I., Schrepp, W., and Walther, H.,** Speciation of actinide ions in aqueous solution by laser induced pulsed spectrophotoacoustic spectroscopy, *Appl. Phys. B,* 34, 203, 1984.
28. **Fang, H. L. and Swofford, R. L.,** The thermal lens in absorption spectroscopy, in *Ultrasensitive Laser Spectroscopy,* Klinger, D. S., Ed., Academic Press, New York, 1983, 175.
29. **Berthoud, T. and Mauchien, P.,** Determination of low level of U(VI) in water solution by means of the laser induced thermal lensing effect, *Anal. Chim. Acta,* 153, 265, 1983.
30. **Moulin, C., Delorme, N., Berthoud, T., and Mauchien, P.,** Double beam thermal lens spectroscopy for actinide detection and speciation, *Radiochim. Acta,* 44/45, 103, 1988.

APPENDIX 6.1

EXPRESSIONS FOR THE GRAPHICAL REPRESENTATION OF A MICROCOMPONENT SOLUTION

An explanation of the curves in Figure 6.1 is given below (autoprotolysis of water is neglected).

1. In sulfuric acid media (see Figure 6.1A).
For any medium, the only equilibrium considered is

$$HSO_4^- \rightleftarrows H^+ + SO_4^{2-}, \quad K$$

Omitting charges and setting $[H^+] = h$,

$$C_H = h + [HSO_4] \qquad C_{SO_4} = [SO_4] + [HSO_4]$$

The concentration of sulfuric acid is C. Hence,

$$C_H = 2\,C + \epsilon \qquad C_{SO_4} = C$$

where ϵ is the initial $[H^+]$ concentration, i.e., $\approx 10^{-5}\,M$. In practice, for $h > \epsilon$, C is a function of h and $[SO_4]$ and becomes

$$C = \frac{h\,(K + h)}{2K + h} \quad \text{(Curve 1)} \quad C = \frac{[SO_4]\,(K - [SO_4])}{K - [SO_4]} \quad \text{(Curve 2)}$$

Curve 3, representing log C as a function of log h or log $[SO_4]$, is constructed from curves 1 and 2 by changing h or $[SO_4]$ stepwise by a factor of 10, taking $K = 10^{-1}$. The equation of curve 3 is

$$h = \frac{2\,K\,[SO_4]}{K - [SO_4]}$$

C as a function of h and $[HSO_4]$ becomes

$$C = \frac{h\,(K + h)}{2\,K + h} \quad \text{(Curve 4)}$$

$$C = [HSO_4]\left(1 + \frac{K}{1 + \dfrac{[HSO_4]}{2} + \sqrt{1 + \dfrac{8\,K}{[HSO_4]}}}\right) \quad \text{(Curve 5)}$$

As before, acceptable pairs of C and h, or C and $[HSO_4]$ are selected and curve 6 is drawn.

If sulfate salts or $HClO_4$ are added to the solution in order to vary independently h and $[SO_4]$, the concentrations C_H and C_{SO_4} are changed.

2. *In phosphoric and citric acid media (see Figure 6.1B).*

The following equilibria are considered:

$$H_3PO_4 \rightleftarrows H^+ + H_2PO_4^-, \quad K_{11}$$

$$H_2PO_4^- \rightleftarrows H^+ + HPO_4^{2-}, \quad K_{12}$$

$$HPO_4^{2-} \rightleftarrows H^+ + PO_4^{3-}, \quad K_{13}$$

$$2\,H_3PO_4 \rightleftarrows H_6P_2O_8, \quad K_{20}$$

$$H_6P_2O_8 \rightleftarrows H^+ + H_5P_2O_8^-, \quad K_{21}$$

Omitting charges, the balance of H and PO_4 reads

$$C_H = 3\,[H_3PO_4] + 2\,[H_2PO_4] + [HPO_4] + 6\,[H_6P_2O_8] + 5\,[H_5P_2O_8] + h \qquad (1)$$

$$C_{PO_4} = [H_3PO_4] + [H_2PO_4] + [HPO_4] + 2\,[H_6P_2O_8] + 2\,[H_5P_2O_8] \qquad (2)$$

Setting C equal to the total H_3PO_4 concentration,

$$C_H = 3\,C + \epsilon$$

$$C_{PO_4} = C$$

In this case the variables C and h or C and $[H_3PO_4]$ cannot be separated because (1) and (2), as a function of the K values, lead to two relations of type $C = f(h, [H_3PO_4])$. After elimination of C, and for $h > \epsilon$, one obtains

$$a\,[HPO_4]^2 + b\,[H_3PO_4] + c = 0$$

with $a = K_{21}\,K_{20}\,h^2$, $b = K_{11}\,(h^2 + 2\,K_{12}\,h + 3\,K_{13}\,K_{12})$, $c = -h^4$

The solution

$$[H_3PO_4] = \frac{-\,b + \sqrt{b^2 + 4\,ac}}{2a}$$

is represented by curves 1 or 2.

The values of C for given pairs h, $[H_3PO_4]$ are computed from

$$C = [H_3PO_4] + K_{11}\frac{[H_3PO_4]}{h} + K_{11}K_{12}\frac{[H_3PO_4]}{h^2} + K_{11}K_{12}K_{13}\frac{[H_3PO_4]}{h^3} + \ldots$$

$$\ldots + 2\left\{K_{20}[H_3PO_4]^2 + K_{20}K_{21}\frac{[H_3PO_4]^2}{h}\right\}$$

For the ionic strength $\mu = 0.2$, the following K values are used:

$$K_{11} = 1.12 \times 10^{-2};\ K_{12} = 1.26 \times 10^{-7};\ K_{13} = 3.2 \times 10^{-13};$$

$$K_{20} = 1.17 \times 10^{-1};\ K_{21} = 1.9 \times 10^{-2}$$

Citric acid media are simpler since polymers need not be considered ($K_{20} = K_{21} = 0$) and

$$b\,[H_3Cit] + c = 0 \quad \text{(Curve 3)}$$

It follows, for instance, that

$$C = -\frac{c}{b}\left\{1 + \frac{K_{11}}{h} + \frac{K_{11}K_{12}}{h^2} + \frac{K_{11}K_{12}K_{13}}{h^3}\right\}$$

When h and $[H_3Cit]$ are varied independently, C_H and C_{PO_4} are modified as before. For $\mu = 0.1$, $K_{11} = 2 \times 10^{-3}$; $K_{12} = 1.26 \times 10^{-4}$; $K_{13} = 10^{-5}$.

3. Carbonate media (see Figure 6.1C).
The equilibria involved are

$$CO_2(g) \rightleftarrows CO_2\ (aq), \qquad K_0$$

or

$$CO_2(g) \rightleftarrows H_2CO_3\ (aq), \qquad K_0$$

$$H_2CO_3(aq) \rightleftarrows H^+ + HCO_3^-, \qquad K_1$$

$$HCO_3^- \rightleftarrows H^+ + CO_3^{2-}, \qquad K_2$$

$$CO_2(g) + H_2O \rightleftarrows 2\,H^+ + CO_3^{2-}, \qquad K = K_0K_1K_2$$

As before, it follows that

$$\log h = \frac{1}{3}(\log K + \log P_{CO_2})$$

$$\log [CO_3] = 0.3 + \log h$$

This is represented by curve 1.

In an open system, for which C is the concentration of dissolved carbonate and $P_{CO_2} = p$,

$$C_H = h + [HCO_3] + 2\,[H_2CO_3]$$

$$C_{CO_3} = [CO_3] + [HCO_3] + [H_2CO_3]$$

$$C_H = 2\,C_{H_2CO_3} + \epsilon = 2\,K_0\,p + \epsilon$$

$$C_{CO_3} = C_{H_2CO_3} + C = K_0\,p + C$$

Here, $\epsilon = 10^{-7}\,M$ for pure water. The set of equations leads to

$$C = K_0\,p\left[\frac{2B}{A} - 1\right] + \frac{B}{A}(\epsilon - h)$$

with

$$A = \frac{h}{K_2} + 2\frac{h^2}{K_1K_2} \qquad B = 1 + \frac{h}{K_2} + \frac{h^2}{K_1K_2}$$

Log C can be calculated stepwise as a function of log h. Now $[CO_3]$ corresponding to the pairs C, h is obtained from

$$[CO_3] = \frac{K_0\,p + C}{B}$$

and curve 2 is constructed using the following values: $K_0 = 7.11 \times 10^{-2}$, $K_1 = 3.02 \times 10^{-7}$, $K_2 = 2.4 \times 10^{-10}$ and $P_{CO_2} = 3.16 \times 10^{-4}$ atoms, which corresponds to the amount of CO_2 in the atmosphere.

It may be noted that all relations C_H and C_L as a function of C_i are equivalent to the expression of electroneutrality.

CHAPTER 7

CHEMISTRY AT THE TRACER LEVEL

I. EXPERIMENTAL SYSTEMS

Radiochemistry is the realm of tracer chemistry. The reactions involved are those of type 2A and 2B between species $E_i(m)$ of a microcomponent m and species $E_i(M_j)$ or $E_i(L_j)$ of macrocomponents M_j. They also include 3A and 3B reactions between species of two microcomponents $E_i(m_j)$, as described in Chapter 6.

A given chemical system at the tracer level is always defined by two sets of extensive parameters pertaining respectively to the micro- and macrocomponent(s). The first set, specific for the microcomponent, includes the radionuclide concentration, which is calculated from the known specific activities or from measurement of the relative or absolute activities. If the specific activities are unknown or if measurements of the number N_m of radionuclides cannot be performed, one must rely on *arbitrary* relative concentrations. However, at least the concentration *range* can be estimated from the conditions under which the radionuclide and the solution have been prepared. Microcomponents also comprise the concentrations of nonradioactive impurities, which can neither be determined nor accurately estimated.

It should be recalled that *carrier-free* for short-lived radionuclides of a stable element really means that the concentration of the element is determined by the fortuitous (and obviously unknown) impurity concentration of that element. In this case, the specific activity is less than the theoretical value and it cannot be quantified.

The parameters for macrocomponents include the actual and equilibrium concentrations $E_i(L_j)$ or $E_i(M_j)$, which are usually known, and the equilibrium concentration of H^+ ions in the case of aqueous solutions.

The second set of parameters can be treated independently, since consumption of the macroreagents during the course of reactions is not perceptible. In particular, equilibrium concentrations of the $E_i(L_j)$ entities can always be calculated as a function of pH and analytical concentrations of the reagents introduced into the system. In natural systems, concentrations of the latter can be determined from measurements performed following completion of the reaction, but this generally requires *destructive* analysis.

The specific parameters for microcomponents which participate in 3A and 3B reactions are not independent. The mutual reactions between $E_i(m_j)$ radioactive microcomponents, as well as interactions of a microcomponent with an impurity, must be taken into account in the usual manner.

II. SOLUTIONS CONTAINING RADIOCOLLOIDS

At the tracer level, a solution of an element is not necessarily truly homogeneous and thermodynamically stable, but may be rather a polydispersed system containing not only simple and complex ions and molecules, but also microparticles of variable size or aggregates. Formation of the latter depends on the nature of the element and on many, sometimes unsuspected and largely uncontrollable, parameters. This phenomenon was discovered by radiochemists because the methods of radioactivity are particularily suitable for detecting small amounts of matter. First mention of the existence of these aggregates or *radiocolloids* was made by Paneth in 1913, who found that in neutral solutions Bi and Po form colloids, but not Pb. At that time, the colloidal behavior of stable matter at concentrations below 10^{-9} *M* was still unknown.

Radiocolloids in aqueous solutions have been amply investigated and comprehensive reviews are available. It will suffice here to mention only a few problems that may be encountered when dealing with radioactive matter at the tracer level.

The presence of radiocolloidal species in a solution of radionuclides can be inferred from the following criteria:[1-3]

1. The radioactive species can be separated from the solution by a physical method such as filtration, ultrafiltration, centrifugation, or ultracentrifugation.
2. The species do not obey laws for the behavior of true solutions when a chemical gradient (diffusion, dialysis, isotopic exchange) or an electric gradient (electrophoresis, electrolysis, electrodialysis) is applied.
3. Adsorption on solid surfaces and spontaneous deposition differ from those observed for species in true solutions.
4. Autoradiography reveals the formation of aggregates of radioactive atoms.

The early work of Paneth was followed by numerous investigations using improved methods, of which a recent one is based on correlated diffusion of laser light.[4-6] The terms *real* or *true* colloids and *pseudocolloids* have appeared in the literature;[7-10] however, the exact nature of radiocolloids is still open to discussion.

The behavior of radiocolloids may be compared to that of ordinary colloids. The latter are defined in two different ways. The first is *operational* and considers colloids as polynuclear particles ranging in size from about 1 nm to 0.5 μm and dispersed in a medium. Because of their small size, these particles remain in suspension since the size limit for spontaneous sedimentation is a few micrometers.[11-18]

The second definition is more *conceptual* and considers the *nature* of colloids. Lyophobic colloids are either inorganic, e.g., polysilicic acids and their salts, or very sparingly soluble hydroxides like those of Fe and Mn, or organic acids originating from vegetal matter, such as humic or fulvic acids and similar materials. These colloids are found in nearly neutral artificial aqueous solutions and even in extremely pure water, but are much more abundant in natural aquatic systems.

The appearance of these colloids may result from more or less continuous and rapid condensation of monomeric hydrolyzed species of M or M_j linked by oxo or hydroxo bridges.[19,20] Another condensation mechanism may involve electrostatic interactions due to polarization of OH groups. The self-condensation rate increases with a high power of the concentration of monomeric species. The development of aggregates continues until eventually microphases are formed; the latter can again recreate monomeric forms by equilibrium reactions. With this picture, colloids have a finite lifetime and their number grows until a stationary concentration is reached. An illustration of this scheme is provided by the evolution of Fe(III) in solution:

$$\text{Fe(III)} \rightarrow \text{Fe}_x\text{O}_y \rightarrow \alpha\text{-FeOH (goethite) or a Fe}_2\text{O}_3\text{ (hematite)} \rightarrow \text{Fe(III)} \rightarrow \ . \ . \ .$$

It is an important property of these colloids that they possess:

1. A superficial diffusive electric, positive or negative charge, depending on the composition of the bulk solution
2. Localized electric charges
3. Superficial sites suitable for exchange, such as OH groups of $M_j(OH)$ or $X_j(OH)$ species
4. Sites suitable for binding by formation of inner- or outer-sphere complexes

Carrier-free radioactive tracers have only two ways of attaining a colloidal state, namely by sorption on a preexisting colloidal impurity in the solution or by polycondensation of a radioactive monomeric species. It is this duality which led to the definitions of *pseudo* radiocolloids and *real* radiocolloids.

Owing to the superficial properties of colloids, the C_0^z entities are readily fixed by physical adsorption, chemisorption, ion exchange, or surface complexation.[21-24] In this case, the radionuclide is carried by the colloidal support and behaves *like* the host colloid. The observed behavior reflects that of the support and not really that of the radionuclide. Formation of radiocolloids in this way increases with the concentration of the supporting material and should expectedly be impeded in solutions previously purified by ultrafiltration. The reversibility of pseudocolloids is determined by chemical reactions of the radionuclide at the surface of the host or by further condensation processes of the macroelement M.

The origin of true radiocolloids is still uncertain. They are composed of aggregates of 10^3 to 10^7 radioactive atoms of an element at tracer concentration (e.g., Ce, Eu, Pm, Ru, Zr, Nb, Bi, Po, Th, Am) and appear in solutions in which the presence of colloidal impurities is ruled out. But one may wonder if such solutions really exist.

A correlation has been observed between the appearance of radiocolloids and hydrolysis of the corresponding element. At present, it would be vain to attempt to quantify the two phenomena because the mechanism of hydrolysis is very complicated and probably not yet fully understood. Qualitatively, formation of colloids seems to be favored for heavy elements (Au, Po, Pa, Pu) in relatively high oxidation states.

The formation of true radiocolloids can hardly proceed in a homogeneous medium because of kinetic hindrance. The seed for condensation must be an aggregate of entities; its formation would imply the encounter of two hydrolyzed radioactive species, which is highly improbable at tracer and sub-tracer levels (reaction of type 3A); moreover, such a process would be extremely slow. Since true radiocolloids do exist and develop within short periods of time, it must be concluded that local heterogeneous zones of very limited extent can be present, in which for some reason the concentration of radionuclides is much higher than in the bulk of the solution. In this manner, the kinetics of certain processes may be favored, as will be discussed later.

Radioactivity may intervene in the mechanism. A radionuclide fixed on the surface of a colloidal impurity is ejected by recoil and a few neighboring atoms are detached simultaneously by sputtering. This small group of atoms may serve as a seed for further condensation. The same process is plausible in decay of radionuclides fixed on the surface of vessel walls or on the solid phase of a radioelement.

An increase in total activity has little effect on the formation of radiocolloids unless the radiolytic products actually participate in the reaction.[25]

Solutions of radiocolloids are not in thermodynamic equilibrium or, alternatively, the time required for establishment of a steady state is very long. For these reasons, their behavior can neither be described nor predicted with precision.[26]

Hereinafter, chemistry at very low concentrations will be assessed by considering truly homogeneous solutions. This ideal situation is best approximated when the medium is sufficiently acidic or contains complexing agents which ensure that the radiotracer exists as C_l^z entities.[27]

In practice, a solution which initially contained radiocolloids at the tracer level can be considered homogeneous after ultrafiltration on filters with pores of a few nanometers (the size of C_0^z or C_i^z species), and provided the activity does not change with time.

III. KINETICS OF REACTIONS AT VERY LOW CONCENTRATION

Quite often, even at ordinary concentrations, the mechanism of a reaction with a given stoichiometry is unknown. Hence, derivation of kinetics of the same reaction under conditions for which one or more of the reacting species is present at the tracer scale concentration is not straightforward.[28]

In the following it will be assumed that the reaction is controlled by a slow rate-determining step involving some entity E.

If the kinetics of this elementary reaction are first order with respect to E, the latter disappears at the same rate, whatever its concentration. Hence, the overall rate will not change upon dilution. This situation is not common in liquid phase.

The case of interest is that of an entity E composed of several elements. It is formed in the direct, or at least close encounter of two entities of these elements, which exchange energy, charge, or atoms. The probability of this event depends on concentration, and discussion will be focussed on a second-order elementary reaction between two entities E_1 and E_2 and its change in rate with dilution. The same treatment applies to two species of a same element.

The most general expression of this reaction, assuming that all stoichiometric coefficients are equal to one, is

$$E_1 + E_2 \rightleftarrows E_3 + E_4$$

with a forward rate constant k_+ and a backward rate constant k_-. These are overall rate constants that include the diffusion-controlled mutual approach (E_1 and E_2, E_3 and E_4) and the probability of chemical reaction between the pairs E_1–E_2 and E_3–E_4. In the following discussion, knowledge of the nature of the rate-determining step is not required and it can be viewed as a ''black box''. The important point is that the probability of encounter between E_1 and E_2 decreases with concentration and, correspondingly, the reaction rate will also decrease.

If the reactions that precede the slow rate-determining step are instantaneous or very fast, the initial concentrations of E_1 and E_2 will rapidly attain their equilibrium values, i.e., $[E_1]_0$ and $[E_2]_0$. The latter are of the same order of magnitude as the concentrations of the elements that they embody (see Appendix 7.1).

The half-time τ of the reaction can be calculated as a function of these concentrations and of the rate constants k_+ and k_-. It is assumed that at time zero, which coincides in practice with the instant of mixing of the precursors of E_1 and E_2, the concentrations of products $[E_3]$ and $[E_4]$ are zero.

The general case, which would correspond to $k_+ \neq k_-$ and $[E_1]_0 \neq [E_2]_0$ leads to a complicated expression of τ (see Appendix 7.2). For the simpler case with $k_+ >> k_-$ and $[E]_0 = [E_1]_0 = [E_2]_0$, the half-time of the reaction becomes

$$\tau = \frac{1}{k_+ [E]_0}$$

A 10^n-fold decrease in concentrations of species E_1 and E_2 (or of the incorporated element) increases τ in the same proportion. Hence, the slow rate-determining step can hinder the reaction for a high enough value of n and a reaction which is fast at ordinary concentrations may become slow at tracer concentrations. Similar conclusions hold for situations intermediate between the general case and the previous simple conditions. (see Appendix 7.2).

As a general rule, if the concentration of a species E_i in the experimental macroscopic rate law appears at a power different from unity, the reaction kinetics of the radionuclide will differ from that of the element, and possibly also the mechanism of the reaction. Quite commonly, the reaction rate decreases with C_M. Cases for which the concentration is at a power less than unity and reactions involving dimers should be considered specifically.

For further insight on the effect of dilution on kinetics, a simple case will be considered for which $k_+ > k_-$ and the reaction is separated into two steps:

$$E_1 + E_2 \rightleftarrows (E_1 - E_2), \quad k_{12}, k_{21}$$
$$(E_1 - E_2) \rightleftarrows E_3 + E_4, \quad k_c$$

k_{12} and k_{21} are the rate constants of mutual approach and separation, respectively, of the entities E_1 and E_2; k_c is the rate constant of the chemical reaction.

Setting $(E_1 - E_2) \equiv E$, and considering steady state conditions for which $d[E]/dt = 0$, it follows that

$$\frac{d[E_3]}{dt} = k_c \left(\frac{k_{12}}{k_{21} + k_c}\right) [E_1] [E_2]$$

Comparing with

$$\frac{d[E_3]}{dt} = k_+ [E_1] [E_2]$$

it results that

$$k_+ = \frac{k_c\, k_{12}}{k_{21} + k_c}$$

If $k_c << k_{21}$, i.e., diffusion proceeds faster than the chemical step,

$$k_+ = k_c\, k_{12}/k_{21} = k_c\, K$$

The reaction is controlled by the reaction of E_1 with E_2 and an equilibrium state is established between the species E_1, E_2 and (E_1-E_2). If $k_c >> k_{21}$, the limiting case is $k_+ = k_{21}$. The reaction cannot be faster than the diffusion, which is the controlling step. Rate constants of diffusion-controlled reactions are of the order of 10^{10} L mol^{-1} s^{-1}.

Absolute values of k_c are not available for reactions controlled by the chemical step. On the other hand, theoretical expressions for k_{12} constants have been established.[29-30] The simplest approach is based on the Fick diffusion laws and for the case where the chemical reaction occurs at each encounter and $k_{12} > k_{21}$,

$$k_{12} = 4\pi\, \rho\, D$$

D is the sum of the diffusion coefficients of the entities E_1 and E_2, and ρ the capture distance of the two entities, which depends on the size of E_1 and E_2. Alternatively, ρ may represent the distance at which the entities undergo interaction at long range. Any factor which changes ρ and D also affects the reaction rate.

For macrocomponents, the diffusion coefficients D_{E_i} may depend on the concentrations of M_1 and M_2, particularly when the latter determine the ionic strength of the medium, since

$$D_{E_i} = D^0_{E_i}\,(1 - \alpha\sqrt{\mu})$$

At the tracer level and for zero ionic strength, $D^0_{E_i}$ is the diffusion coefficient D_{E_i} at infinite dilution, and is a constant of the order of 10^{-5} to 10^{-6} cm^2 s^{-1}. The values of ρ are independent of the concentrations C_{M_1} and C_{M_2} and lie at the 100 pm level.

Apart from these rudimentary cases and hypotheses, theoretical expressions of rate constants are complicated and lead to time- and distance-dependent diffusion coefficients.

The time required for an encounter between E_1 and E_2 can be estimated by simple reasoning. In an isotropic medium, the dispersion law of a population of $N_{E_i}(0)$ entities from a surface of 1 cm^2 in a direction x perpendicular to the surface is the Gaussian

$$N_{E_i}(x,t) = \frac{N_{E_i}(0)}{\sqrt{\pi Dt}} \exp\left(-\frac{x^2}{4\,Dt}\right)$$

The mean value of x is

$$\bar{x} = 2\left(\frac{Dt}{\pi}\right)^{1/2}$$

and the standard deviation (i.e., the mean free path) is

$$\sigma = (2\ Dt)^{1/2}$$

In a solution, after a time of 1 ns, σ is about 1 nm. The distance between entities E_i in a volume V is proportional to $C_i^{-1/3}$. For encounter by diffusion, the entities must travel over a distance of the order of $\bar{x}$ or σ, which requires a time proportional to $C_i^{-2/3}$. Hence, the reaction time increases with decreasing concentration.

The reaction rate of a given number N_m of atoms of the micro component decreases with the spatial expansion of the system in which they are contained.

From the preceding, it is obvious that kinetics play an important role at the tracer level. At some critical concentration, kinetics prevail over thermodynamics in a reaction.

A. Reactions 3A and 3B between Microcomponents

Reactions at the tracer level may involve two species $E_1(m)$, $E_2(m)$ of an element 1, or two species of elements 1 and 2, i.e., $E_1 = E_1(m_1)$ and $E_2 = E_2(m_2)$, or the element 1 and an impurity corresponding to an entity $E_i(L_j)$. The concentrations C of the unique element or C_1 and C_2 of elements 1 and 2 or the concentration of element 1 and the species $E_i(L_j)$ are now decreased by a factor 10^n until all reaction partners become microcomponents. It is assumed that the macrocomponents are not affected by dilution, i.e., remain in macroamounts.

In the case of two elements, the order of magnitude of the initial concentrations $[E_1]_0 + [E_2]_0$ is now about 10^{-n} C_1 and 10^{-n} C_2 because, as will be shown later, the decrease in initial concentrations *may not be exactly the same* for elements 1 and 2.

Two conclusions can be drawn: (1) two radionuclides at the tracer level or a radionuclide and a microcomponent impurity cannot react during the time generally available for performing an experiment (Reaction 3A). Accordingly, they coexist and remain in the form in which they were introduced into the solutions. Of course, the situation may be different for 3B reactions if the half-lives of the radionuclides are sufficiently long. Each case must be examined separately. The same applies for two species of a unique element. (2) All tracer level entities containing radionuclides are monomeric, with the possible exception of natural entities of very long-lived radionuclides.

This is exemplified by the various species of fallout Pu in the different (3 to 6) oxidation states of the element. Fallout Pu has been released during the last 40 years and is encountered in large aqueous reservoirs at concentrations of the order of $10^{-17 \pm 1}$ *M*. Disproportionation of Pu(V) and redox reactions between various oxidation states of Pu may not yet have reached thermodynamic equilibrium, which might be established in some distant future.

Two reactions are of particular interest in the present context: isotopic exchange and disproportionation.[28]

1. Isotopic Exchange at the Tracer Level

Isotopic exchange involves two entities E_1 and E_2 of the same element M composed of two isotopes $M(A_1)$ and $M(A_2)$, where the A's are mass numbers. In the absence of isotope effect, exchange of the isotopes between the two entities is characterized by the equilibrium

$$E_1[M(A_1)] + E_2[M(A_2)] \rightleftarrows E_1[M(A_2)] + E_2[M(A_1)]$$

In many cases, exchange is an elementary second-order reaction with $k_+ = k_- = k$ and an equilibrium constant $K = k_+/k_- = 1$. The half-time of the exchange reaction takes the simple form

$$\tau = \frac{\ln 2}{k\,([E_1]_0 + [E_2]_0)}$$

Hence, at the tracer scale (i.e., when the macrocomponent M becomes a microcomponent) the isotopic exchange reaction of type 3A cannot be observed, but equilibrium can possibly be established over geological times, i.e., for type 3B reactions.

Many isotopic exchange reactions, and occasionally the mechanism of the process, are known.[31,32] They occur in homogeneous or heterogeneous systems, sometimes accompanied by isotope effects. Kinetically, these reactions are not always of second order.

The extent of an isotope exchange reaction is usually expressed in terms of the exchange rate R defined as the number of moles of the isotope $M(A_1)$ transferred from E_1 to E_2 per unit volume and time. The simplest case is that of one type of atom exchanged without isotope effect. The exchange rate is

$$R = \frac{C_1^0\,C_2^0 \ln 2}{\tau\,[C_1^0 + C_2^0]}$$

where C_1^0 and C_2^0 are the initial concentrations of E_1 and E_2, i.e.,

$$C_i^0 = C_{E_iM(A_1)} + C_{E_iM(A_2)}$$

It is found experimentally that R is given by

$$R = k\,[C_1^0]^n\,[C_2^0]^m$$

in which m and n are positive integers or fractions. The n- or m-dependence of the rate is obvious.

The exchange rate of U between U(IV) and U(VI) is highly dependent on experimental conditions, particularly on the pH.[31] In 0.1 *M* HCl and $C^0_{U(IV)} = C^0_{U(VI)} = 2.5 \times 10^{-2}$ *M*, the half-time τ is about 2 h and the coefficients n = m = 1. It follows that k = 0.116 mol^2 L^{-2} min^{-1} and R = 0.73 × 10^{-4} mol min^{-1}.

In a 10^{-10} *M* solution of ^{230}U(IV) (i.e., 6 × 10^{13} radioactive species L^{-1}) and, say, total uranium concentration 10^{-5} *M*, the half-time τ would be of the order of 400 d at constant $C_{U(IV)}$. In fact, the radionuclide disintegrates with a 20.8 d half-life and the exchange rate R, which has an initial value of 7 × 10^9 species L^{-1} min^{-1}, tends toward zero.

An increase in half-time with dilution has been reported for the exchange of I between C_2H_5I and NaI in alcohol (m = n = 1).[31] The τ value is 1 h for $C_1^0 = 0.9$ *M* and $C_2^0 = 0.135$ *M*. It increases to 53 h for $C_1^0 = 0.044$ *M* and $C_2^0 = 2.5 \times 10^{-4}$ *M*. More complex cases occur which must be treated specifically.

At the tracer level, radionuclides are often adsorbed on walls of vessels. The exchange occurs in a heterogeneous system and depends on the transfer rate of the macroscopic partner to the walls.

2. Disproportionation

A simple disproportionation reaction of a tracer level element M in oxidation state N is characterized by the equilibrium

$$2E_1[M(N)] \rightleftarrows E_3[M(N+1)] + E_4[M(N-1)]$$

It is a particular case of a reaction between two microcomponents with $E_1 = E_2$ and $[E_1]_0 = [E_2]_0$. The expression of τ is rather complicated (Appendix 7.2) and leads to the following conclusions: (1) disproportionation of a radionuclide at the tracer level is precluded, unless the reaction time t is extremely long with respect to τ, and provided the half-life of the nuclide is sufficiently long; and, (2) oxidation states that are unstable for radionuclides in ponderable amounts may be observed in solutions at tracer level concentrations. ^{237}Np(V) and other pentavalent actinides disproportionate in acidic medium. However, ^{239}Np(V) at the tracer level, e.g., 10^5 Bq L^{-1} (≈ μCiL^{-1}) is stable with respect to disproportionation.

3. Other Cases

The situation is more complicated when the reaction mechanism involves several second-order slow steps that are in competition with one another. An example is provided by the set of reactions

(a) $E_1 + E_2 \rightarrow E_3 + E_4$
(b) $E_1 + E_5 \rightarrow E_6 + E_7$

If the rate constants k_{+a} and k_{+b} of the forward reactions largely exceed those of the backward reaction and $k_{+a} >> k_{+b}$, and for the conditions $[E_2]_0 = [E_5]_0$, the initial yield of products $E_3 + E_4$ will be dominant at ordinary concentrations. Conversely, if $[E_2]_0 << [E_5]_0$, $[E_2]_0$ being constant, products $E_6 + E_7$ will dominate. The *nature* of the reaction products changes with concentration of the element in the entities E_2, as in the preceding cases.

B. Reactions 2A and 2B between Micro- and Macrocomponents

Reactions between a microcomponent species $E_1(m)$ and a macrocomponent species E_2 are not kinetically hindered because the concentration of one of the reagents, say $[E_2]_0$, largely exceeds that of the other. A reaction between two entities that is allowed at usual concentrations is also observable if one of the species is at the tracer level. For any reaction, elementary or not, of the type

$$E_1(m) + \sum E_i(L_j) \rightarrow \dots$$

consumption of the microcomponent follows the expression

$$\frac{d[E_1(m)]}{dt} = -\ k_+[E_1(m)] \prod [E_i(L_j)]^{x_i}$$

in which the powers in x_i are the experimental partial orders of the reaction with respect to $E_i(L_j)$ entities. Since the latter are not consumed in the reaction, they can be grouped together with the rate constant k_+ in a single conditional constant λ with the result

$$\frac{d[E_1(m)]}{dt} = -\ \lambda[E_1(m)]$$

Here, the proportionality constant is represented by λ in order to maintain a formal analogy with the radioactive decay law.

All reactions of type 2A and 2B between micro- and macrocomponents are of pseudo first order with respect to the microcomponent. This result has important consequences, as will be shown in Chapter 8.

IV. THERMODYNAMICS OF TRACER LEVEL REACTIONS

At the tracer level the number of entities $E_i(m_j)$ is still very high and the law of mass action can be applied without restriction to all reactions of interest to the radiochemist. The standard free energy of a reaction

$$\sum \nu_i E_i = 0$$

can be expressed in the usual way as a function of the equilibrium concentrations $[E_i]$ or of the thermodynamic activities a_i of entities E_i

$$\Delta G^o = -RT \ln \prod a_i^{\pm\nu_i} = -RT \ln \prod [E_i]^{\pm\nu_i} = -RT \ln K$$

It should be noted that the activity coefficients of $E_i(M_j)$ entities with concentrations much lower than those usually encountered in chemistry, are not necessarily equal to unity (see Appendix 8.3).

The thermodynamical behavior of an element does not depend on concentration, at least in the domain in which formation of polymers is precluded. However, the behavior may be affected owing to the very different consumptions of the $E_i(L_j)$ entities at the tracer scale and at usual concentrations. This will be illustrated by two examples of acid-base and redox reactions.

A. Acid-Base Equilibrium

Entities $E_i(M,L)$ which are represented by complexes C_l^z will be considered. In its most general expression, the successive formation of two complexes is represented by

$$C_l^{z_1} + H_p L \rightleftarrows C_{l+1}^{z_2} + (z_1 - z_2) H^+$$

Retaining the simple case for which $p = 1$, equilibrium can be expressed (omitting the charges) in the usual manner

$$ML_l + L \rightleftarrows M_{l+1}, \quad K$$

The ratio of concentrations of the two complexes

$$R = \frac{[ML_{l+1}]}{[ML_l]} = K[L]$$

depends directly on the ligand concentration [L] at equilibrium. The latter is a complicated function of C_L, C_M, K, and l. It can be shown (see Appendix 7.3) that the concentration [L] is always a decreasing function of C_M, and that the decrease is faster for higher values of l. It follows that, for given conditions, element M at the tracer level is incorporated in the complex of order $l + 1$, whereas at higher concentrations the complex ML_l will be predominant. This behavior is enhanced for higher l values. For example, in 1 *M* HCl macroscopic amounts of Pb(II) precipitate as $PbCl_2$, whereas 10.6 h ^{212}Pb is present in the form of anionic chloro complexes. The behavior of M thus depends on the concentration range, although the thermodynamic law valid for M is the same in both cases.

B. Redox Equilibrium

This example will illustrate the necessity of examining each system separately before any conclusion can be drawn with respect to the existence of a specific oxidation state at a tracer concentration. The equilibrium concentrations of species involved in the redox reaction between two elements M_1 and M_2, i.e.,

$$n_2\ ox_1 + n_1\ red_2 \rightleftarrows n_1\ ox_2 + n_2\ red_1$$

[in which the usual ox and red symbols stand for species $E_i(M_j,L_j)$] are expressed, for instance, as the fraction of total concentration: $[red_1] = \alpha_1 C_1$ and $[red_2] = \alpha_2 C_2$. The α terms are further denoted α_i and α_e for the initial and equilibrium conditions, respectively. For the simple case with $n_1 = n_2 = 1$, the ratio $\gamma = C_2/C_1$ of the total concentration of M_1 and M_2 takes the form (see Appendix 7.4):

$$\gamma = \frac{\alpha_{2e}[K(1 - \alpha_{1i}) + \alpha_{1i}] - \alpha_{1i}}{(\alpha_{2e} - \alpha_{2i})[\alpha_{2e}(1 - K) - 1]}$$

With this expression, α_{2e} can be computed as a function of the initial conditions and of K, and in particular as a function of the ratio γ of the concentration of the two elements. When the equilibrium constant K of the redox reaction is less than one, i.e., $E_1^\circ < E_2^\circ$, the reaction proceeds from right to left. The couple ox_2/red_2 is oxidizing with respect to the couple ox_1/red_1. At first sight, it would appear that the concentration of the reduced form M_2 should increase, which signifies that α_{2e} would also increase. With a set of values: $\gamma = 0.1$, $K = 0.1$, $\alpha_{1i} = 0.1$, $\alpha_{2i} = 0.9$, it is found that $\alpha_{2e} = 0.6$, i.e., α_2 has decreased!

Such situations may be encountered for ponderable as well as for tracer amounts of M_1 and M_2. Consider a system of the microcomponent (subscript 1) Fe and the macrocomponent U. In ordinary chemistry, $\gamma << 1$ and the effect of the redox couple 1 is not perceptible or is neglected. The amount of U(IV) oxidized by traces of Fe(III) is not measurable.

At the tracer level, if element (1) is the macrocomponent, $\gamma >> 1$, and element (2) is virtually in a single oxidation state. In acidic medium, ^{230}U(IV) is quantitatively oxidized to U(VI) in the presence of 10^{-8} *M* Fe(III) under conditions for which the kinetics is appreciable.

Intermediate cases at the tracer level for which $\gamma \approx 1$ must be considered individually. This is again illustrated by U(IV) in presence of Fe(III). The oxidation of U(IV) with Fe(III) ($n_1 = 1$; $n_2 = 2$) at pH 5.5 in a noncomplexing medium and in absence of O_2 is represented by the reaction

$$U(OH)_4 + 2\ Fe(OH)_2^+ + 3\ H^+ \rightleftarrows UO_2OH^+ + 2\ Fe^{2+} + 5\ H_2O,$$
$$K' = 10^{28}$$

The value of K′ is computed using the following data:

$$E^o(UO_2^{2+}/U^{4+}) = 0.254\ V;\ E^o(Fe^{3+}/Fe^{2+}) = 0.77\ V;$$
$$\log \beta_4(U^{4+}) = -5.13;\ \log \beta_1(UO_2^{2+}) = -5.24;$$
$$\log \beta_2\ (Fe^{3+}) = -5.67$$

This reaction may occur in a closed system of UO_{2+x}, a U(IV) oxide with an altered nonstoichiometric surface, in contact with an aqueous solution containing Fe(III) impurities at a concentration $\approx 10^{-8}$ M. Typical values are $x = 0.2$, $\gamma = 0.1$, $C_U = 10^{-7}$ M, $\alpha_{2e} = 0.2$. Taking $\alpha_{1i} = 0$ and considering $K = K'[H^+]^3 >> 1$ it results, as expected for a complete reaction (see Appendix 7.4), that

$$\alpha_{2e} = \alpha_{2i} - 2\gamma$$

This relation gives the amount of U(IV) ($\alpha_{2i} = 0.4$) which would have been found in the solution without the presence of Fe(III). Under other pH conditions, for which K is close to unity or less, the full expressions derived in Appendix 7.4 should be used.

Iron impurities can hardly be avoided in practice, since a unique aerosol particle of Fe_2O_3 with a size of 1 μm leads in 10 mL of water to $C_{Fe} \approx 10^{-8}$ *M*.

V. THE EXPERIMENTAL APPROACH TO TRACER LEVEL CHEMISTRY

As pointed out before, chemical information on a system must be obtained in a way that does not alter the true forms of the various species present. For this purpose, an external perturbation is applied to the system, which then responds with a signal carrying the chemical information. At concentrations

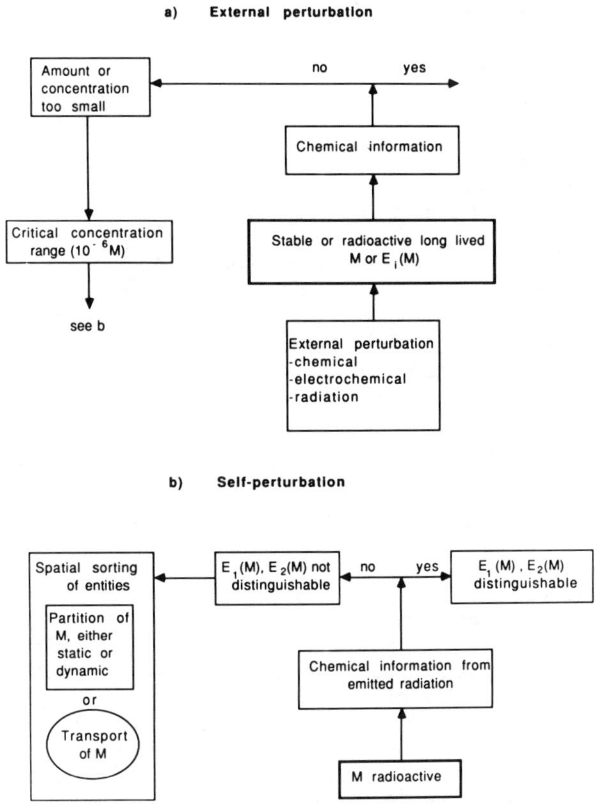

FIGURE 7.1. Strategies for speciation of trace elements.

below 10^{-6} *M*, the system becomes transparent to any form of perturbation and speciation is no longer possible.

However, if the species are radioactive, the system is self-perturbed because it spontaneously emits a signal in the form of radiation. This signal, in principle, can be exploited in two different ways: (1) in a direct manner, which would also be the most appealing, since the radiation itself would carry the chemical information; or, (2) in cases where this is not possible, the radiations are only indicators of the radionuclides present. Since all species of a given element emit the same signal, indirect procedures are required to sort out spatially the various forms of a given radionuclide.

For this purpose, the system must be perturbed. This can be achieved chemically, by opposing to the system a second phase and following the distribution of the radionuclides between the two phases, either in static or dynamic experiments. A second approach is perturbation by the application of an electric potential or a chemical gradient, i.e., utilization of transport methods.

Figure 7.1 provides an overview of strategies that can be used in determining the chemical behavior of radionuclides at the tracer scale.

GENERAL REFERENCES

Atkins, P. W., *Physical Chemistry,* 2nd ed., Oxford University Press, 1982.

Baes, C. F., Jr. and Mesmer, R. E., *The Hydrolysis of Cations,* John Wiley & Sons, New York, 1976.

Benes, P. and Majer, V., *Trace Chemistry of Aqueous Solutions. General Chemistry and Radiochemistry,* Elsevier, Amsterdam, 1980.

Evans, E. A. and Muramatsu, M., Eds., *Radiotracer Techniques and Applications,* Marcel Dekker, New York, 1977.

Hiemenz, R. C., *Principle of Colloid and Surface Chemistry,* Marcel Dekker, New York, 1986.

Hunter, R. J., *Foundation of Colloid Science,* Vol. 1 and 2, Clarendon Press, Oxford, 1989.

McKay, H. A. C., *Principles of Radiochemistry,* Butterworths, London, 1971.

Shaw, D. J., *Introduction to Colloid and Surface Chemistry,* Butterworths, London, 1966.

Wahl, C. A. and Bonner, N. A., Eds., *Radioactivity Applied to Chemistry,* John Wiley & Sons, New York, 1958.

REFERENCES

1. **Griffiths, P. J. F. and Park, G. S.,** Diffusion, in *Radiotracer Techniques and Application,* Vol. 1, Evans, E. A. and Muramatsu, M., Eds., Marcel Dekker, New York, 1977, chap. 12.
2. **Kepak, F.,** Behavior of carrier free radionuclides, in *Radiotracer Techniques and Application,* Vol. 1, Evans, E. A. and Muramatsu, M., Eds., Marcel Dekker, New York, 1977, 339.
3. **Benes, P. and Majer, V.,** *Tracer Chemistry of Aqueous Solutions. General Chemistry and Radiochemistry,* Elsevier, Amsterdam, 1980, 62.
4. **Triay, I. R., Hobart, D. E., Mitchell, A. J., Newton, T. W., Ott, M. A., Palmer, P. D., Rundberg, R. S., and Thomson, J. L.,** Size determinations of plutonium colloids using autocorrelation photon spectroscopy, *Radiochim. Acta,* 52/53, 127, 1991.
5. **Imae, T., Otani, W., and Oka, K.,** Electrophoretic light scattering of surfactant micelle colloids, *J. Phys. Chem.,* 94, 853, 1990.
6. **Goff, J. R. and Luner, P.,** Measurement of colloid mobility by laser Doppler electrophoresis: the effect of salt concentration on particle mobility, *J. Colloid Interface Sci.,* 99, 468, 1984.
7. **Kim, J. I.,** Actinide colloid generation in groundwater, *Radiochim. Acta,* 52/53, 71, 1991.
8. **Gutierrez, M. G., Bidoglio, G., Avogadro, A., Mingarro, E., and D'Alessandro, M.,** Experimental investigations of radionuclide transport through cored granite samples, *Radiochim. Acta,* 52/53, 213, 1991.
9. **Ramsay, J. D. F.,** The role of colloids in the release of radionuclides from nuclear waste, *Radiochim. Acta,* 44/45, 165, 1988.
10. **Nitsche, H.,** Basic research for assessment of geologic nuclear waste repositories: what solubility and speciation studies of transuranium elements can tell us, in *Materials Research Society Symposium,* Vol. 212, Materials Research Society, Pittsburgh, PA, 1991.
11. **Lieser, K. H., Ament, A., Hill, R., Singh, R. N., Stingl, U., and Thybusch, B.,** Colloids in groundwater and their influence on migration of trace elements and radionuclides, *Radiochim. Acta,* 49, 83, 1990.
12. **Barth, H. G.,** Particle size analysis, *Anal. Chem.,* 63, 1R, 1991.
13. **Maiti, C. T., Smith, M. R., and Saul, J. C.,** Colloid formation study of U, Th, Ra, Pb, Po, Sr, Rb, and Cs in briny groundwaters, *Nucl. Technol.,* 84, 82, 1989.
14. **Dearlove, J. P. L., Longworth, G., Ivanovich, M., Kim, J. I., Delakowitz, B., and Zeh, P.,** A study of groundwater-colloids and their geochemical interactions with natural radionuclides in Gorleben aquifer systems, *Radiochim. Acta,* 52/53, 83, 1991.

15. **Degueldre, C., Baeyens, B., Goerlich, W., Riga, J., Verbist, J., and Stadelmann, P.,** Colloids in water from a subsurface fracture in granitic rock, Grimsel test site, Switzerland, *Geochim. Cosmochim. Acta,* 53, 603, 1989.
16. **Degueldre, C. A. and Wernli, B.,** Characterization of the natural inorganic colloids from a reference granitic ground water, *Anal. Chim. Acta,* 195, 211, 1987.
17. **Vilks, P., Cramer, J. J., Shewchuk, T. A., and Larocque, J. P. A.,** Colloid and particulate matter studies in the Cigar Lake natural-analog program, *Radiochim. Acta,* 44/45, 305, 1988.
18. **Dellis, T. and Moulin, V.,** Isolation and characterization of natural colloids, particularly humic substance, present in a groundwater, in *Water Rock Interaction,* Miles, E., Ed., Balkema, Rotterdam, 1989, 197.
19. **Combes, J. M., Manceau, A., Calas, G., and Bottero, J. Y.,** Formation of ferric oxydes from aqueous solutions: a polyhedral approach by X ray absorption spectroscopy. Hydrolysis and formation of ferric gels, *Geochim. Cosmochim. Acta,* 53, 583, 1989.
20. **Thiyagarajan, P., Diamond, H., Soderholm, L., Horwitz, E. P., Toth, L. M., and Felker, L. K.,** Plutonium (IV) polymers in aqueous and organic media, *Inorg. Chem.,* 29, 1902, 1990.
21. **Schindler, P. W. and Stumm, W.,** The surface chemistry of oxides, hydroxides and oxide minerals, in *Aquatic Surface Chemistry,* Stumm, W., Ed., John Wiley & Sons, New York, 1987, 83.
22. **Dzombak, D. A. and Morel, F. F. M.,** *Surface Complexation Modelling: Hydrous Ferric Oxide,* John Wiley & Sons, New York, 1990.
23. **Tessier, A. and Campbell, P. G. C.,** Partitioning of trace metals in sediments, in *Metal Speciation: Theory, Analysis and Application,* Kramer, J. R. and Allen, H. E., Eds., Lewis Publishers, Chelsea, MI, 1988, 183.
24. **Murray, J. W.,** The surface chemistry of hydrous manganese dioxide, *Colloid Interface Sci.,* 46, 357, 1990.
25. **Newton, T. W., Hobart, D. E., and Palmer, P. D.,** The formation of Pu(IV)-colloid by alpha-reduction of Pu(V) or Pu(VI) in aqueous solutions, *Radiochim. Acta,* 39, 139, 1986.
26. **Meier, H., Zimmerhackl, E., Zeitler, G., and Menge, P.,** Correlations of electrokinetic properties of sorption data of radionuclides in site-specific samples, *Radiochim. Acta,* 52/53, 195, 1991.
27. **Choppin, G. R. and Allard, B.,** Complexes of actinides with naturally occurring organic compounds, in *Handbook on the Physics and Chemistry of the Actinides,* Freeman, A. J. and Keller, C., Eds., North-Holland, Amsterdam, 1985, 407.
28. **Guillaumont, R., Adloff, J. P., and Peneloux, A.,** Kinetic and thermodynamic aspects of tracer scale and single atom chemistry, *Radiochim. Acta,* 46, 169, 1989.
29. **Noyes, M.,** Effects of diffusion rates on chemical kinetics, *Prog. Reaction Kinetics,* 1, 129, 1961.
30. **North, A.,** Diffusion controlled reactions. *Q. Rev. Chem.,* 20, 421, 1966.
31. **Myers, E. O. and Prestwood, R. J.,** Isotopic exchange reactions, in *Radioactivity Applied to Chemistry,* Wahl, A. C. and Bonner, N. A., Eds., John Wiley & Sons, New York, 1951, 349.
32. **Johnston, J.,** Isotopic exchange processes, in *Radiotracer Techniques and Application,* Vol. 1, Evans, E. A. and Muramatsu, M., Eds., Marcel Dekker, New York, 1977, 405.

APPENDIX 7.1

EQUILIBRIUM CONCENTRATIONS

Equilibrium concentrations of the entities of element M reacting in solution can always be calculated as a function of the analytical concentrations *C* of the reagents and on the basis of equilibrium constants K. The expressions are classical, but will be recalled because of their frequent use when dealing with microcomponents. In the following, equilibrium constants are expressed in terms of equilibrium concentrations, in brackets, and the total concentrations of exchangeable atoms, groups of atoms and of the element are denoted by C.

A simple case is represented by the reactions

$$M^{n+} + HL \rightleftarrows ML^{(n-1)+} + H^+, \quad K_1$$

$$HL \rightleftarrows L^- + H^+, \quad K_L$$

Omitting charges and replacing activities by concentrations, the following five relations for five unknowns are established:

$$(1)\ K_1 = \frac{[ML]\,[H]}{[M]\,[HL]} \quad (2)\ K_L = \frac{[L]\,[H]}{[HL]} \quad (3)\ C_H = [H] + [HL]$$

$$(4)\ C_L = [L] + [HL] \qquad (5)\ C_M = [M] + [ML]$$

For illustration, let us suppose that the solution is prepared with $HClO_4$, HL, NaL and ML_n. The analytical concentrations are C_{HClO_4}, C_{HL}, C_{NaL}, and C_{ML_n}. Then the concentrations

$$C_H = C_{HClO_4} + C_{HL}; \quad C_L = C_{HL} + C_{NaL} + n\,C_{ML_n}; \quad C_M = C_{ML_n}$$

are experimental data corresponding to the preparation of the solution. In a given *natural* solution, the analytical concentrations are determined by elemental analysis.

Following measurement of pH, the four remaining unknowns are obtained from combinations of the previous expressions. From (2) and (4),

$$C_L = [L] + \left(1 + \frac{[H]}{K_L}\right) + [ML]$$

From (1) and (2),

$$K_1 K_L^{-1} = \frac{[ML]}{[M]\,[L]}$$

These last two relations yield

$$C_L = \frac{[ML]}{K_1 K_L^{-1} [M]} \left(1 + \frac{[H]}{K_L}\right) + [ML]$$

Introducing now (5),

$$C_L = \frac{C_M - [M]}{K_1 K_L^{-1}} \left(1 + \frac{[H]}{K_L}\right) + C_M - [M]$$

With this last relation, [M] and all other unknown concentrations can be calculated. [M] as a function of C_L, C_M, [H], K_1 and K_L is obtained by solving a quadratic equation. Cases other than the simple example given here would lead to complicated calculations. The complex ML may be the entity E_1 considered in Appendix 7.2.

APPENDIX 7.2

REACTION HALF-TIMES

Consider the elementary reaction

$$E_1 + E_2 \rightleftarrows E_3 + E_4$$

with the forward rate constant $k_1 = k_+$ and the backward rate constant $k_2 = k_-$. The initial equilibrium concentrations $[E_1]_0 = a$ and $[E_2]_0 = b$ are calculated as shown in Appendix 7.1. At time t, $[E_3] = [E_4] = x$. At any time, $[E_1] = a - x$ and $[E_2] = b - x$. It follows that

$$\frac{dx}{dt} = k_1 (a - x)(b - x) - k_2 x^2$$

Rearranging,

$$\frac{dx}{dt} = (k_1 - k_2) x^2 - k_1(a + b) x + k_1 ab = \alpha x^2 + \beta x + \gamma$$

The time t is obtained by integration of this equation using the initial conditionsgiven above. The solution depends on the sign of the parameter $q = 4\alpha\gamma - \beta^2$. Taking $q < 0$,

$$t = \frac{1}{\sqrt{-q}} \ln \frac{-2\alpha x - \beta + \sqrt{-q}}{2\alpha x + \beta + \sqrt{-q}} \left(\frac{\beta + \sqrt{-q}}{-\beta + \sqrt{-q}} \right)$$

The equilibrium value of x, i.e., x_e, is obtained from

$$x_e^2 = \frac{k_1}{k_2} (a - x_e)(b - x_e)$$

with the result

$$x_e = k_1 \frac{(a + b) \pm \sqrt{(a - b)^2 + 4 k_2/k_1}}{2 (k_1 - k_2)}$$

The + sign corresponds to $ab \leq k_2/k_1$ and the − sign to $ab > k_2/k_1$. The reaction half-time τ is by definition the time for which $x = x_e/2$. Its value is

$$\tau = \frac{1}{\sqrt{-q}} \ln \frac{\sqrt{-q} - (\pm\, 2\, k_1 \sqrt{(a-b)^2 + 4\, k_1/k_2}}{\sqrt{-q} + (\pm\, 2\, k_1 \sqrt{(a-b)^2 + 4\, k_1/k_2}} \cdots$$

$$\cdots \frac{\sqrt{-q} - k_1\,(a+b)}{\sqrt{-q} + k_1\,(a+b)}$$

This general expression will now be applied to several specific cases:

1. The backward reaction is negligible with respect to the forward one, i.e., $k_1 >> k_2$ and $a \neq b$. In this case, $\alpha = k_1$, $\beta = -k_2(a + b)$, $\gamma = k_1ab$ and the parameter q becomes

$$q = 4\, k_1^2\, ab - k_1^2\, (a+b)^2 = -\, k_1^2\, (a-b)^2 < 0$$

Now the expression for t takes the form

$$t = \frac{1}{k_1\,(a-b)} \ln \frac{-b}{a} \left(\frac{a-x}{x-b}\right)$$

The reaction proceeds until $x_e = a$ if $a > b$, and $x_e = b$ if $a < b$. In the former case,

$$\tau = \frac{1}{k_1\,(a-b)} \ln \frac{-b}{a-2b}$$

2. The backward reaction is still negligible, and $a = b$. Since $\alpha = k_1$, $\beta = -2k_1$ and $\gamma = k_1a^2$, the parameter q is zero and the expression of t is indeterminate. The kinetics is defined by the equation

$$t = \frac{1}{k_1\,(a-b)} \ln \frac{-b}{a} \left(\frac{a-x}{x-b}\right)$$

which leads to

$$x = \frac{a^2\, k_1\, t}{1 + a\, k_1\, t}$$

The reaction half-time τ is the time for which $x \approx a/2$, i.e.,

$$\tau \approx \frac{1}{k_1\, a}$$

3. The rate constants are equal: $k_1 = k_2 = k$ and $a \neq b$.
The coefficient α vanishes and

$$\frac{dx}{dt} = k(a - x)(b - x) - k\,x^2 = k[ab - x\,(a + b)]$$

The product ab of the initial concentrations is obtained from the equilibrium equation

$$x_e^2 = (a - x_e)(b - x_e)$$

which yields

$$x_e = \frac{a\,b}{a + b}$$

The differential equation is now written

$$\frac{dx}{x_e - x} = k\,(a + b)\,dt$$

which leads to

$$\ln\left(1 - \frac{x}{x_e}\right) = -\,k\,(a + b)\,t$$

In this case the half-time of the reaction corresponds to $x = x_e/2$, from which

$$\tau = \frac{\ln 2}{k\,(a + b)}$$

4. $k_1 \neq k_2$ *and* $a = b$: this situation is characteristic of a disproportionation reaction. Here, $\alpha = k_1 - k_2$, $\beta = -2k_1a$ and $\gamma = k_1a^2$. The parameter $q = -4\,k_1k_2\,a^2$ is negative. Hence, from the general expression

$$t = \frac{1}{2\sqrt{k_1\,k_2}\,a} \ln \frac{a - 1 + \sqrt{k_2/k_1}}{a + 1 + \sqrt{k_2/k_1}}$$

the half-time of the reaction becomes

$$\tau = \frac{1}{2\,\sqrt{k_1\,k_2}\,a} \ln \frac{a - 2 + 2\,\sqrt{k_2/k_1}}{a + 2 + 2\,\sqrt{k_2/k_1}}$$

APPENDIX 7.3

ACID-BASE EQUILIBRIUM AT THE TRACER LEVEL

The following set of relations in which the charges are omitted, is considered:

$$[ML_{l+1}] = K[L]\,[ML_l] \tag{1}$$

$$C_L = [L] + l\,[ML_l] + (l + 1)\,[ML_{l+1}] \tag{2}$$

$$C_M = [ML_l] + [ML_{l+1}] \tag{3}$$

From (2) and (3)

$$[ML_l] = (l + 1)\,C_M - C_L - [L] \tag{4}$$

$$[ML_{l+1}] = C_L - [L] - l\,C_M \tag{5}$$

The derivatives of expressions (1), (2), and (3) with respect to C_M give:

$$\partial[ML_{l+1}]/\partial C_M = K[L]\,\partial[ML_l]/\partial C_M + K[ML_l]\,\partial[L]/\partial C_M \tag{6}$$

$$0 = \partial[L]/\partial C_M + l\,\partial[ML_l]/\partial C_M + (l + 1)\,\partial[ML_{l+1}]/\partial C_M \tag{7}$$

$$1 = \partial[ML_l]/\partial C_M + \partial[ML_{l+1}]/\partial C_M \tag{8}$$

From expressions (7) and (8),

$$\partial[ML_{l+1}]/\partial C_M = -\,l - \partial[L]/\partial C_M \tag{9}$$

$$\partial[ML_l]/\partial C_M = 1 + l + \partial[L]/\partial C_M \tag{10}$$

Terms from expressions (5), (9), and (10) are inserted in (6), with the result that

$$\frac{\partial\,[L]}{\partial\,C_M} = -\,\frac{1 + K[L](l + 1)}{1 + K[L] + K[(l + 1)C_M - C_L - [L]]}$$

Because all terms are positive, the derivative is always negative and its absolute value increases with l.

APPENDIX 7.4

REDOX EQUILIBRIUM

Consider the redox couples of two elements M_1 and M_2 with apparent standard reduction potentials E_1^o and E_2^o:

$$ox_1 + n_1\, e^- \rightarrow red_1$$

$$ox_2 + n_2\, e^- \rightarrow red_2$$

and the overall reaction

$$n_2\, ox_1 + n_1\, red_2 \rightleftarrows n_1\, ox_2 + n_2\, red_1$$

for which the equilibrium constant K is given by

$$-RT \ln K = n_1\, n_2\, F(E_2^o - E_1^o)$$

Setting for the initial state

$$C_{red_1} = x_{1i} = \alpha_{1i}\, C_1 \qquad C_{red_2} = x_{2i} = \alpha_{2i}\, C_2$$

and for the equilibrium

$$[red_1] = x_{1e} = \alpha_{1e}\, C_1 \quad [red_2] = x_{2e} = \alpha_{2e}\, C_2$$

it follows from the exchange of electrons that

$$(x_{1e} - x_{1i})\, n_2 = (x_{2i} - x_{2e})\, n_1 \qquad (1)$$

Expressing the equilibrium constant in the usual way, one obtains

$$\ln \frac{x_{1e}^{n_2}\,(C_2 - x_{2e})^{n_1}}{(C_1 - x_{1e})^{n_2}\, x_{2e}^{n_1}} = \frac{n_1\, n_2\, F}{RT}\,(E_1^o - E_2^o)$$

From expression (1), it results that

$$x_{1e} = x_{1i} + \frac{n_1}{n_2}\,(x_{2i} - x_{2e}) \qquad (2)$$

which for the simplest case $n_1 = n_2 = 1$, yields the ratio $\gamma = C_1/C_2$ as

$$\gamma = \frac{\alpha_{2e}[K(1 - \alpha_{1i}) + \alpha_{1i}] - \alpha_{1i}}{(\alpha_{2e} - \alpha_{2i})\,[\alpha_{2e}\,(1 - K) - 1]} \tag{3}$$

With this last expression, α_{2e} can be first calculated in terms of α_{2i}, α_{1i}, γ and K, by solving a quadratic equation; α_{1e} is evaluated afterwards from (2), with the result

$$\alpha_{1e} = a_{1i} + \frac{1}{\gamma}(\alpha_{2i} - \alpha_{2e})$$

For the case in which $n_1 = 1$ and $n_2 = 2$,

$$\gamma = \frac{\alpha_{2i}[K(1 - \alpha_{1e})^2 + \alpha_{1e}^2] - \alpha_{1e}^2}{2\,(\alpha_{1e} - \alpha_{1i})\,[K(1 - \alpha_{1e})^2 + \alpha_{1e}^2]} \tag{4}$$

With this relation α_{1e} is first derived in terms of α_{2i}, α_{1i}, γ, and K; next α_{2i} is obtained from

$$\alpha_{2e} = \alpha_{2i} + 2\gamma\,(\alpha_{1i} - \alpha_{1e}) \tag{5}$$

The special case $\alpha_{1i} = 0$ and $K >> 1$ yields

$$2\gamma\,\alpha_{1e}^3[\alpha_{2i} + 4\gamma]\,\alpha_{1e}^2 + 2(\alpha_{2i} + \gamma)\,\alpha_{1e} - \alpha_{2i} = 0 \tag{6}$$

for which $\alpha_{1e} = 1$ is an obvious solution. Thus equation (6) can be rewritten

$$(\alpha_{1e} - 1)[2\gamma\,\alpha_{1e}^2 - (\alpha_{2i} + 2\gamma)\,\alpha_{1e} + \alpha_{2i}] = 0 \tag{7}$$

of which the two solutions are $\alpha_{1e} = \alpha_{2i}/2\gamma$ and again $\alpha_{1e} = 1$. Equation (6) eventually takes the form

$$(\alpha_{1e} - 1)^2\left(\alpha_{1e} - \frac{\alpha_{2i}}{2\gamma}\right) = 0$$

According to (5), the solution $\alpha_{1e} = \alpha_{2i}/2\gamma$ implies $\alpha_{2e} = 0$. However, $\alpha_{1e} < 1$ and $\alpha_{2e} < 1$, from which it follows that $\gamma > 0.5$. The solution $\alpha_{1e} = 1$ yields $\alpha_{2e} = \alpha_{2i} - 2\gamma$, but since $\alpha_{2e} < 1$, it is required that $\alpha_{2i} < 1 - 2\gamma$. The values of α_{2e} can be calculated for different α_{2i} as a function of γ.

CHAPTER 8

RADIOCHEMICAL METHODS. ADVANTAGES AND LIMITATIONS

PART ONE — CHEMICAL INFORMATION CARRIED BY RADIATION

I. INTRODUCTION

In order to determine the *chemical state*, i.e., the oxidation state and bond properties, of an element in entities constituting radioactive or nonradioactive matter, it is necessary to use a probe that is sensitive to the organization of the atomic electrons. Such a probe is offered by the radiations emitted *in situ*, provided their properties depend on the electronic environment. Alternatively, a positron emitter can be introduced into the system and the properties of the annihilation radiation can be investigated. In a sense, the positron can be considered as a ''radioactive'' tracer of the electron.

Radiations emitted by radionuclides have their origin mainly in the nucleus. However, the electron orbitals intervene also in the two decay modes corresponding to electron capture (EC) and internal conversion (IC). The dependence of nuclear properties (and consequently of the emitted radiations) on the electron environment is termed *hyperfine interaction*. It results from the coupling of a nuclear parameter with other parameters characteristic of the electronic environment and involves the energy, the wave function, and the degeneracy of nuclear states (see Appendix 8.1). The practical consequence of this is a change in *energy* or in *intensity* or in both properties of the emitted radiations.

In chemistry, a change in the electronic environment is principally produced by a change in the oxidation state. Accordingly, this is the easiest to observe. The establishment of a chemical bond also slightly affects the electron distribution. Because the properties of bonds may differ among various compounds, this provides another potential source of relevant information.

A straightforward interaction between the nucleus and the atomic electrons results from the finite density $[\Psi(0)]^2$ of electrons inside the nuclear volume.

Only s-type electrons and relativistic $p_{1/2}$ electrons in heavy nuclei are concerned. Obviously, the two decay modes that are most sensitive to the electron density are EC and IC. Hence, it would be expected that these two processes should be somewhat dependent on the chemical properties that affect the density of s electrons in the nucleus. This is indeed observed in compounds of light elements such as Be, for which the 2s valence electrons are involved in the formation of bonds, but the effects are much less apparent in heavier elements.

In general, the s-electron density at the atomic nucleus has two contributions, namely from filled s-orbitals of inner shells, and from partially filled valence shells. The valence electron contribution comprises the atom's own electrons together with electrons from surrounding ligands involved in the formation of chemical bonds. Thus, it is very sensitive to chemical influences. Changes in the oxidation state and in bond properties influence the s-electron density at the nucleus in two ways: (1) *directly*, by changing the s and $p_{1/2}$ electron densities in the valence shell; or, (2) *indirectly*, by shielding the s electrons by electrons from p, d, and f subshells. An increase in population of the corresponding orbitals causes a weaker attraction of the s-electron cloud by the nuclear charge and thus decreases $[\Psi(0)]^2$. Conversely, delocalization of the p or d electrons increases the electron density.

Calculations show that the absolute changes in s-electron density with electronic configuration are small. For example, electron densities at the nucleus for free-ion configurations obtained from Dirac-Fock calculation increase from 6 965 162 (in atomic units) for Np^{3+} ($5f^4$) to 6 966 057 for Np^{7+} ($5f^0$).[1]

Accordingly, it is evident that the changes in nuclear properties with variations in the electronic environment are extremely small. Their observation therefore requires very precise measurements of energy and intensity of the emitted radiations. Moreover, the number of convenient radionuclides is very limited, and the choice is even further restricted by chemical considerations, since, as said, the best candidates are multivalent elements.

II. INFORMATION FROM ENERGY CHANGE OF RADIATION WITH THE ELECTRONIC ENVIRONMENT

Chemical information cannot be inferred from the energy of α particles, because the electronic environment is not directly involved in the emission; its role is merely that of a screen between the nucleus and the atomic environment.

The prognosis is more complicated for β decay, which implies the creation of an $e^- - \bar{\nu}$ or $e^+ - \nu$ pair. The characteristics of the disintegration are fixed by the electronic wave functions of the initial and final states. The energy of the emitted radiation depends on the oxidation state of the nuclide and on the decay probability as will be discussed in a later section. The chemical shift of the β ray endpoint energy will in any event be extremely weak, but it is most pronounced for the soft β^- emitters ^{63}Ni, ^{187}Re, ^{106}Ru, and ^{214}Pu.[2,3]

Much more important energy shifts may appear in the emissions originating from the electronic environment, i.e., in X-rays and Auger electrons following EC and IC. It should be recalled that, in the former process, the message conveyed is characteristic of the *daughter* atom.

The energy of X-rays and Auger electrons is determined by the differences in energies between inner atomic levels. The energy shift of these levels with a change in the valence electron configuration is about 0.1 eV, which is an appreciable value. However, all electron orbitals are shifted by about the same amount and the net effect is very small.

IC of the energy available in a nuclear transition releases an electron as in the photoelectric effect. Since the nuclear energy is highly monochromatic, the kinetic energy of the conversion electron

$$E_e = E(\text{nuclear}) - E(\text{binding})$$

is also sharply defined. The same property is shared by the X-rays and Auger electrons arising from IC. Spectroscopy of Auger electrons with a resolution better than 0.1 eV is excluded, but as in the case of XPS, information on the chemical state can be gained from the energy of the photoelectron.

The energy of both X-rays and γ photons is slightly dependent on the chemical state of a radioactive emitter. The γ photons originate from a transition between nuclear levels. By virtue of the hyperfine interactions, the nominal energy of the nuclear levels may change with the electronic environment by a magnitude close to the natural width of the radiation, which is of the order of 10^{-8} eV. Such a small shift can be determined by Mössbauer spectroscopy.

Finally, the energy of annihilation radiation from positons may provide information about the characteristics of the site at which the electron–positron pair has interacted.

A. Internal Conversion Electron Spectroscopy

At present, speciation of a radioelement based on the energy of conversion electrons has been reported for only one case, i.e., technetium. Although unique, this example is quite informative.[4,5] Fortuitously, ^{99m}Tc possesses an appropriate nuclear transition and forms a large variety of compounds in which the oxidation state of the element may have values between +1 and +7. Many of these compounds are used in nuclear medicine. The properties of ^{99m}Tc thus furnish an ideal case study.

This nuclide decays initially by a 2172.6 eV transition to a state involving a spin change $\Delta I = 3$ for which γ emission is strongly prohibited. Internal conversion is highly effective and is energetically allowed in the 3p and 3d *valence* shells and sub-shells. The $3p_{3/2}$ and $3p_{1/2}$ levels from spin-orbit coup-

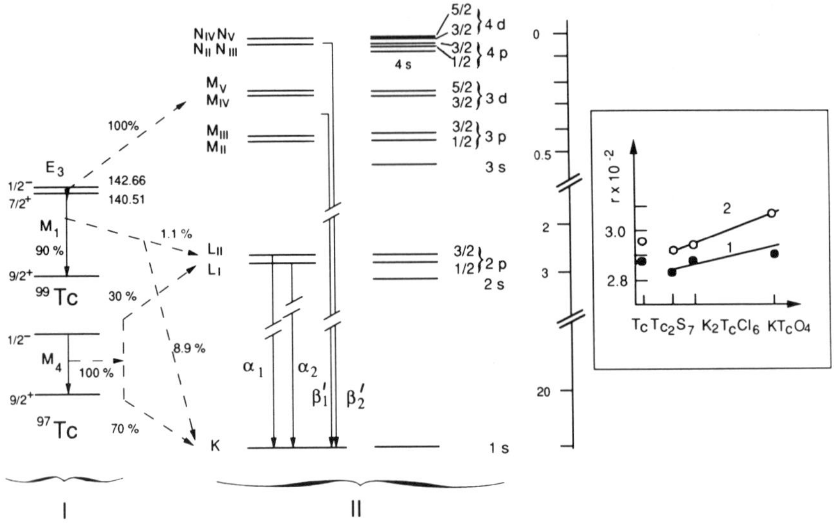

FIGURE 8.1. Nuclear states (I) of ^{97}Tc and ^{99}Tc and atomic levels (II) of technetium involved in chemical bonds. The energy of nuclear levels is given in keV. ^{99m}Tc (T = 6.006 h) decays to ^{99}Tc (T = 3.13 × 10^5 years) and ^{97m}Tc (T = 90 d) to ^{97}Tc (T = 2.6 × 10^6 years). Percentages refer to IC coefficients. Energies (in keV) of atomic states involved in X-ray emission: K: 21.044; L_{II} 2.793; L_{III} 2.677; M_{II} 0.045; M_{III} 0.425; M_{IV} 0.257; M_V 0.253; $N_{II,III}$ 0.039; $N_{IV,V}$ 0.002. Energies (in keV) and relative intensities of Tc X-rays in Tc(0): $K_{\alpha 2}$ 18.251 (52.6%); $K_{\alpha 1}$ 18.367 (100%); $K_{\beta' 1}$ 20.61 (26.2%); $K_{\beta' 2}$ 21.01 (4.3%); ω_K = 0.779. Insert: Variation of the $K_{\beta' 2}/K_{\alpha}$ intensity ratio r for various compounds of Tc in different oxidation states: (1) ^{97m}Tc; (2) ^{99m}Tc. (Adapted from Yoshihara, K., *Topics in Current Chemistry*, Vol. 157, Springer-Verlag, Berlin, 1990, 7.)

ling are split by about 20 eV and are well resolved in the electron conversion spectrum. A change of one unit in the oxidation state of Tc causes a change in electron binding energy of 1 eV (Figure 8.1). All compounds of the element in a given oxidation state exhibit the same chemical shift within ± 0.2 eV, which makes the determination of the valence state of Tc unambiguous. Only picogram amounts of compounds are required for the analysis, whereas microgram quantities are necessary for the same information by the XPS technique.

B. Mössbauer Emission Spectroscopy

In a classical, i.e., *absorption* Mössbauer experiment, the source is a radionuclide in tracer amounts embedded in a nonmagnetic, highly symmetric solid matrix, in such a way that a monochromatic *recoil-free* γ photon is emitted. This photon is resonantly reabsorbed without energy loss in the ground state isomer, bound in a macroscopic compound (Figure 8.2). The

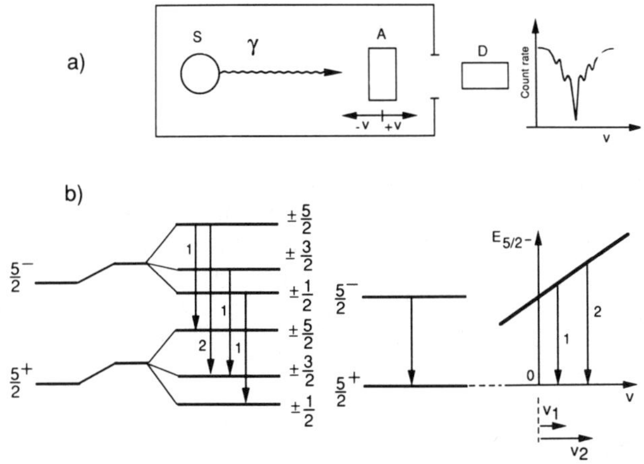

FIGURE 8.2. Principle of Mössbauer emission spectroscopy of ^{237}Np. (a) Simplified scheme of experimental setup. The source is a compound of ^{241}Am (T = 433 years). This nuclide decays via α emission to the 59.6-keV Mössbauer level of ^{237}Np. The absorber is a neptunium compound with ^{237}Np nuclei in the ground state (T = 2.41 × 10^6 years). Absorber and/or source may be cooled to liquid He temperature, 4.2 K. The energy of the source photons is scanned by moving the absorber. The Mössbauer emission spectrum is an intensity plot registered by the detector D as a function of absorber velocity. Resonant absorption is observed as a decrease in intensity. (b) Mössbauer transition in ^{237}Np. The quadrupole moment of the ^{237}Np nucleus is 4.1 b in the $5/2^+$ ground state and also in the 59.6-keV $5/2^-$ excited state. The mean lifetime of the excited state is 90.7 ns and the natural linewidth Γ = 7.24 × 10^{-9} eV, equivalent to a Doppler velocity of 0.034 mm s^{-1}. The absorber is usually NpO_2 because the local symmetry around the Np atom precludes the splitting of the nuclear levels; electric quadrupolar and magnetic dipolar interactions vanish. For simplification, the magnetic interaction in the *source* is omitted. The electric quadrupole interaction separates the 5/2 level into three components in the ground and excited states. The selection rule ΔM = 0, ± 1 allows 7 transitions, but in general only 5 lines can be resolved in the spectrum. One ΔM = 0 transition [labeled (1)] and one ΔM = 1 transition [labeled (2)] are shown on the figure. The centroid of the pattern is the isomer shift of the source photons with respect to the absorber.

photon from the source serves as a probe for measuring the hyperfine interactions in the absorber. For this purpose, the energy of the photon is monitored in order to match the precise energy required for absorption in the various hyperfine states in the absorber. The photon energy is scanned over the energy range of hyperfine splittings, typically of the order of 10^{-8} eV, by imposing a relative linear motion between source and absorber. The nominal energy E_γ of the source radiation, upon displacement with a velocity v, becomes

$$E_s = E_\gamma(1 \pm v/c)$$

and the relative energy change by Doppler shift is

$$\Delta E = |E_s - E_\gamma| = \frac{v}{c} E_\gamma$$

For a photon $E_\gamma = 10$ keV, a displacement with a velocity $v = 1$ mm s^{-1} is equivalent to an energy change of 3×10^{-8} eV, i.e., the order of magnitude of the hyperfine interactions. As a further condition, to serve as a chemical probe, the width of the γ-lines should permit the right resolution, i.e., about 10^{-8} eV or less. The width Γ of the line is related to the mean lifetime of the nuclear level by the uncertainty relation $\Gamma\tau = h/2\pi$, which sets the limits of the acceptable values of τ. The intensity of the Mössbauer effect increases with decreasing energy E_γ, with decreasing IC of the transition and with decreasing temperature.

Despite the nuclear restrictions, the Mössbauer effect has been observed for about 75 transitions in isotopes of 44 elements. In practice, however, the phenomenon is conveniently exploitable for chemical investigation only for some ten or so elements. Radioelements which possess an isotope suitable for absorption Mössbauer spectroscopy include Tc, Th, Pa, U, Np, Pu, and Am. With the exception of Np and Am, Mössbauer experiments with these elements have been limited to the demonstration of the existence of the effect.

In Mössbauer *emission* spectroscopy, the absorber contains the ground state isomer in a nonmagnetic, symmetric, environment which precludes hyperfine splitting of the nuclear level. In this case, spectroscopy reveals splittings at the site of the source nuclei, determined by the electronic environment of a *tracer* nuclide, and accordingly the source photon carries a chemical message. The source is a radionuclide whose half-life may range from several hours to hundreds of years, and which should have activity in the MBq range, decaying via α, β, EC, or IT to a nuclear level which emits photons suitable for the Mössbauer effect.

The following conditions must be fulfilled in Mössbauer emission spectroscopy:

1. In order to ensure recoil-free emission, the emitter must be firmly bound in a lattice and the photon energy must neither exceed 100 keV nor be excessively converted.
2. The mean lifetime of the excited level should lie between 1 ns and 100 ns to ensure sufficient resolution.
3. A sufficient amount, say 100 mg, of absorber cooled at liquid N_2 or He temperature must be available; this requirement is not straightforward when dealing with radioelements. The cross section for resonant absorption of the incoming photons exceeds by at least five orders of magnitude that of absorption by the photoelectric and Compton effects.

Mössbauer emission spectroscopy furnishes information on the chemical environment of the nuclide in the excited state at the instant the photon is emitted. With the exception of cases for which the nuclear level is fed by IT, this information is pertinent to the *daughter nuclide* in an excited state. It does not necessarily reflect the normal chemical state of the daughter because of the *after-effects* that follow decay of the parent atom and which include recoil, shake-off, and possibly Auger electron emission in the case of EC decay or IC. In the solid, the recoiling species comes to rest in a time which is much shorter than the nuclear lifetime. On the other hand, for very short lifetimes of the nuclear level, the ionization and excitation effects may not have attained relaxation at the instant of the emission, thus providing a time-dependent pattern of the Mössbauer spectra.

Indeed, Mössbauer emission spectroscopy is much more appropriate for the investigation of properties of nucleogenic atoms in a solid than for speciation of nuclides at the tracer level.[6]

The chemical information contained in a Mössbauer spectrum appears in three parameters:

1. The isomer shift, which is defined as the position of the centroid of a Mössbauer line on the energy (or velocity) axis and is related to the oxidation state of the emitter and to the degree of covalency of its bonds
2. The quadrupole splitting of the emission line, which reveals asymmetry in the electronic environment of the emitter and provides information on the chemical spin state and intensity of the ligand field
3. The magnetic splitting, indicating the presence of a magnetic field at the Mössbauer nucleus, and which has remained constant for a time exceeding the lifetime of the Mössbauer transition

Mössbauer emission spectroscopy experiments have been described extensively for $^{57}Co(EC)^{57}Fe$ and occasionally for the following decays: $^{119m}Sn(IT)^{119g}Sn$, $^{83}Br(\beta^-)^{83}Kr$, $^{99}Rh(EC)^{99}Ru$, $^{119}Sb(EC)^{119}Sn$, $^{119m}Te(EC)^{119}Sn$, $^{125}Sb(\beta^-)^{125}Te$, $^{125}I(EC)^{125}Te$, $^{151}Gd(EC)^{151}Eu$, $^{161}Tb(\beta^-)^{161}Dy$, $^{193}Os(\beta^-)^{193}Ir$, $^{197}Pt(\beta^-)^{197}Au$, $^{237}U(\beta^-)^{237}Np$, and $^{241}Am(\alpha)^{237}Np$. Data have been obtained on the chemical state of the *daughter* atoms. However, no really pertinent information is provided with respect to the chemical state of the parent atom. Even in the simplest case of an isomeric transition, aliovalent charge states (with respect to the oxidation state of the parent) are observed. Spectra of ^{119m}Sn-labeled tetravalent tin compounds point to the presence of divalent ^{119}Sn species and, conversely, the formation of $^{119}Sn^{4+}$ ions has been observed in $^{119m}Sn(II)$ compounds.

Within the chemistry of the radioelements, the most interesting case is that of ^{241}Am, which decays by α emission to the 59.6 keV/68 ns Mössbauer level of ^{237}Np.[7] The fact that the Mössbauer effect is observed here proves that the daughter atom has reached an equilibrium rest site at the moment of photon emission, despite its initial kinetic energy of 93 keV. Owing to very favorable nuclear properties, Np exhibits an exceptionally wide range of isomer shifts

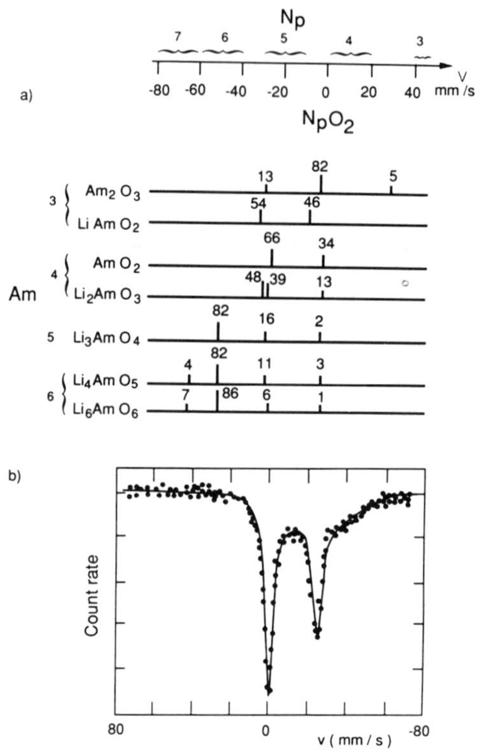

FIGURE 8.3. Mössbauer emission spectroscopy of ^{237}Np. (a) Distribution in percent of the oxidation states of Np formed by α decay of ^{241}Am in various oxides, together with isomer shifts of the Np entities with respect to NpO_2 at 77 K. (b) Mössbauer emission spectrum of $^{241}AmO_2$ at 4.2 K, interpreted as the superposition of two single lines of Np(IV) and Np(V) and a quadrupole pattern. (Adapted from Friedt, J. M., *Radiochim. Acta*, 32, 105, 1983.)

as a function of its oxidation state, extending from +34 mm s^{-1} for Np^{3+} to −77 mm s^{-1} for Np^{7+} relative to $NpAl_2$. This fact allows precise speciation of nucleogenetic ^{237}Np, but again the correlation with the state of the parent is not unequivocal. In most Am compounds, Np is found in several oxidation states, with a propensity for sharing the oxidation state of the parent (Figure 8.3).[7]

C. Positron Annihilation

The positron (e^+) is stable *in vacuo*. In matter, however, it annihilates with an electron, either directly or after forming a bound state with an electron (e^-), the *positronium* Ps (Figure 8.4). The latter is an *exotic* atom which can be viewed as a light isotope of hydrogen. If the spins of the two partners e^+ and e^- are antiparallel, the singlet state *para*-Ps (p-Ps) is formed; the parallel

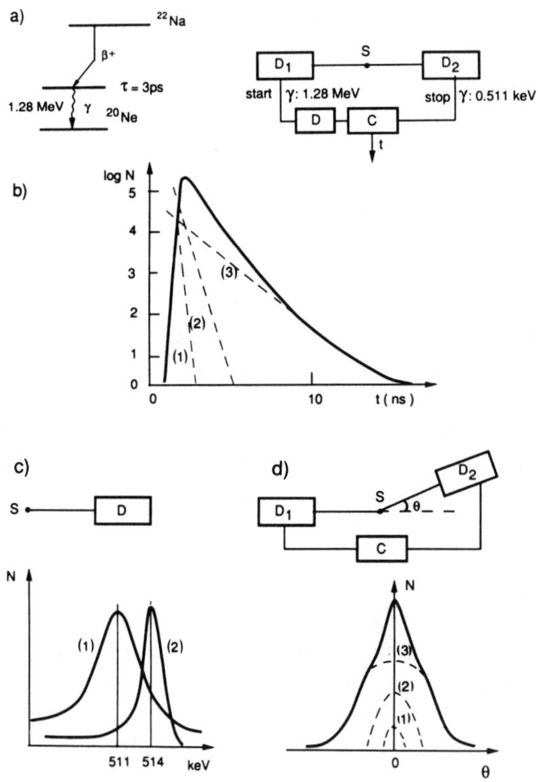

FIGURE 8.4. Principle of positron annihilation spectroscopy. (a) Lifetime spectroscopy. The positron source is ^{22}Na (T = 2.605 years) which decays by emission of a positron with the nearly simultaneous emission of a γ photon which serves as start signal and is registered by detector D_1. The positron is rapidly thermalized and disappears by annihilation with the emission of 511keV photons which are registered by detector D_2. The time interval t between the two signals measured through the delay (D) and coincidence circuit (C) is an indication of the various events in which the positron has been involved. (b) The lifetime spectrum is an intensity plot as a function of time elapsed since emission of the positron. The spectrum is decomposed into three components: para-Ps (1), free e^+ (2) and ortho-Ps (3) with lifetimes of the order of 0.1, 0.5, and 1.8 ns in condensed medium such as water. The relative proportions of the three components is also computed from the spectrum. (c) Measurement of Doppler broadening of the annihilation line. When at the instant of annihilation the pair e^-/e^+ is not at rest, the energy of the photons is slightly and isotropically changed from 511 keV. The width of the γ line (1) measured with a high resolution detector is broadened and after proper deconvolution yields the relative proportions of the various components. The width of the 514 keV γ line (2) of ^{85}Sr with a fwhm of 1.55 keV serves as reference. (d) Measurement of the angular correlation between the two 511 keV annihilation photons. This method is complementary to the previous one. The two annihilation photons are emitted at an angle θ with respect to the emission at 180°. The number N of photons in coincidence recorded by the detectors D_1 and D_2 as a function of the angle θ via a coincidence circuit (C) yields a curve which is deconvoluted into the contributions of para-Ps (1), ortho-Ps (2), and free e^+ (3).

orientation of the spins is characteristic of the triplet state *ortho*-Ps (o-Ps). The singlet p-Ps annihilates intrinsically (i.e., with its bound electron), whereas o-Ps essentially annihilates with an electron of the surrounding medium by a *pick-off* process. Because of various conservation rules (kinetic and angular momentum, energy), the most probable mode of annihilation is, in all cases, creation of two photons.

If, at the instant of annihilation, the center-of-mass of the e^+/e^- pair is at rest, the two photons are emitted in opposite directions, each carrying the same energy of 0.511 MeV. For free e^+ and o-Ps, however, this condition is not fulfilled. Even if the positrons are thermalized, they still have kinetic energy and the annihilation involves orbital electrons belonging to a molecule of the medium. In consequence, the kinetic momentum P of the center-of-mass of the pair is not nil and its kinetic energy amounts to a few electronvolts (eV). From the ensuing Doppler effect, it results that the energies of the photons deviate from 0.511 MeV, and the annihilation γs are not emitted in strictly opposite directions.

Two independent techniques can give access respectively to the transverse and to the longitudinal components of P; these are the angular correlation method, which measures the angular distribution of the annihilation photons, and Doppler broadening spectroscopy, which measures their energy distribution.[8,9] In principle, the measured projected distributions are characteristic of each state of the positron, thus opening the way to analytical applications.

Both e^+ and Ps provide useful information inasmuch as they are able to react with specific solutes. Ps chemistry is complex and involves the possibility of bound-state formation with solutes, but this has not yet found analytical applications.[10] Free e^+ particles also react with various solutes such as halide and pseudo halide ions, but the analytical implications in the present context are meager because the ions must be present at high concentration.[11] Very fast successive reactions have been found in the spur of positrons in polar solvents on a picosecond time scale, opening a promising route to the assessment of existence and nature of very labile species present at tracer concentration.[12] However, these applications are more relevant to radiation chemistry than to radiochemistry. At present, the properties of annihilation photons are of little (if any) concern in the speciation of radionuclides.

III. INFORMATION FROM CHANGE IN RADIATION CHARACTERISTICS WITH THE ELECTRONIC ENVIRONMENT

A. Change in Half-Lives

Attempts to observe a change in half-life of the radionuclides ^{226}Ra and ^{210}Po with the chemical state using external stimuli such as temperature, pressure, application of electric and magnetic fields, or even deliberate change in geographical location were as in vain as they were numerous in the early history of radiochemistry.[13,14] In principle, the half-life of any radioelement, whatever its decay mode, could possibly alter with the electronic environment of the nucleus. This is not unexpected for EC, which depends directly on the 1s-electron density at the nucleus, nor for IC, which may also involve s-type electrons.

Changes in half-lives are expressed by the ratio $\Delta\lambda/\lambda$, which is the relative change of the decay constant for a radionuclide in two different chemical environments. The most pronounced effect has been measured for compounds of ^{7}Be, an EC-decaying nuclide with a half-life of 53 d. Be is a very favorable element, since 2s electrons are directly involved in the formation of chemical bonds. The value $\Delta\lambda/\lambda$ of $(11.3 \pm 0.58) \times 10^{-4}$ obtained for the pair BeO–BeF_2 is a good indicator of the magnitude of the effect and the precision attained in the measurement. Even in this ideal case, systematic speciation of ^{7}Be based on half-life is ruled out. Minor changes in the decay rate are also found for ^{51}Cr [between Cr(III) and Cr(VI)] and ^{54}Mn [between Mn(II) and Mn(VII)]

The same conclusion holds for variation in the decay rate of a converted isomeric transition. Values reported for ^{99m}Tc compounds are of the same order of magnitude. Slight changes in λ have also been reported for ^{117m}Sn in Sn(II) and Sn(IV) compounds and for ^{65}Zn in Zn(s) and Zn(II) compounds. An odd case is 24-s ^{90m}Nb, which emits a soft 2.38 keV γ ray converted in the L_I sub-shell, for which the electron binding energy is 2.37 keV. Sufficient perturbation of the sub-shell population may impede the conversion and considerably affect the half-life.

The rate of α decay may be influenced by the Coulomb barrier around the nucleus, which can be slightly modified by the density of 1s electrons.

In the case of β^- disintegration, the decay constant may, in principle, may be affected by the oxidation state and the chemical bonds of the emitter. Besides the emission of the β^- particle in the continuum, two opposite electronic rearrangements may intervene: (1) orbital electron *creation*, in which the particle remains bound in the electronic cloud, a process that is obviously oxidation-state dependent, and (2) a quantum mechanical exchange process with production of a bound electron, combined with ejection into the continuum of the previously occupying electron. Theory indicates that the probability of the two processes is of the order of 10^{-4}, but the two contributions appear generally to be of *opposite* sign so that the overall changes in $\Delta\lambda/\lambda$ are very small. For ^{106}Ru ($T = 372.6$ d, $E_{\beta-}(\max) = 0.0394$ MeV) the total decay half-life changes by about 0.1% for a shift in one unit of the oxidation number.[2,3] The effect should be more pronounced for soft β^- emitters and large differences in the orbital electron creation, especially when comparing decay for the element in a low oxidation state compound with that of the element in a high oxidation state.

The chemical perturbation of α and β decays has still to be demonstrated, since published results which favor this effect appear rather dubious. In any event, speciation based on this phenomenon is excluded.

B. Change in the Relative Intensity of X-Rays

X-ray emission follows EC decay and IC. The relative intensity of these X-rays depends on the population of electronic levels involved in the transition, which in turn is related to the chemical state of the atom, its oxidation state and bonding with ligands. Transition between deeper shells, such as 2p → 2s, should be only slightly affected, but larger effects occur with X-rays originating from the valence shells. Thus, for the first row of transition elements, it is convenient to quote the relative intensities of X-rays from the 3p → 1s to the 2p → 1s transition, in other words, the ratios of intensities of the $K_{\beta 1,3}$ (or $K_{\beta' 1}$) and $K_{\alpha 1,2}$ lines. Experimental data are available for various compounds of the EC-decay nuclide ^{51}Cr. The K_β/K_α intensity ratio in the vanadium daughter increases slightly with the oxidation state of the parent from 0.12 in Cr(II) compounds to 0.14 in chromates, Cr(VI). However, for each oxidation state the ratio covers a range of values, thus reducing the accuracy of any conclusion with respect to the bonding involved. It should be noted that correlation with the oxidation state of the parent is meaningful only if the immediate after-effects of the EC decay have relaxed at the moment of X-ray emission. Chemical effects for the same X-ray ratio have also been measured for ^{55}Fe-labeled compounds; the K_β/K_α intensity ratio of the Mn daughter depends on the chemical form of iron, but no clear difference is observed between Fe(II) and Fe(III) compounds.

Tc offers a number of possibilities in the search for chemically-sensitive X-ray ratios following IC.[16] This element possesses three isomeric pairs, ^{99m}Tc, ^{97m}Tc, and ^{95m}Tc, of which the latter also decays by EC. For this second-row transition element, conversion takes place in the 3d and 4d shells. The greatest effects are found for the $K_{\beta' 2}/K_{\alpha 1,2}$ intensity ratio, i.e., the 4p → 1s vs. the 2p → 1s transition. For the ^{95m}Tc isomer, the value is 0.0267 for Tc_2S_7, 0.0270 for K_2TcCl_6 and 0.0280 for the pertechnetate $KTcO_4$. The ratio is not simply correlated to the valence state, but shows some dependence on the ionic character of the Tc bonds (see Figure 8.1).

Characteristic X-rays are also emitted when electron shell vacancies are produced by bombardment with external beams of electrons, protons, and heavy ions. More data on chemical shifts can be gleaned from such X-rays than from radioactive decay. The reported K_β/K_α intensity variations are in agreement with recent atomic models.[15]

C. Change in the Relative Intensity of Conversion Electrons

The intensity of conversion electrons is in principle related to the electronic population of the shells and sub-shells in which the nuclear energy is converted. This effect has been measured occasionally for the purpose of standardizing isomer shifts in Mössbauer spectroscopy. Owing to the high absorption of such electrons in matter, this procedure is unsuitable for routine speciation. Moreover, the information obtained is relevant to the daughter and not to the parent, except for isomeric transitions.

Finally, as interesting as they may be for fundamental research, these various chemical shifts are not suitable for reliable chemical speciation of radionuclides at the tracer scale.

D. Change in the Angular Correlation of γ Photons

When a nucleus decays by emission of two successive γ photons, the angle between the directions of the photons is not random, but has a distribution which depends on the properties of the nuclear levels and photons (Figure 8.5). The photons are *correlated*. This distribution may be modified by the chemical environment; in this case, the angular correlation is said to be *perturbed*. Applications of perturbed angular correlation (PAC) in chemistry, and more restrictively for the speciation of radiotracers, are still very limited.[14,16,17]

The emission cascade involves three levels with angular momentum I_1, I_2, and I_3, and two photons γ_1 and γ_2. The emission probability of a photon in a direction $\vec{k}$ relative to the spin axis depends on the angle $(\vec{I}, \vec{k})$. For a group of randomly oriented nuclei with a statistical distribution of the spin axes, all angles are equally probable and the emission is isotropic. It is possible to select the photons γ_1 which are emitted in a given direction simply by recording these photons with a detector D_1. The photons γ_2 are then measured as a function of the position of detector D_2 with respect to D_1. A plot of intensity registered by D_2 as a function of the angle θ between the two detectors represents the angular correlation of the two photons. For this purpose, the second transition from the intermediate level to the ground level must occur very rapidly after the emission of γ_1, so that the spin I_2 has "no time" to change its orientation; in other words, the lifetime τ of the intermediate level should be very short. Since, in addition, the two radiations must emanate from the same nucleus, the second photon is recorded in *fast coincidence* with γ_1.

The angular correlation function, in a simplified form and for the more commonly encountered cascades, is written

$$W(\theta) = 1 + A_2P_2(\cos\theta) + A_4P_4(\cos\theta)$$

$W(\theta)$ is the probability of emission of the photon γ_2 at an angle θ with respect to the direction of the first photon. $W(\theta) = 1$ for $I_2 = 0$ and 1/2. Coefficients A_2 and A_4 contain nuclear parameters of the two transitions and are known

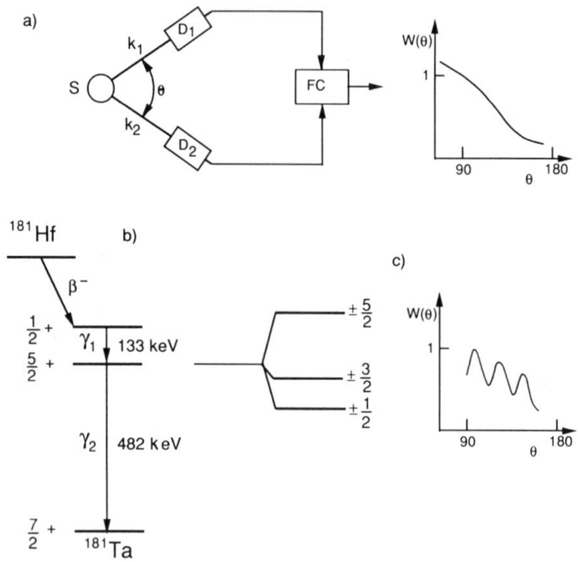

FIGURE 8.5. Principle of angular correlation spectroscopy. (a) Simplified scheme of the experimental setup. The two photons emitted in cascade by the source S are registered by detectors D_1 and D_2 placed at an angle θ, and analyzed by a fast-coincidence circuit (FC). The curve shows the normalized angular correlation function. (b) Perturbed angular correlation of the 133 to 482 keV γ-γ cascade of ^{181}Ta. The source is ^{181}Hf (T = 42.5 d). Coupling of the quadrupole moment (Q = 2.5 b) of ^{181}Ta in the intermediate level with the electronic environment produces a precession of spin during the 15.2 ns mean lifetime of the 482 keV ($5/2^+$) level. The perturbation results in a time-dependent change of the angular correlation. (c) Differential perturbed angular correlation spectroscopy yields a pattern which is characteristic of the electric field gradient acting at the site of the nucleus. The electric field gradient is determined by bonding symmetry and properties and may be altered by aftereffects of the β^- decay which feeds the γ-γ cascade. (Adapted from Adloff, J. P., *Radiochim. Acta*, 25, 57, 1978.)

for all cases of interest to the chemist. The P(cosθ) functions are Legendre polynomials.

As in Mössbauer spectroscopy, chemical effects on the angular correlation function have their origin in the coupling of momenta of the nucleus in the intermediate level with extranuclear electromagnetic fields produced by the electronic environment. This coupling is equivalent to a precession of the intermediate spin, with the result that its orientation changes in the time interval between the emissions of γ_1 and γ_2. The angular correlation function is perturbed provided τ is long enough (between 10^{-11} and 10^{-6} s) to allow the intermediate level to "feel" the perturbation. It is now written

$$W(\theta, t) = 1 + A_2G_2(t) + A_4G_4(t)$$

The perturbation functions G(t) give full information on the type, intensity, and direction of interaction and lead to quadrupole frequencies, electric field gradients, and correlation times.

The experimental procedure is denoted as the "time differential perturbed angular correlation" method. On the whole, it is more difficult to handle and is more time-consuming than Mössbauer spectroscopy. A serious limitation, as far as speciation is concerned, is the absence of an electric monopole interaction and equivalent of isomer shifts. On the other hand, a major advantage is the removal of restrictions based on temperature or solid state. High temperature phase transitions, solution chemistry of metal complexes and rotational diffusion of macromolecules represent topics that are open to investigation by this technique.

The perturbed angular correlation method has a point in common with Mössbauer spectroscopy in that it demands a radionuclide that must decay to a suitable γ-γ cascade. In practice, the choice is *very* limited and most of the reported experiments deal with the excited states of ^{181}Ta and ^{111}Cd populated by the decay of corresponding isobars: ^{181}Hf, ^{111}In, ^{111m}Cd, and ^{111}Ag. The list of potential candidates includes a dozen other nuclides, among which several isotopes decay to interesting cascades in multivalent elements. Examples of such cascades are found in ^{60}Ni, ^{75}As, ^{106}Pd, ^{188}Os, ^{187}W, and ^{100}Rh.

It is important to realize that, as in Mössbauer emission spectroscopy, speciation of a radioactive tracer is actually that of the excited *daughter* atom and not of the parent. A prerequisite for chemical application of perturbed angular correlation is the relaxation of after-effects of nuclear decay before the emission of the first photon. In this case, the chemical state of the daughter is identical to that at the initial level of the cascade, but extrapolation to the state of the parent is not straightforward.

This point is illustrated by the most commonly used nuclide, ^{181}Hf. In hafnium compounds, the element is normally in its tetravalent state and bonded to ligands L such that the charge balance following β^- decay should read

$$\mathrm{HfL_4} \rightarrow [\mathrm{TaL_4}]^+ + \beta^-$$

From all cases so far investigated, it appears that the 5d° configuration of Hf(IV) is preserved in the daughter, which acquires its normal pentavalent state prior to γ emission. However, the tantalum atom must adapt itself to the structure of the hafnium host, which may lead to unusual configurations for Ta.

If the after-effects of decay have not relaxed at the moment of emission of γ_1, the method is suitable for an *in situ* investigation of these effects. Perturbed angular correlation, like Mössbauer emission spectroscopy, is more relevant to the chemistry of nuclear transformations than to speciation of radionuclides at the tracer level.

Application of perturbed angular correlation to biological substances serves a different purpose. The radionuclides acts merely as a tracer for the study of the rotational motions of macromolecules and biomolecules.

PART TWO — PARTITION AND TRANSPORT METHODS

I. GENERAL CONSIDERATIONS

Owing to the limited extent of chemical information which can be obtained from the characteristics of the decay and from the properties of radiations, the latter serve generally as mere witnesses of the presence and of the behavior of radionuclides.

In this chapter, the main objective is to study the *speciation* of radionuclides. Even at the *macroscopic* level, the determination of the chemical state of elements is often problematic, and the task of speciation becomes all the more acute at the tracer level since reference entities for comparison are generally not available.

The most common methods are based on the *partition* of a radioactive tracer between two phases and on the *displacement* of a radioactive tracer within a single phase. Both methods rely on radiation detection. As will be shown, these procedures lead to first-hand quantitative information on the various E_i species among which a microcomponent m_j is distributed. It is recalled that these species are denoted $E_i(m_j)$.

Let S_1 be the system under investigation, which for the time being includes a single type m of radionuclide at the tracer level. In any partition method, a second system S_2 is opposed to S_1 such that part of the microcomponent is transferred from S_1 into S_2 (Figure 8.6). It is assumed that: (1) the reagents which constitute S_2 are present in sufficient concentration to behave as macrocomponents with respect to the tracer. This condition will ensure convenient reaction rates between micro- and macrocomponents; (2) the macrocomponents present in S_1 are not affected by S_2, and conversely. This requirement may require special care in the preparation of S_2.

The entities which are extracted into S_2 are the species initially present in S_1, or the species which are formed in the two-phase combination. In order to identify the species which pass into S_2, the latter are overlined: $\overline{E_i(m)}$. After separation of the two phases, the systems S_1 and S_2 are recovered, the only modification being the distribution of M. With this technique, the species of the microcomponent that have different chemical behavior in the two-phase system are sorted out spatially as a function of the time elapsed after the two systems are brought into contact. From a simple measurement of the activities of each phase, the concentration C_M of all $E_i(m)$ in S_1 and $\overline{C_M}$ of all $\overline{E_i(m)}$ in S_2 are obtained. They represent the *only measurement* which can be performed on the radionuclide, and all further information has to be deduced from these data.

The latter belong to the $E_i(m)$ species in the two-phase system and one may ask if they still reflect the properties of the species in the initial system S_1 prior to perturbation by S_2. In other words, the chemical perturbation which

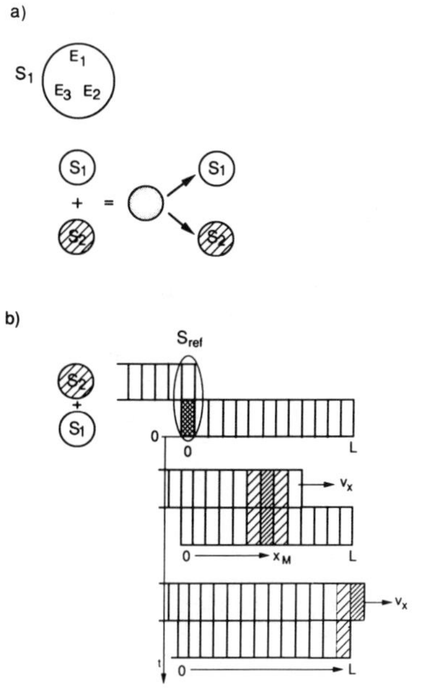

FIGURE 8.6. Schematic representation of partition systems. (a) Static partition: the entities $E_i(i = 1, 2, 3, \ldots)$ are distributed among the systems S_1 and S_2. (b) Dynamic partition: S_{ref} is the static reference system, v_x the relative velocity of the two phases and x_M the average displacement of a radionuclide. For $x_M < L$, x_M is a function of time; for $x_M > L$, x_M is a function of the elution volume. The successive positions of S_{ref} are shown on the figure.

is required in the partition method should not be so strong that it could alter the information that is really sought. This concern is rather general in chemistry, but is more acute when dealing with tracer amounts.

By combination of solid, liquid, and gaseous phases, five types of biphasic partition systems are conceivable (Figure 8.7). Since the composition of the perturbing S_2 system can be varied *ad libitum*, partition methods can be applied in a number of ways. In practice, the most widely used biphasic systems are liquid-liquid, liquid-solid, and solid-gas.

Depending on each specific case and the information sought, the partition experiment may be performed by following either a static or a dynamic mode. In the first instance, a unique partition equilibrium is set up and the two phases are generally immobile. A dynamic or chromatographic partition can be viewed as a succession of static partitions which may or may not be identical, and the two phases are in relative motion.

In the transport method, a chemical or electric gradient is imposed on S_1 and the microcomponent is displaced towards the lower gradient. The displacement velocity and position of the $E_i(m)$ entity are the *only experimental* parameters which can be measured, and all information of interest must be derived from these data. Any perturbation of the system S_1, if present, will always be of less importance than in partition methods.

In the following, selected examples will be presented to illustrate the potentialities of the two methods. Further applications are discussed in Chapter 10.

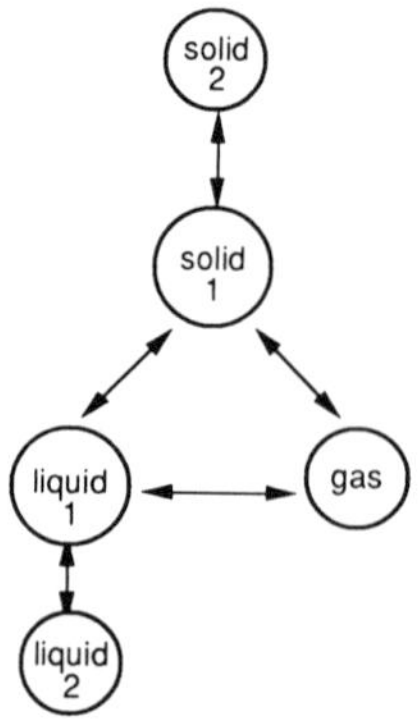

FIGURE 8.7. Various types of biphasic systems. Liquid systems may be aqueous/organic, fused salts/organic, or fused salts/fused metal.

II. STATIC PARTITION

A. Distribution Coefficients

The distribution of a radionuclide between two phases is expressed by the experimental dimensionless distribution coefficient

$$D = \frac{\overline{C_M}}{C_M}$$

where the C_M refer to the concentration of M in the two phases. In a liquid-liquid biphasic system, the value of D is easily obtained from the measurement of the activities of equal aliquot parts of the two phases (keeping in mind that the detection efficiency is generally different for the two phases). This practice is to be preferred to indirect determination, such as the measurement of the activity decrease in the initial phase, which may lead to substantial errors if the tracer is adsorbed on the walls of the container.

However, in the case of a liquid-solid system, this latter procedure cannot be circumvented. The activity of the solution is measured prior to the introduction of the solid and again at the end of the experiment. The partition is characterized by a coefficient K_d in units of $cm^3\ g^{-1}$

$$K_d = \frac{\overline{q_M}}{C_M}$$

in which $\overline{q_M}$ is the amount of M per unit weight of the dry solid and C_M the concentration of M in the liquid phase. If a weight m of the solid with a swelling coefficient δ is brought into contact with a volume V of the liquid phase,

$$K_d = \frac{V}{m}(D + 1 - \delta)$$

with

$$D = \frac{A_0 - A}{A}$$

A_0 is the initial activity of the liquid phase and A the activity following contact with the solid.

The measured D values are meaningful only if the two phases are correctly separated. This condition may not be easily satisfied when one of the phases is a solid or when colloids or pseudocolloids appear in the two-phase system. In all cases, completeness of phase separation is essential, and requires great care. The balance of the introduced activity should always be verified.

For a given system, D varies with the time elapsed after the two phases are brought into contact and tends towards an equilibrium value, D_e. This latter value corresponds to the thermodynamic equilibrium in which the chemical potentials of the $E_i(m)$ and other species involved in the partition are equal in both phases. The value of D_e is controlled by alternating the phase in which the radionuclide is introduced.

The equilibrium coefficient D_e depends on the various intensive or extensive parameters $\mathbf{P}_i$ which define the biphasic system: P, T, pH, equilibrium concentrations $[E_i(L_j)]$ of ligands, and thermodynamic constants of the equilibria involved in the two phases and in the partition equilibrium.

An unequivocal relation always exists between D and D_e or between D_e and the derivative dD/dt extrapolated to time zero. This is a remarkable property of a partition system at the tracer level. Hence, when working with radionuclides under these conditions, it may be sufficient to measure solely D, or its variations during the first instants of the experiment in order to derive the thermodynamic data which, in principle, are contained in D_e. Observation of a system that is not in equilibrium for the purpose of describing its equilibrium state would appear somewhat paradoxical, but this approach holds at the tracer level.

B. Kinetics of Transfer

The transfer rate of M depends on the mixing conditions of the phases and on the kinetics of the partition process, in which the crossing of the interface may be an important step. Diffusion in the bulk is generally slow for solid phases. The partition may proceed by one or several successive slow steps. An example is provided by the transfer of UO_2^{2+} aqua ions into an organic phase containing chelating and solvating reagents. The first step may involve a complexing reaction in the aqueous phase,

$$UO_2(H_2O)_6^{2+} + 2\ HA \rightleftarrows UO_2A_2(H_2O)_2 + 2\ H^+ + 4\ H_2O \qquad \text{(ue10)}$$

in which HA is the chelating agent, followed by the transfer into the organic phase by a solvation process

$$UO_2A_2(H_2O)_2 + \overline{S} \rightleftarrows \overline{UO_2A_2.2S} + 2\ H_2O$$

These steps are generalized according to

$$E_1(m) + X \rightleftarrows E_2(m)$$
$$E_2(m) + Y \rightleftarrows \overline{E_3(m)}$$

in which $E_1(m)$ is an aqua ion of M and X and Y are complexing or solvating species of the type $E_i(L_j)$. For the sake of simplicity, it will be assumed that the rate-controlling step is the transfer

$$E_2(m) \rightarrow \overline{E_3(m)}$$

between the two phases. Since M is a microcomponent, the direct and reverse reactions are of pseudo-first order with respect to $E_2(m)$ and $E_3(m)$. To maintain a formal analogy with the time evolution equation of nuclides in a radioactive filiation, the transfer rate is written

$$\frac{d[E_2(m)]}{dt} = \lambda_{32}[\overline{E_3(m)}] - \lambda_{23}[E_2(m)]$$

The λ_{ij} parameters depend on the rate constants, on the known and constant concentrations $[E_i(L_j)]$ of the species involved in the partition reaction, and, more generally, on the parameters $\mathbf{P}_i$. If the latter are constant — which will be assumed — the λ_{ij} are also constant. The equation is solved as follows.

Setting $[E_2(m)] = y$ and $[\overline{E_3(m)}] = y_0 - y$, an equivalent form is

$$\frac{dy}{dt} = -\lambda_{23}\,y + \lambda_{32}(y_0 - y) = c + fy$$
$$\text{with } c = \lambda_{32}\,y_0 \text{ and } f = -(\lambda_{23} + \lambda_{32}) < 0$$

Integration leads to

$$c + fy = e^{ft}e^{fK}$$

The constant K and the parameter c are combined, taking into account the initial and final conditions: $y = y_0 = [E_2(m)]_0$ for $t = 0$ and $y = y_e = [E_2(m)]_e$ for $t = \infty$. Hence,

$$y = y_e + ([E_2(m)]_0 - y_e)\, e^{ft}$$

and

$$[E_2(m)] = [E_2(m)]_e + ([E_2(m)]_0 - [E_2(m)]_e)\, e^{ft}$$

$[E_1(m)]$ is related to $[E_2(m)]$, but their dependency involves only concentrations of macrocomponents that remain constant. The time variation of $[E_1(m)]$ thus corresponds to that of $[E_2(m)]$:

$$[E_1(m)] = [E_1(m)]_e + ([E_1(m)]_0 - [E_1(m)]_e)\, e^{ft}$$

It results that

$$C_M = [E_1(m)]_e + [E_2(m)]_e + \{[E_1(m)]_0 + [E_2(m)]_0 - ([E_1(m)]_e + [E_2(m)]_e)\}\, e^{ft}$$

which is equivalent to

$$C_M = C_e + (C_0 - C_e)\, e^{ft}$$

With

$$D = \frac{C_0 - C}{C} \quad \text{and} \quad D_e = \frac{C_0 - C_e}{C_e}$$

one obtains

$$D = \frac{1 - e^{ft}}{\dfrac{1}{D_e} + e^{ft}}$$

Introducing the equilibrium half-time τ for which

$$y = y_e + \frac{[E_2]_0 - y_e}{2}$$

it results that

$$\tau = \frac{-0.7}{f}$$

so that

$$D = \frac{1 - e^{-0.7t/\tau}}{\frac{1}{D_e} + e^{-0.7t/\tau}}$$

It is deduced that

$$\frac{dD}{dt} = \frac{0.7/\tau \; e^{-0.7t/\tau} (1 + 1/D_e)}{(1/D_e + e^{-0.7t/\tau})^2} > 0$$

The second derivative

$$\frac{d^2D}{dt^2} = \frac{(0.7/\tau)^2 (1 + 1/D_e) \; e^{-0.7t/\tau} (e^{-1.4t/\tau} - 1/D_e^2)}{(1/D_e + e^{-0.7t/\tau})^4}$$

is zero for

$$t = \frac{\tau}{0.7} \ln D_e$$

Hence, D is always increasing and has an inflexion point (Figure 8.8). Further,

$$\left(\frac{dD}{dt}\right)_{t=0} = \frac{0.7}{\tau} \left(\frac{D_e}{1 + D_e}\right)$$

Variations of D as a function of time are shown in Figure 8.8 for various situations. The corresponding expressions for the more general case are derived in Appendix 8.2.

The previous relations between D and D_e may be verified fortuitously for macroscopic amounts of M.

C. Thermodynamics of Transfer

In order to derive a more general expression of D_e it is assumed, for the time being, that the thermodynamic activities at equilibrium can be replaced by equilibrium concentrations (see Appendix 8.3). Of various equivalent ways of expressing D_e, the equilibria can be formulated arbitrarily by specifying the entities of M in one phase and assuming that they will be transferred into the other phase. In a partition between an aqueous and an organic phase, this procedure would consist in writing the equilibria in the aqueous phase involving the aqua ion C_0^z and assuming that among the various possible entities $C_{l,q}^z$ only the neutral ones $C_{\bar{l},\bar{q}}^0$ are extracted into the organic phase.

The resulting expression for D_e is

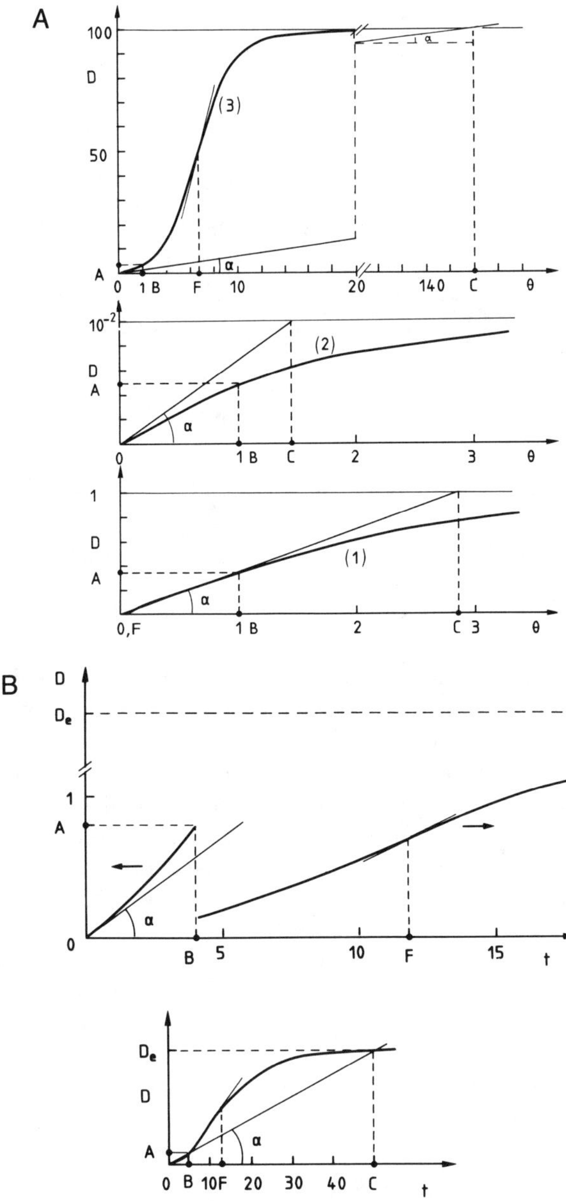

FIGURE 8.8. Theoretical variations of D with time (see text). (A) Variation of D as a function of the reduced time $\theta = t/\tau$ for: (1) $D_e = 1$; (2) $D_e = 10^{-2}$; (3) $D_e = 10^2$. Points marked in the figure have the following meanings: $OA = 0.5\, D_e/(1 + 0.5\, D_e)$ is the D_e value for $OB = \theta = 1$. $OC = (1 + D_e)/0.7$ is the abscissa of the intersection of the tangent at the origin with D_e. $OF = \log D_e/0.7$ is the abscissa of the inflexion point for $D_e \geq 1$. $\alpha = 0.7\, D_e/(1 + D_e)$ is the slope of the tangent at the origin. (B) Variation of D as a function of time with the following parameters: $D_e = 8$; $\alpha = 0.155\ \text{min}^{-1}$; $OA = 0.8$; $OB = 4$ min; $OC = 51$ min; $OF = 11.9$ min. (Note the different scales).

$$D_e = \frac{\sum [\overline{C^0_{l,\bar{q}}}]}{\sum [C^z_{l,q}]}$$

The summation is extended over all possible values of l and q which define the order of the complexes. This expression will be developed in a later section.

A similar fundamental relationship can be established for all static partition systems. In principle, information on the species involved in the partition can be easily derived provided that the species C^0_{l},q are effectively present in *one* phase. This may not be the case if these species are dissociated in an organic phase.

In practice, however, derivation of quantitative information on the species present is not always straightforward and the limitations of the procedure must be kept in mind.

For a given partition system, the values of D_e are characteristic for any element in a given oxidation state, provided, of course, that S_2 does not perturb S_1 to the extent that this oxidation state may be changed.

III. DYNAMIC PARTITION

A dynamic partition in which one of the phases is mobile will be described in simple terms rather than using the elaborated expressions applied, for instance, in chromatography. Several reference static partition systems S_{ref} are considered (see Figure 8.6). They represent imaginary, contiguous systems and are all identical if the parameters $\mathbf{P}_i$ remain constant along the length L of the stationary phase of a chromatographic column. On the other hand, they differ if some of the parameters $\mathbf{P}_i$ change with the distance x ($0 < x < L$) starting from the extremity of the stationary phase. This is equivalent to the application of a gradient $d\mathbf{P}_i/dx$ to the parameter $\mathbf{P}_i$.

When the mobile phase moves starting from the point $x = 0$, an increasing number of reference systems S_{ref} participate in the transfer of M from one phase to the other. At any given time, the displacement of M is characterized by the values of x_M and dx_M/dt and the partition of M in each system S_{ref} is fixed by its distribution coefficient D. The latter depends on the equilibrium value D_e in the system S_{ref} and on the time Δt during which the two phases are in contact, or, equivalently, on the velocity v_x of the mobile phase (which is also the velocity of S_{ref} at distance x).

The position x_M and the displacement velocity dx_M/dt of the microcomponent are easily recorded from radioactivity measurements. The derivation of the expression of the distance x_M as a function of D_e, v_x, and $d\mathbf{P}_i/dx$, which contains the required chemical information on M, is a more difficult task.

In the absence of a gradient imposed on $\mathbf{P}_i$, if the velocity $v_x = v$ is constant, the contact time Δt between each phase of the identical reference

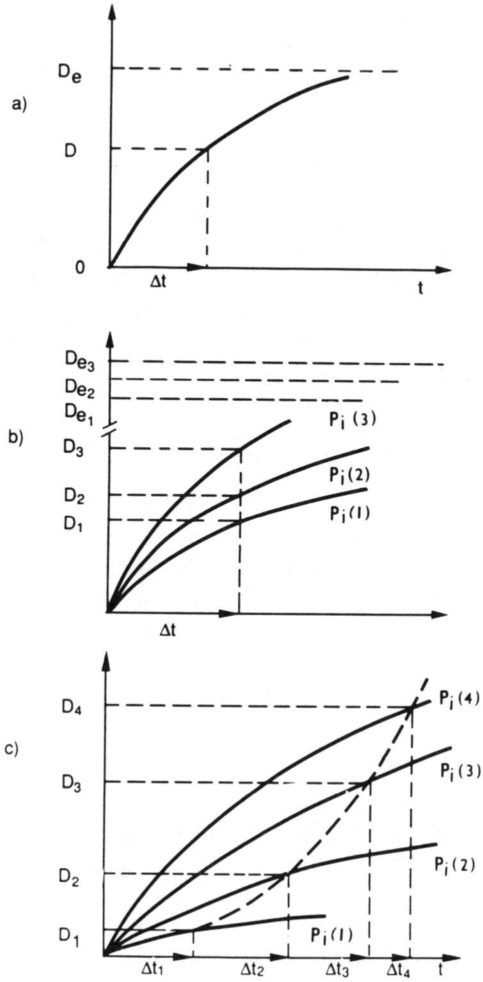

FIGURE 8.9. Variation of D with the contact time Δt between two phases in a reference system. The relative velocity of the phases is v_x. Terms in $\mathbf{P}_i$ are macroscopic parameters which define the system. (a) v_x and $\mathbf{P}_i$ are constant. (b) v_x is constant, one parameter $\mathbf{P}_i$ is x-dependent. The curves correspond to different values of $\mathbf{P}_i$ at distances x_1, x_2, x_3 . . . (c) v_x and one parameter $\mathbf{P}_i$ vary. The curves correspond to different values $\mathbf{P}_i$ at distances x_1, x_2, x_3, x_4 . . .

systems S_{ref} is a constant, as is also the distribution coefficient $D = \overline{C_M}/C_M$ (Figure 8.9). Let $\overline{V}$ and V be the volumes of the imaginary phases of the reference system and f the fraction of M in the mobile phase:

$$\frac{dx_M}{dt} = fv = v\,\frac{\overline{V}\,\overline{C_M}}{V\,C_M + \overline{V}\,\overline{C_M}} = v\,\frac{D}{A + D}$$

From this it follows that

$$x_M = v \frac{D}{A + D} t$$

A is a constant (unknown) characteristic of the geometry, size, and nature of the chromatographic system. This description holds for an ideal situation that does not take into account a possible linear diffusion of M along one or both phases. It should be pointed out that dx_M/dt remains measurable only as long as the value of D is not excessively large or small.

When a gradient is imposed on $\mathbf{P}_i$, v_x being constant and equal to v, Δt is constant but the distribution coefficient D is a function of x_M and

$$\frac{dx_M}{dt} = v \frac{D(x_M)}{A + D(x_M)}$$

If the velocity of the mobile phase v_x is not constant, the contact time Δt between the two phases depends on x_M between the two limiting values of the velocity, and D is still a function of x_M via the gradient $d\mathbf{P}_i/dx_M$:

$$\frac{dx_M}{dt} = v_x(x_M) \frac{D(x_M)}{A + D(x_M)}$$

In both cases the distribution coefficient $D(x_M)$ must be known in order to permit the integration.

The partition methods used in radiochemistry, apart from their static or dynamic nature, are further classified as a function of the property which is measured:

The equilibrium distribution coefficient D_e
The distribution coefficient D
The derivative dD/dt extrapolated to time $t = 0$
The displacement x_M of the microcomponent
The displacement velocity dx_M/dt
The parameters $\mathbf{P}_i$ subjected to a gradient, $d\mathbf{P}_i/dL$

This classification, together with the designation of the related methods, is recapitulated in Table 8.1. The dynamic partition methods, such as chromatography, are extremely important in radiochemistry because they can be applied to a very small number of atoms, and at the limit to a single atom. The latter case exhibits a *time-dependent* statistical behavior during the transfer from one reference system S_{ref} to another, identical to the statistical *time-independent* behavior of a system of many atoms which are necessarily required for a unique static partition.

Table 8.1
Classification of Radiochemical Partition Methods

Property measured	Static partition	Partition method
Equilibrium distribution coefficient, D_e		Solvent extraction ion exchange on resin syncrystallization
Distribution coefficient, D (dD/dt) extrapolated at time zero		Radiopolarography Radiocoulometry

Property measured	Dynamic partition Parameters dv/dx	dP_i/dx	Partition method
Displacement x_M and dx_M/dt	0	0	Resin chromatography
	—	—	Extraction chromatography
	0	$\neq 0$	Same with gradient
	$\neq 0$	$\neq 0$	Thermochromatography

IV. FORMULATION OF THE SPECIES INVOLVED IN PARTITION METHODS

Partition methods have a serious limitation because, as will be shown, they lead only to *differences* in the behavior and properties of the species present in the two phases. This is a delicate point which requires further precision in the formulation of complex species $E_i(M_j)$ of the microcomponents and $E_i(M_j)$ of the macrocomponents.

A. Aqueous Phase

To outline the problem, let us consider an aqueous solution of U(IV) in the presence of a complexing agent H_pL such as H_3PO_4 (see Chapter 6). Partition methods do not distinguish the following couples:

$C_0^+ \equiv U(OH)_3^+$ and $UOOH^+$
$C_2^- \equiv U(OH)_3\ (H_2PO_4)_2^-$ and $UO(HPO_4)\ (H_2PO_4)^-$ or $U\ PO_4\ (HPO_4)^-$
$C_2^0 \equiv UPO_4\ (H_2PO_4)$ and $U(HPO_4)_2$

More generally, it is not possible to distinguish between two species $E_i(m) = C_l^z$ which differ by the number of coordinated water molecules or the nature of the ligands derived from a same complexing reagent H_pL. It should be recalled that the equilibrium concentrations $[H_{p-1}L^-]$, . . . $[L^{p-}]$ of the complexing ligands are not independent.

On the other hand, the order l and the charge z of the complexes are known unambiguously. For a clearer identification of the species and for developments below, the following symbolic notation is introduced instead of C_l^z.

$$C_{l,y}^z = M(H_pL)_l H_{-y}^{(n-y)}$$

In this expression, n is the charge of the aqua ion. If the oxidation state N of the microcomponent M is ≤ 4, then $n = N$ and the aqua ion is M^{N+}; if $N > 4$, the aqua ion is generally an oxo ion, for instance $M^{n+} = MO_x^{(N-2x)}$ and $n = N - 2x$, like pentavalent MO_2^+ and hexavalent MO_2^{2+} actinyl aqua ions. The charge z is equal to $n - y$.

Together with this notation, the following conventions are adopted:

$$H_{-y} \equiv y\ OH \text{ if } y > 0$$

$$H_{-y} \equiv y\ H \text{ if } y < 0$$

i.e., $H_{-1} \equiv OH$, $H_{-2} \equiv 2\ OH$, $H_1 \equiv H$, $H_2 \equiv 2H$, etc.

Accordingly, one can write

$$H_{-1} + H_{+1} \equiv H_2O$$

$$H_{-1} + H_{-1} \equiv O + H_2O$$

To illustrate the notation,

$$C_0^+ = C_{0,3}^+ = U(OH)_3^+ = UOOH^+ = UH_{-3}^+$$
$$C_4^{2-} = C_{4,6}^{2-} = U(H_3PO_4)_4H_{-6}^{2-}$$
$$C_2^0 = C_{2,4}^0 = U(H_3PO_4)_2H_{-4}$$

This manner of representing the complexes may appear somewhat abstruse but it has several advantages:[18-22]

1. All possible complexes of order l that differ by the number of water molecules or by the ligands derived from the complexing agent H_pL are represented by the symbol $C_{l,y}^z$.
2. The charge $z = n - y$ of the complexes appears in the unique coefficient y, which is related to H, and accordingly to the solvated protons in the aqueous phase; in other words it is associated with the pH.
3. The formation of the complexes C_l^z can be rationalized directly from the aqua ion. Hence, the equilibrium

$$M^{n+} + l\,HpL \rightleftarrows M(H_pL)_l\,H_{-y}^{(n-y)+} + yH^+, \quad K_{l,y}$$

replaces

$$C_0^n + l\,H_pL \rightleftarrows C_l^z + (n - z)H^+, \qquad K_l$$

or

$$C_{0,0}^n + l\,H_pL \rightleftarrows C_{l,y}^z + y\,H^+, \qquad K_{l,y}$$

The formation of the complexes is characterized by the conditional equilibrium constants $K_{l,y}$. This will be illustrated for two complexes of $M^{3+} \equiv C_0^{3+}$ with phosphoric acid, $H_3PO_4 \equiv H_3L$. The complexes considered are of type $C_2^- = M(H_3PO_4)_2H_{-4}$. The formation of one of these complexes may be represented by the equilibrium

$$M^{3+} + 2\,H_2L^- + 2\,OH^- \rightleftarrows M(H_2L)_2\,(OH)_2^-, \quad \beta$$

Together with the equilibrium constants

$$K_w = \frac{[H]\,[OH]}{[H_2O]}$$

and

$$K_a = \frac{[H_2L]\,[H]}{[H_3L]}$$

in which the charges are omitted for the sake of simplicity, one obtains

$$\beta = \frac{[M(H_2L)_2(OH)_2]}{[M]\,[H_3L]^2\,[H_2O]^2\,[H]^{-4}}\,\frac{1}{K_a^2K_w}$$

The concentration $[H_2O]$ is a constant in dilute solutions, and *a fortiori* at the tracer level. Hence, the formation constant of the complex can be represented by

$$K_{24} = \beta\,[H_2O]^2\,K_a^2K_w = \frac{[M(H_2L)_2(OH)_2] \equiv [M(H_3L)_2\,H_{-4}]}{[M]\,[H_3L]^2\,[H]^{-4}}$$

This constant describes the equilibrium

$$M^{3+} + 2\,H_3L + 4\,H_{-1} \rightleftarrows C_2^-$$

or

$$C_0^{3+} + 2\,H_3L \rightleftarrows C_2^- + 4\,H^+$$

A second example refers to the complexing reaction

$$M^{3+} + HL^{2-} + H_2L^- + OH^- + H_2O \rightleftarrows$$
$$M(HL)(H_2L)\,OH^-,\ H_2O,\ \beta'$$

The dissociation constant of H_2L is

$$K_a' = \frac{[HL]\,[H]}{[H_2L]}$$

from which

$$K_a K_a'\,[H_3L] = [HL]\,[H]^2$$

and

$$\beta' = \frac{[M(HL)(H_2L)OH, H_2O]}{[M]\,[H_3L]^2\,[H]^{-4}}\,\frac{1}{K_a^2 K_a' K_w [H_2O]^2}$$

From $\beta' K_a' = \beta$, it follows that

$$K_{24} = \beta'\,K_a'\,K_a^2\,K_w [H_2O]^2 = \frac{[M(H_3L)_2\,H_{-4}]}{[M]\,[H_3L]^2\,[H]^{-4}}$$

K_{24} pertains to all C_2^- complexes formed according to

$$C_0^{3+} + 2\,H_3L \rightleftarrows C_2^- + 4\,H^+$$

4. The notation can be extended to mixed complexes when several complexing reagents compete: H_pL, $H_{p'}Q$, $H_{p''}R$ and their formation constants are expressed by equilibrium constants $K_{l,q,r,y.}$
5. Finally, with this special notation, a unitary treatment of the various radiochemical methods can be established without the assumptions on the species involved in the reactions which are usually introduced, in particular for aqueous solutions.

B. Non-Aqueous Phases

In an organic phase of low dielectric constant, the species (e.g., molecules, ion pairs) are neutral and are characterized by specific values of l, q . . . which are denoted $\bar{l}$, $\bar{q}$ . . . , and y = n. These species will be written $C^0_{\bar{l},\bar{q},\ldots n}$ It is assumed that the transfer of these species from the aqueous phase, in which they are formed, to the organic phase is characterized by a partition coefficient $P_{\bar{l},\bar{q},\ldots n}$.

For illustration, consider an organic solution in which:

1. The chelate of U(VI), $\overline{C}^{0}_{\bar{2},2} = UO_2A_2$, is written $UO_2(HA)_2H_{-2}$
2. The ion pair $\overline{C}^{0}_{6,\bar{2},\bar{4}} = UCl_6(R_3NH)_2$ of U(IV) between UCl_6^{2-} and R_3NH^+ is written $U(HCl)_6(R_3N)_2H_{-4}$

In a gaseous phase, all species are neutral; in a solid the various species, if not molecular, may be associated arbitrarily in ion pairs and the preceding considerations are still valid.

In a solvent with a low dielectric constant, the neutral ion pair may dissociate in anionic and cationic species $C_{\bar{l},\bar{q},\bar{y}}$, depending on $\overline{C}_M$ or on C_M. For instance, the chlorocomplex of U(IV) dissociates according to

$$\overline{UCl_6(R_3NH)_2} \rightleftarrows \overline{UCl_6^{2-}} + 2\,\overline{R_3NH^+}$$

which is also represented as

$$\overline{U(HCl)_6(R_3N)_2\,H_{-4}} \rightleftarrows \overline{U(HCl)_6\,H_{-6}} + 2\,\overline{R_3NH^+}$$

or

$$\overline{C^{0}_{6,\bar{2},4}} \rightleftarrows \overline{C_{\bar{6},\bar{0},\bar{6}}} + 2\,\overline{R_3NH^+}$$

The expressions of the partition coefficients will now be established for the principal methods used in radiochemistry. Experimental problems related to the utilization of the methods will not be discussed, and no attempt will be made to describe the mechanisms and the microscopic aspect of the reactions, because many of them have not been elucidated.

V. DETERMINATION OF THE EQUILIBRIUM DISTRIBUTION COEFFICIENT

Radiochemical methods based on measurement of the equilibrium distribution coefficient are by far the most numerous and most commonly used. They are not all amenable to a quantitative treatment because quite often the complexity of partition phenomena precludes adequate modeling.

A. Solvent Extraction

The advantage of the solvent extraction method lies in its versatility. A radionuclide present in a natural or intentionally prepared aqueous system is extracted into an organic phase after complexation, solvatation, or association with an ion of opposite charge. The composition of the organic phase and the nature of complexing or other binding agents can be varied *ad libitum* and the number of practical combinations of both parameters is unlimited.

If the objective of the study is speciation rather than separation of radionuclides, the most simple reagents, whose behavior is well known in the two phases, should be selected. Furthermore, it is recommended to choose a system for which the species remain undissociated in the organic phase and for which the ionic strength of the aqueous phase can be kept constant.

Here it is not intended to discuss the numerous aspects of solvent extraction; only the points of interest with respect to speciation will be emphasized.

1. Extraction of Complexes

The usual complexing agents are chelating compounds, derivatives of organophosphoric acids, and organic acids dissolved in an inert solvent.[23-29] They all behave like weak acids; for simplification they are considered monoprotic and are represented as an acid HA with an acidity constant pK_a. Partition of HA between the organic and aqueous phases is characterized by the coefficient

$$P_{HA} = \frac{[\overline{HA}]}{[HA]}$$

Knowing the values of P_{HA} and pK_a and, in appropriate cases, the polymerization constants of HA in the organic phase, the concentrations $[\overline{HA}]$ and $[A^-]$ can be calculated as a function of C_{HA} and the pH of the aqueous phase:

$$\log\,[A^-] = \log C_{HA} - \left[1 + [H^+]\frac{(1 + P_{HA})}{K_a}\right]$$

$$\log\,[\overline{HA}] = \log C_{HA} - \log\left[1 + \frac{1}{P_{HA}}\left(1 + \frac{K_a}{[H^+]}\right)\right]$$

The pH must be checked because the dissociation of HA may change the initial concentration $[H^+]_0$ of the protons. The actual value of $[H^+]$ is obtained from

$$C_{HA} + [H^+]_0 = [H^+] + [\overline{HA}] + \frac{K_a}{[H^+]}[A^-]$$

When choosing the complexing agent HA, the sum $pK_a + \log P_{HA}$ should be considered, since for $P_{HA} > 1$ (which is generally satisfied) it determines the complexing power of the chelating agent in the aqueous phase. The concentration $[A^-]$ is lower for high values of the sum, as is also the eventual perturbation of the system for a given C_{HA}.

First, it is assumed that the aqueous phase contains only one reagent H_pL and the organic phase one chelating agent HA. The latter separates to some extent in the biphasic system, with the result that mixed complexes $C_{l,q,y}^{(n-y)+}$ can be formed owing to competition between H_pL and HA.

The equilibria of interest concerning the aqua ion in the aqueous phase are those by which charged and neutral species soluble in the organic phase are formed:

$$M^{n+} + l\,H_pL + q\,HA \rightleftarrows C_{l,q,y}^{(n-y)+} + y\,H^+, \qquad K_{l,q,y}$$

$$M^{n+} + \bar{l}\,H_pL + \bar{q}\,HA \rightleftarrows C_{\bar{l},\bar{q},n}^{0} + n\,H^+, \qquad K_{\bar{l},\bar{q},n}$$

$$C_{\bar{l},\bar{q},n}^{0} \rightleftarrows \overline{C_{\bar{l},\bar{q},n}^{0}}, \qquad P_{\bar{l},\bar{q},n}$$

The equilibrium distribution coefficient is written

$$D_e = \frac{\sum\limits_{\bar{l},\bar{q}} [\overline{C_{\bar{l},\bar{q}}^{0}}]}{\sum\limits_{l,q} [C_{l,q}^{(n-y)+}]}$$

Summation is extended over all allowed values of the indices starting from zero up to the maximum value. With the preceding equilibria the distribution coefficient becomes

$$D_e = \frac{\sum\limits_{\bar{l},\bar{q}} P_{\bar{l},\bar{q},n}\,K_{\bar{l},\bar{q},n}\,[H_pL]^{\bar{l}}\,[HA]^{\bar{q}}\,[H^+]^{-n}}{\sum\limits_{l,q,y} K_{l,q,y}\,[H_pL]^{l}\,[HA]^{q}\,[H^+]^{-y}}$$

This expression contains the equilibrium concentrations of the non-dissociated forms of the complexing agents H_pL and HA, which, as said, can be calculated from the concentrations C_L, C_A, and the pH. Although they may often be devoid of physical meaning, they appear of interest here since D_e can be derived with respect to the independent parameters [HA], $[H_pL]$, and $[H^+]$. For instance, setting $[H_pL] = u$ and regrouping all constants terms as coefficients a and b,

$$D_e = \frac{a\sum\limits_{\bar{l}} K_{\bar{l},\bar{q}}\,u^{\bar{l}}}{b\sum\limits_{l} K_{l,q}\,u^{l}} = \frac{aA}{bB}$$

Further,

$$\frac{\partial \ln D_e}{\partial \ln u} = \frac{\partial \ln A}{\partial \ln u} - \frac{\partial \ln B}{\partial \ln u}$$

$$\frac{\partial \ln A}{\partial \ln u} = \frac{\partial \ln K_{\bar{l},\bar{q}}\, u^{\bar{l}}}{\partial \sum K_{\bar{l},\bar{q}}\, u^{\bar{l}}} \frac{\partial \sum K_{\bar{l},\bar{q}}\, u^{\bar{l}}}{\partial u} \frac{\partial u}{\partial \ln u}$$

$$\frac{\partial \ln A}{\partial \ln u} = \frac{1}{\sum K_{\bar{l},\bar{q}}} K_{\bar{l},\bar{q}} \sum \bar{l} u^{(\bar{l}-1)}\, u = \frac{\sum \bar{l}\, K_{\bar{l},\bar{q}}\, u^{\bar{l}}}{\sum K_{\bar{l},\bar{q}}\, u^{\bar{l}}}$$

This expression defines the mean value of the order of the complexes extracted into the organic phase

$$\langle \bar{l} \rangle = \frac{C_{H_pL} - [H_pL]}{C_M}$$

In the same way for the aqueous phase,

$$\langle l \rangle = \frac{\partial \ln B}{\partial u}$$

from which

$$\frac{\partial \ln D_e}{\partial \ln [H_pL]} = \langle \bar{l} \rangle - \langle l \rangle$$

The other expressions are obtained in the same way:

$$\frac{\partial \log D_e}{\partial \log [H^+]} = -(n - \langle y \rangle) \quad \text{or} \quad \frac{\partial \log D_e}{\partial\, pH} = n - \langle y \rangle$$

and

$$\frac{\partial \log D_e}{\partial \log [\overline{HA}]} = \frac{\partial \log D_e}{\partial \log [HA]} = \langle \bar{q} \rangle - \langle q \rangle$$

It follows that the charge of the complexes in the aqueous phase can always be obtained from the variation of log D_e with log $[H^+]$, but, on the other hand, only the *difference* between the average values of the order of the complexes present in the two phases is accessible. For this purpose, the change in D_e with one independent parameter which defines the biphasic system must be investigated, all other parameters being kept constant. It is important to vary each $\mathbf{P}_i$ within the largest possible limits. When the slopes $\partial \log D_e / \partial \log [\mathbf{P}_i]$ have integer values, the mean values are also the exact ones.

It is seen that only three relations are available for the determination of five coefficients. This is not a major problem, since in practice experimental conditions can be arranged for which some of the indices can reasonably be expected to be zero; the number of unknowns can then be reduced to three. As an example if, $\langle \bar{l} \rangle = \langle q \rangle = 0$, $\langle \bar{q} \rangle$ and $\langle l \rangle$ are known, then the charge as well as the order of the complexes can be obtained. For this purpose, the formation of mixed complexes in the two phases should be avoided, and that of $C^z_{l,0,y}$ in the aqueous phase and $\overline{C^0_{0,q,n}}$ in the organic phase should be favored. A hydrophobic HA should be preferred and the ligand L can be labeled with an appropriate radionuclide to check that it is not extracted into the organic phase in the form of a $\overline{C^0_{\bar{l},\bar{q}}}$ species.

In general, experiments are conducted in several steps.[30,31] First the pH and C_{H_pL} are kept constant, and $[\overline{HA}]$ is varied to verify that the extracted species always remain the same (constant $\bar{q}$). Next the pH is varied for different values of C_{H_pL} at constant $[\overline{HA}]$. Finally, the variations of D_e thus obtained are used to plot log D_e as a function of pH at constant log $[H_pL]$ and as a function of log $[H_pL]$ at constant pH. The slope of the curves gives the limiting values of l and z. The case for which $l = 0$ provides information on the extent of hydrolysis.

A relatively simple example is provided by the partition of U(VI) (as ^{230}U or ^{233}U) in the biphasic system:
HDEHP(diethylhenyl phosphoric acid)-C_6H_6/H_3PO_4, $HClO_4$, $LiClO_4$, ionic strength $\mu = 0.2$, T = 25°C with $\mathbf{P}_i$ parameters in the following ranges: $10^{-4} < \overline{C_{HA}} < 0.5\ M$, $C_{H_3PO_4} \leq 4.2\ M$ and $-4 < \log [H^+] < -0.7$.

The aqua ion UO_2^{2+} and the phosphoric complexes of U(VI) are all extracted from $HClO_4$ 0.2 *M* and complexing media with a slope $\partial \log D / \partial \log [\overline{HA}] \leq 2$, depending on $[\overline{HA}]$. From the experimental points of the curve with slope 2 (Figure 8.10a), it is deduced that $\langle \bar{q} \rangle - \langle q \rangle = \bar{q} - q = 2$. This difference is the maximal one compatible with UO_2^{2+} and it can be reasonably assumed that q = 0. It follows that the extracted species is $\overline{UO_2A_2}$.

D_e values obtained for $[\overline{HA}]$ such as $\bar{q} = 2$, may be normalized with respect to an arbitrary value $[\overline{HA}]_n$:

$$\log D_n = \log D_e + 2\,[\log[\overline{HA}] - \log[\overline{HA}]_n]$$

The variations of log D_n for different values of $C_{H_3PO_4}$ are shown in Figure 8.10b. The concentration $[H_3PO_4]$ is calculated for each value of log $[H^+]$ and $C_{H_3PO_4}$ to draw the curves shown in Figure 8.10c and d, which lead to $\langle l \rangle \leq 3$ and $-2 < \langle z \rangle < 2$. The experimental partition coefficients D_e are correctly fitted assuming the presence of seven different complexes. The predominating form in H_3PO_4 4.2 *M* (log $[H^+]$ = 0.7) is $C^+_{3,1}$.

This treatment is particularly simple in the ideal case with $\bar{q} = n$, $\bar{l} = 0$, q = 0, and y = lp, since the aqueous phase then contains the complexes ML_l and the species extracted into the organic phase is $\overline{MA_n}$. On the other hand, complications arise when more than one complexing agent is present

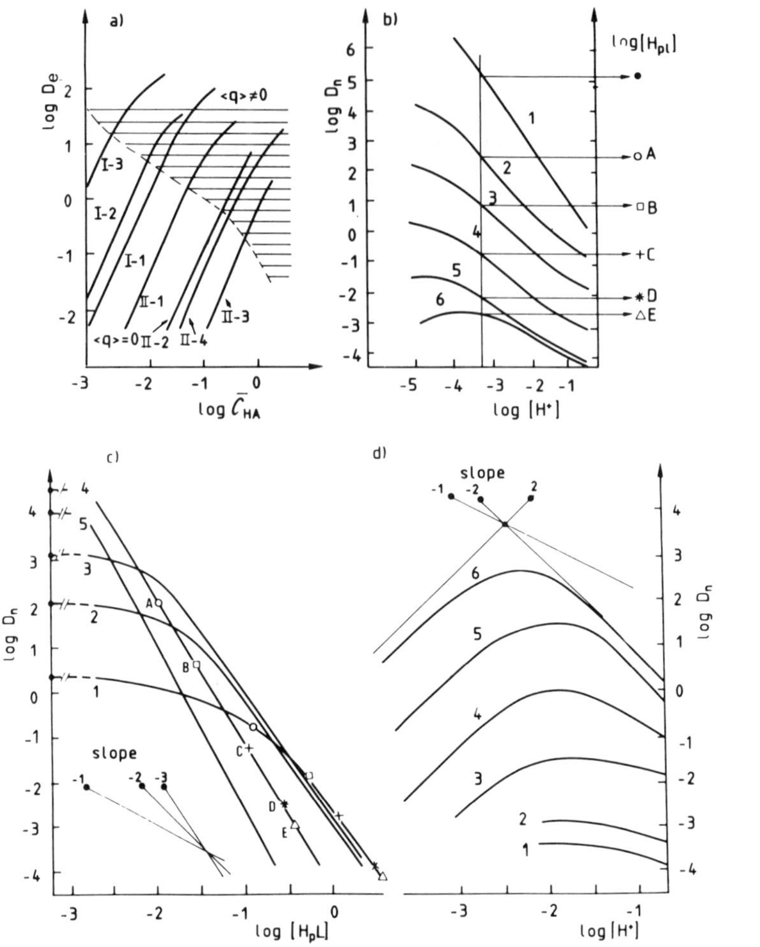

FIGURE 8.10. Partition of $^{230}U(VI)$ in the two phase system HDEHP $- C_6H_6/$ (H^+,$\overline{Li^+}$,ClO_4^-,H_3PO_4), $\mu = 0.2$. All concentrations are in mol L^{-1}. (a) Variation of D_e with $\overline{C_{HA}} \equiv \overline{[HA]}$, (HA is HDEHP). I. $C_{H_3PO_4} = 0$ and $[H^+] = 0.2$ (curve marked I-1), 10^{-1} (curve I-2), 10^{-2} (curve I-3). II. $C_{H_3PO_4} = a$ and $[H^+] = b$; a, b = 0.1, 0.2 (curve II-1); 1, 0.2 (curve II-2); 3, 0.3 (curve II-3); 4.2, 10^{-2} (curve II-4). In the shaded zone, $\langle q \rangle \neq 0$. Corresponding D_e values are ignored in the following; $\langle q \rangle = 0$ in the nonshaded zone. (b) D_n values normalized for $\overline{C_{HA}} = 10^{-2}$ *M*. Variation with log $[H^+]$ for different values of $C_{H_3PO_4}$ (M): (1) 0; (2) 0.1; (3) 0.3; (4) 1; (5) 3; (6) 4.2. (c) For a given pH and the different values of C_{H_pL} in figure (b), the values A, B, C . . . of $[H_pL]$ are calculated and curves drawn for the following values of log $[H^+]$: (1) -0.7; (2) -1.5; (3) -2; (4) -3; (5) -3.7. (d) Variation of D_n with log $[H^+]$ for different $[H_pL]$ calculated from figure (c). $[H_pL] = 2$ (curve 1); 1 (curve 2); 0.3 (curve 3); 0.1 (curve 4); 3×10^{-2} (curve 5); 10^{-2} (curve 6).

in the aqueous phase. For one additional complexing agent, the number of $\mathbf{P}_i$ increases only by one unit with respect to the simple case, but the number of experiments that must be performed is considerably greater.

It may be noticed that the expression of D_e has been established independently of a partition equilibrium. The choice of the latter is arbitrary because its actual partition mechanism is generally not known.

If errors on activity measurements are minimized, fluctuations of D_e values in repetitive trials are due to nonreproducibility in pH, $\overline{C_{HA}}$ and C_{H_pL}. They increase with n, $\langle\overline{q}\rangle$ and $\langle l\rangle$. For instance, in the case of a complex of order 2 formed by M^{3+} ion and extracted with 3 moles of HA,

$$\frac{\Delta D_e}{D_e} = 3\,\frac{\Delta[H^+]}{[H^+]} + 3\,\frac{\Delta[\overline{HA}]}{[\overline{HA}]} + 2\,\frac{\Delta C_{H_pL}}{C_{H_pL}} + 2\,\frac{\Delta V}{V}$$

If the relative errors on the volumes of the two phases used for measurement of the activities, on C_{HA} and on C_{H_pL} are 1%, and if ΔpH is 0.02 unit, fluctuations of D_e values may be as high as 15% and are mainly due to the pH effect ($\Delta[H^+]/[H^+] = 2.3\ \Delta pH$). A better precision requires many determinations, which is actually made in practice. If *relative* D_e values for two or more radionuclides are to be measured, it is expedient to study both simultaneously.

2. Extraction of Solvates

The procedure utilizes ethers, ketones, or esters of organophosphoric acids, either pure or dissolved in an inert solvent.[23-29,32,33] These reagents, which will be denoted S, have the property of extracting to some extent the acids present in the aqueous phase. For this reason, the organic phase should be saturated prior to extraction with the aqueous phase in the absence of M. This ensures a constant composition of macrocomponents in the aqueous phase.

If the aqueous phase contains only one complexing agent H_pL, the species in the organic phase are the $C^0_{\bar{l},\bar{s},n} = M(H_pL)_{\bar{l}}\,S_{\bar{s}}\,H_{-n}$. The complexes $C^{(n-y)+}_{l,y} = M(H_pL)^z_l$ which cannot be solvated by S are present in the aqueous phase. The following equilibria then hold:

$$M^{n+} + l\,H_pL \rightleftarrows C^{(n-y)+}_{l,y} + y\,H^+, \qquad K_{l,y}$$

$$M^{n+} + \bar{l}\,H_pL + \bar{s}\,S \rightleftarrows C^0_{\bar{l},\bar{s},n} + n\,H^+, \qquad K_{\bar{l},\bar{s},n}$$

$$C^0_{\bar{l},\bar{s},n} \rightleftarrows \overline{C^0_{\bar{l},\bar{s},n}}, \qquad P_{\bar{l},\bar{s},n}$$

The distribution coefficient takes the form

$$D_e = \frac{\sum_{\bar{l},\bar{s}} [\overline{C^0_{\bar{l},\bar{s},n}}]}{\sum_{l,y} [C^{(n-y)+}_{l,y}]} = \frac{\sum_{\bar{l},\bar{s}} P_{\bar{l},\bar{s},n}\,K_{\bar{l},\bar{s},n}\,[H_pL]^{\bar{l}}\,[S]^{\bar{s}}\,[H^+]^{-n}}{\sum_{l,y} K_{l,y}\,[H_pL]^{l}\,[H^+]^{-y}}$$

The latter expression leads to

$$\frac{\partial \log D_e}{\partial \log [S]} = \langle \bar{s} \rangle$$

$$\frac{\partial \log D_e}{\partial \log [H^+]} = -(n - \langle y \rangle) \text{ or } \frac{\partial \log D_e}{\partial pH} = n - \langle y \rangle$$

$$\frac{\partial \log D_e}{\partial \log [H_pL]} = \langle \bar{l} \rangle - \langle l \rangle$$

The remarks of the preceding section concerning the number of unknowns also apply here.

A classical solvating reagent is *tributylphosphate*, TBP. The extraction of Pa(V) from 3 *M* $HClO_4$

$$PaOOH^{2+} + 2\ ClO_4^- + 3\ \overline{TBP} \rightleftarrows \overline{PaO(OH)(ClO_4)_2.3\ TBP}$$

follows the preceding expression with $\langle \bar{s} \rangle = 3$.[34] The *Purex* process for reprocessing of spent nuclear fuels relies on TBP, because of the high resistance of the reagent toward radiolysis.

TBP can solvate chelates and trigger synergistic effects. The latter may be useful for determining the number of water molecules in chelates, as illustrated by the extraction of UO_2^{2+}. The extraction with thenoyltrifluoroacetone, (TTA≡HA) follows the equilibrium

$$UO_2(H_2O)_x^{2+} + 2\ \overline{HA} \rightleftarrows \overline{UO_2A_2(H_2O)_{X-4}} + 2\ H^+$$

Addition of TBP to the organic phase may increase the extraction by a factor of 10^4, indicating that the desolvation of the chelate is probably complete. The extraction of U(VI) is represented by

$$UO_2(H_2O)_x^{2+} + 2\ \overline{HA} + 2\ \overline{TBP} \rightleftarrows \overline{UO_2A_2.2\ TBP} + 2\ H^+$$

It is concluded that $x = 6$, which is in agreement with the hexasolvated aqua ion of U(VI).

Information on the formulation of chelates can be obtained in a similar approach. In $HClO_4$, Pa(V) is extracted with four TTA molecules as a species $Pa(HA)_4H_{-5}$ which may be written $Pa(OH)A_4$, $PaOA_3HA$ . . . In the presence of TBP, synergetic extraction requires 2 moles TTA and 2 moles TBP:

$$PaOOH^{2+} + 2\ \overline{HA} + 2\ \overline{TBP} \rightleftarrows \overline{PaOOHA_2.2\ TBP} + 2\ H^+$$

The composition of the chelate is best represented by $PaOOHA_2$ 2HA.[34] Many examples are provided in Reference 35.

3. Extraction of Ion Pairs

The reagents are salts of amines, or ether-acid or ketone-acid associations soluble in an inert solvent. Only anionic species ($z < 0$) present in the aqueous phase can be extracted. Since complexation occurs in the aqueous phase, high values of C_{H_pL} are required.

A simple case is that in which the aqueous phase contains only HL species and the organic phase the amine salt $R_3NH^+L^-$, written SHL. Complications may arise if the ion pairs dissociate in the organic phase. The neutral species in the organic phase are $C^0_{\bar{l},\bar{y}-n,n} = M(HL)_{\bar{l}}\,S_{\bar{y}-n}\,H_{-n}$ and the complexes in the aqueous phase are $C_l^{(n-y)+} = M(HL)_l^z$ since ion pairs cannot be formed in these phases with amine salts. Following equilibria hold

$$M^{n+} + l\,HL \rightleftarrows C_{l,y}^{(n-y)+} + y\,H^+, \quad K_{l,y}$$

$$M^{n+} + (\bar{l} - \bar{y} + n)\,HL + (\bar{y} - n)\,SHL \rightleftarrows C^0_{\bar{l},\bar{y}-n,n} + nH^+, \quad K_{\bar{l},\bar{y}-n,n}$$

with $\bar{y} \geq n + 1$ and

$$C^0_{\bar{l},\bar{y}-n,n} \rightleftarrows \overline{C^0_{\bar{l},\bar{y}-n,n}}, \qquad P_{\bar{l},\bar{y}-n,n}$$

The distribution coefficient becomes

$$D_e = \frac{\sum_{\bar{l},\bar{y}} [\overline{C^0_{\bar{l},\bar{y}-n,n}}]}{\sum_{l,y} [C_{l,y}^{(n-y)+}]} = \frac{\sum_{\bar{l},\bar{y}} P_{\bar{l},\bar{y}-n,n}\,K_{\bar{l},\bar{y}-n,n}\,[HL]^{\bar{l}-(\bar{y}-n)}\,[SHL]^{\bar{y}-n}\,[H^+]^{-n}}{\sum_{l,y} K_{l,y}\,[HL]^l\,[H^+]^{-y}}$$

The summation in the organic phase begins with indices for which $z \leq -1$. The following relations are deduced:

$$\frac{\partial \log D_e}{\partial \log [SHL]} = \langle\bar{y}\rangle - n$$

$$\frac{\partial \log D_e}{\partial \log [H^+]} = -(n - \langle y\rangle) \quad \text{or} \quad \frac{\partial \log D_e}{\partial \log [pH]} = n - \langle y\rangle$$

$$\frac{\partial \log D_e}{\partial \log [HL]} = \langle\bar{l}\rangle - \langle l\rangle - \langle\bar{y}\rangle + n$$

The only valid information which can be derived for the species in aqueous solution is the charge of the species.

Now the more complicated case involving dissociation of the species $\overline{C^{0}_{\bar{l},\bar{y}-n,n}}$ and $\overline{SHL}$ is considered. At least two additional equilibria must be taken into account:

$$\overline{C^{0}_{\bar{l},\bar{y}-n,n}} \rightleftarrows \overline{C^{(n-\bar{y})}_{\bar{l},0,\bar{y}}} + (\bar{y} - n)\,\overline{SH^{+}}, \qquad \overline{K_{\bar{l},0,\bar{y}}}$$

$$HL + \bar{S} \rightleftarrows \overline{SH^{+}} + \overline{L^{-}}, \quad K_s$$

Together with the electroneutrality relationship

$$[\overline{SH^{+}}] = (\bar{y} - n)\,[\overline{C_{\bar{l},0,\bar{y}}}] + [L^{-}]$$

and the mass balance in the organic phase

$$\overline{C_M} = [\overline{C^{0}_{\bar{l},\bar{y}-n,n}}] + [\overline{C_{\bar{l},0,\bar{y}}}]$$

and setting $\sum_{l,y} [C_{l,y}] = C_M$, the distribution coefficient takes the form

$$(D_e)_{diss} = \frac{1}{C_M}\left[\sum_{\bar{l},\bar{y}} [\overline{C^{0}_{\bar{l},\bar{y}-n,n}}] + \sum_{\bar{l},\bar{y}} [\overline{C_{\bar{l},0,\bar{y}}}]\right]$$
$$= D_e[1 + K_{\bar{l},0,\bar{y}}\,[\overline{SH^{+}}]^{\bar{y}-n}]$$

$[\overline{SH^{+}}]$ is obtained as the root of an equation of order $(\bar{y} - n + 1)$ and, omitting charges, is given by

$$[\overline{SH}] = (\bar{y} - n)\frac{C_M}{1 + \dfrac{[\overline{SH}]^{\bar{y}-n}}{\overline{K_{\bar{l},0,\bar{y}}}}} + \frac{K_s\,[HL]\,[\bar{S}]}{[\overline{SH}]}$$

This last expression shows that $[\overline{SH^{+}}]$ is a function of $\overline{C_M}$ or C_M, as is also the distribution coefficient when all M species are monomeric. The "slope" method does not give a clear identification of the stoichiometric coefficients of interest. At the tracer level, when $\overline{C_M}$ or C_M are very small, or equivalently when

$$[\overline{C_{\bar{l},0,\bar{y}}}] << [\overline{L^{-}}],\ [\overline{SH^{+}}] \quad \text{or} \quad [\bar{S}]$$

$$(D_e)_{diss} = D_e[1 + K_s[HL]\,[\bar{S}]^{(\bar{y}-n)/2}]$$

The distribution coefficient is then no longer dependent on $\overline{C_M}$, but the slope analysis requires particular attention (see Appendix 8.4).

An example of extraction by ion pair formation is provided by Pa(V). In the absence of dissociation in the organic phase, the species is extracted into xylene from an HCl medium $\geq 3\ M$ with two molecules of tri-n-octylamine (TNOA). In 10 *M* or higher HCl concentrations, cationic complexes C_2^{2+} can be inferred from the slopes of the extraction curves as a function of $[Cl^-]$ and $[H^+]$. On the other hand, extraction of Pa(V) with TTA points to the formation of anionic complexes C_3^{3-}.

Partition between the two phases is hampered if the microcomponent forms radiocolloids which cannot be extracted. Several procedures have been described for determination of these forms, e.g., changing the volume of the phases or performing repetitive extraction from a same phase (see Appendix 8.4).

As pointed out, the relations which have been discussed so far should be used with caution, since truly chemically inert diluents do not exist and the thermodynamic activity of each micro- and macrocomponent entity, rather than the concentrations, should appear in the formulae. The influence of activities ought to be examined for each particular partition system, because it depends strongly on the properties of the medium (see Appendix 8.3).

It should be emphasized that, with the exception of ion pairs, the proposed relations are independent of the concentration of the microcomponent and of the presence of nonextractable species. Special cases for which this property does not hold are discussed in Appendix 8.4.

B. Ion Exchange

Partition of the microcomponent between a liquid phase and a cationic or anionic, organic, or inorganic ion exchanger is not used as universally as solvent extraction.[36,37] Several reasons may be invoked, among others the lack of precision in the measurement of D or D_e because of the swelling of resins, and difficulty in the separation of phases.

Ion exchangers are insoluble in the aqueous phase; their exchange capacity is of the order of a few milliequivalents per gram of dry resin, which is always greatly in excess over the amounts of radionuclides present at the tracer level.

The interpretation of D_e is simple in one case only, namely the exchange of complexes $C_{l,y}^{z}$ $(z > 0)$ with a cationic exchanger RH. It is analogous to the solvent extraction of ion pairs, with $C^0_{\bar{l},n-\bar{y},n} = M(HpL)_{\bar{l}}\ (RH)_{n-\bar{y}}\ H_{-n}$ species in the solid phase, but without dissociation.

Since the exchanger RH is insoluble, it is expedient to consider that the following equilibria are determinant:

$$M^{n+} + l\ H_pL \rightleftarrows C_{l,y}^{(n-y)+} + y\ H^+, \quad K_{l,y}$$

$$M^{n+} + \bar{l}\ H_pL \rightleftarrows C_{\bar{l},n-\bar{y}}^{n-\bar{y}} + n\ H^+, \quad K_{\bar{l},n-\bar{y}}$$

with $\bar{y} \leq n-1$ and

$$C^{n-\bar{y}}_{l,n-\bar{y}} + (n - \bar{y})\,\overline{RH} \rightleftarrows \overline{C^{0}_{\bar{l},n-\bar{y},n}} + (n - \bar{y})\,H^{+}, \quad K_e$$

The summation of the indices $\bar{y}$ in the expression

$$D_e = \frac{\sum_{\bar{l},\bar{y}} [\overline{C^{0}_{\bar{l},n-\bar{y},n}}]}{\sum_{l,y} [C^{(n-y)+}_{l,y}]}$$

is performed from n − 1 to 0. The usual expressions are derived:

$$\frac{\partial \log D_e}{\partial \log [RH]} = n - \langle\bar{y}\rangle$$

$$\frac{\partial \log D_e}{\partial \log [H^+]} = -(n - \langle y\rangle) \quad \text{or} \quad \frac{\partial \log D_e}{\partial [pH]} = n - \langle y\rangle$$

$$\frac{\partial \log D_e}{\partial \log [H_pL]} = \langle\bar{l}\rangle - \langle l\rangle$$

Generally, the only species which participates in exchange is M^{n+} because affinity for the resin decreases rapidly with the positive charge of the species. In this case $\langle\bar{l}\rangle = 0$ and the expressions are simplified.

It is a common practice to use a monofunctional cationic exchanger RM′ (where M′ is a nonhydrolyzable, noncomplexing element such as Li^+ or Na^+) rather than RH. The competition between M and M′ in the exchange reaction can be investigated as a function of $C_{M'}$ to obtain $n - \langle\bar{y}\rangle$ and as a function of pH to obtain $n - \langle y\rangle$.

In the same way, when exchange with RH is performed at constant ionic strength μ, complications arise. If cations such as Li^+ or Na^+ are added in order to impose the required value of μ, the resin takes the form RLi or RNa, and the number of equilibria which must be taken into account is much higher.

An anionic resin exchanges only the complexes $C^{z}_{l,y}$ with $z < 0$, but the anions from H_pL, as well as those from added lithium or sodium salts also participate in the exchange. The number of equilibria then becomes still higher and the expression of D_e is rather complicated.

Inorganic exchangers have the advantage of being able to resist high doses of radiation. Hence, they are primarily used in radiochemistry to separate very active species rather than for speciation.

C. Coprecipitation

Coprecipation is the incorporation of a radionuclide in a solid phase formed within a solution and at the tracer level. Hence, this method requires the use of a *carrier*.[38-40] If the latter is a stable or long-lived isotope of the microcomponent, the coprecipitation is similar to a normal precipitation procedure and is then not of interest in the present context, since it does not give additional information on the behavior or properties of the microcomponent. A special case is that of a radioactive tracer for determining the solubility of highly insoluble compounds.

Coprecipitation with a nonisotopic carrier that has chemical properties similar to those of the radionuclide in the same oxidation state proceeds via the two mechanisms *syncrystallization* and *adsorption*.

The purpose of coprecipitation is either to separate or concentrate species at the tracer level, or to obtain information on the chemical properties of $E_i(m)$ species. This latter goal is much more difficult to achieve.

It is obvious that the proper character of tracer scale chemistry is lost since one is now dealing with weighable amounts of matter in the form of a carrier. It is also to be expected that surface phenomena which remain unnoticed at the tracer level may play an important and possibly erratic role in the presence of a bulk material: one monolayer corresponds to about 10^7 species per square centimeter.

The element Ra was first isolated from tons of pitchblende by syncrystallization with chlorides and sulfates of barium; the first separation of Pu from neutron-irradiated U was based on coprecipitation (probably by syncrystallization) with bismuth phosphate. Later, both procedures were extended on an industrial scale.

1. Syncrystallization

Two isomorphous compounds (i.e., of same composition and structure) *syncrystallize* when they are coprecipitated from a liquid phase. In the present case, one of the compounds corresponds to the microcomponent M and is represented by $C^0_{l,n}$, and the second is the carrier written $C'^0_{l,n}$.

Depending on experimental conditions, two types of distribution of the tracer in the carrier can be obtained. In the first, the distribution of M in the lattice of the carrier is homogeneous and thermodynamic equilibrium prevails during the formation of the crystal. In the second case, distribution is heterogeneous and equilibrium exists only on the surface of the crystal during coprecipitation.

Obviously, only the first case can lead to an expression for D_e that will be independent of time and experimental conditions. To achieve proper conditions, crystallization should be slow in order to form small crystallites. Alternatively, it may be performed very rapidly, followed by extended stirring, to permit diffusion of the tracer throughout the crystal. Equilibrium may eventually be established in this way.

In the absence of redox reactions, the equilibria in which a tracer M is involved are of the general type:

$$C_{l,y}^{(n-y)+} \rightleftarrows \overline{C_{\bar{l},n}^{0}} + (\bar{l} - l)\, H_pL + (n - y)\, H^+, \quad K_{l,\bar{l},n}$$

The precipitation of the carrier M′ may *a priori* proceed from complexes different from those of M, with equilibria of the form

$$C'^{(n-y')+}_{l',y'} \rightleftarrows \overline{C'^{0}_{\bar{l},n}} + (\bar{l} - l')\, H_pL + (n - y')\, H^+, \quad K'_{l',\bar{l},n}$$

The equilibrium distribution coefficient D_e becomes

$$D_e = \frac{[\overline{C_{\bar{l},n}^{0}}]}{\sum_{l,y} [C_{\bar{l},n}^{(n-y)+}]} = \sum_{l,y} K_{l,\bar{l},n}\, [H_pL]^{-(\bar{l}-l)}\, [H^+]^{-(n-y)}$$

The charge and order of the complexes may be derived from the variation of D_e as a function of $[H_pL]$ and $[H^+]$ using the expressions

$$\frac{\partial \log D_e}{\partial \log [H^+]} = -\,(n - \langle y \rangle) \quad \text{or} \quad \frac{\partial \log D_e}{\partial\, [pH]} = n - \langle y \rangle$$

$$\frac{\partial \log D_e}{\partial \log [H_pL]} = \langle \bar{l} \rangle - \langle l \rangle$$

The value of $\bar{l}$ is known from the composition of the macroscopic crystalline precipitate. It should always be checked that this composition does not change with the experimental conditions of the precipitation reaction.

It is also possible to measure the partition of the carrier M′ between its saturated solution and the crystal $C_{\bar{l},n}^{0}$ and compare it with the distribution of M. For this purpose, a fractionation coefficient F is defined as

$$F = \frac{D_e}{D'_e}$$

in which D'_e is the distribution coefficient of the macrocomponent. The complete expression of F is

$$F = \frac{\sum_{l,y} K_{l,\bar{l},n}\, [H_pL]^{-(\bar{l}-l)}\, [H^+]^{-(n-y)}}{\sum_{l',y} K'_{l',\bar{l},n}\, [H_pL]^{-(\bar{l}-l')}\, [H^+]^{-(n-y')}}$$

Syncrystallization is said to occur if F is independent of the conditions under which the crystal is formed: this corresponds to the *Berthellot-Nernst* law. It requires fulfillment of the conditions

$$\frac{\partial \log F}{\partial \log [H^+]} = \langle y \rangle - \langle y' \rangle = 0 \qquad \frac{\partial \log F}{\partial \log [H_pL]} = \langle l \rangle - \langle l' \rangle = 0$$

In other words, the same type of complexes of M and M′ should coexist in the aqueous phase; the behavior of both elements must be strictly the same in solution and during formation of the precipitate. Only under these conditions, F will remain constant in the range of $[H_pL]$ and pH explored. It is seen that the species of a *macrocomponent* remains to be identified in order to obtain l' and that the advantages of the tracer scale are lost; in particular, the concentration $[H_pL]$ must be measured.

It may be noted that

$$D_e = F\, D'_e = k\, F \frac{C'_{sat}}{\rho'}$$

in which ρ' is the specific mass of the crystal in g cm^{-3} and C'_{sat} the concentration of M′ in the saturated solution after crystallization. The term k is a correction factor for conversion of units. Determination of D_e in the case of syncrystallization can be based on the measurement of solubility of the macrocomponent as a function of $[H_pL]$ and pH.

If $M = M'$, the fractionation coefficient $F = 1$; the solubility of the crystal labeled with the radioactive microcomponent is

$$S = C_{sat} = \frac{\rho\, D_e}{k}$$

Provided $\bar{l}$ remains constant, the charge and the order of the complexes in equilibrium with the crystal can be derived from the variation of log S with log $[H_pL]$ and pH. In this way, the solubility of $Th_3(PO_4)_4$ has been measured using the ^{227}Th-labeled compound.

What is the smallest quantity of a solid that is sufficient for determining the solubility of a compound? The amount must be large enough to display the normal lattice energy. Thus, for an ionic compound, it should incorporate all ions whose coordinates make a significant contribution to the Madelung constant. The number of ions involved depends on the crystal structure and probably includes more than ten unit cells.

Well-defined coprecipitation by syncrystallization involves ionic crystals formed under conditions for which the species in solution are simple ions M^{n+} and crystals of the carrier M′ are isomorphous with those of the tracer M. In this case,

$$\frac{[\overline{M^{n+}}]}{[M^{n+}]} = F \frac{[\overline{M'^{n+}}]}{[M'^{n+}]}$$

Such coprecipitation has been known for a long time; examples are syncrystallization of chlorides (F = 53), bromides (F = 10), nitrates (F = 1.88), chromates (F = 15), and acetates (F = 0.25) of Ba and Ra, sulfates of Sr and Ra (F = 340), and sulfates of Pb and Ra (F = 11). More recent examples include the syncrystallization of Cf^{2+} (F = 0.026), Es^{2+} (F = 0.15), and Fm^{2+} (F = 1) (prepared by reduction of the trivalent ions with Mg) with $SmCl_2$ in a medium of HCl and C_2H_5OH. The divalent lanthanide is itself obtained by reduction with Mg. The F values are independent of the conditions of precipitation and of the amount of macrocomponent precipitated.[41]

Odd cases are reported for which the distribution of M is homogeneous at the tracer level with constant F values, but the crystals are then not isomorphous. Such crystals are called *anomalous* mixed crystals. Examples are provided by Ag_2CrO_4 and $^{212}Pb^{2+}$ in 10^{-2} *M* HNO_3 (F = 1.4) or $SmCl_2$ and La^{3+} in HCl + C_2H_5OH (F = 0.7). These crystals are not formed for ponderable amounts of M and M′. Hence, syncrystallization should always be interpreted with caution.

Finally, it may happen that the tracer is distributed homogeneously inside the crystal of carrier M′ even if one or both elements are not in the same oxidation state in the two phases. This is not a syncrystallization in the strict sense because F cannot remain a constant independently of the conditions of precipitation. In this case, which resembles a *solid phase extraction*, the following redox equilibrium with a constant $K_{l,\bar{l},n_1,n_2}$ holds for the microcomponent:

$$nC_{l,y}^{(n_1-y)+} + (n_2 - n_1)\text{red} \rightleftarrows n\overline{C_{\bar{l},n_2}^{0}} + n(\bar{l} - l)\,H_pL + n(n_1 - y)\,H^+ + (n_2 - n_1)\text{ox}$$

In this expression, red and ox refer to the two forms of the couple which reduces M by exchange of n or $n_2 - n_1$ electrons. An equivalent equilibrium can be written for M′ (with corresponding parameters denoted by a dash). The fractionation coefficient is now

$$F = \frac{\left(\dfrac{[\text{red}]}{[\text{ox}]}\right)^{(n_2-n_1)/n}}{\left(\dfrac{[\text{red}']}{[\text{ox}']}\right)^{(n'_2-n'_1)/n'}} \; \frac{\sum\limits_{l,y} K_{l,\bar{l},n_1,n_2}\,[H_pL]^{-(\bar{l}-l)}\,[H^+]^{-(n_1-y)}}{\sum\limits_{l',y} K'_{l',\bar{l}',n'_1,n'_2}\,[H_pL]^{-(\bar{l}'-l')}\,[H^+]^{-(n'_1-y')}}$$

This elaborate expression is replaced by the simpler one when the oxidation state of M in solution changes to that of M′ in the crystal: then $n_2 = n_2' = n'$. Each case should be examined specifically.

Two illustrative examples will be described:[41,42]

(α) Fm^{2+} syncrystallizes with $SrCl_2$ in a hydro-alcoholic medium. In the presence of the couple Yb^{2+}/Yb(III), which fixes the redox potential of the system, Fm(III) is reduced according to

$$\mathrm{Fm(III)} + \mathrm{Yb}^{2+} \rightleftarrows \mathrm{Fm}^{2+} + \mathrm{Yb(III)}, \quad \mathrm{K}$$

the partition of Fm between a solution of Fm^{2+} and Fm(III) and $SrCl_2$ is described by

$$\mathrm{Fm}^{2+} + 2\,\mathrm{Cl}^- \rightleftarrows \overline{\mathrm{FmCl}_2}, \quad \mathrm{K}_2$$

Similarily,

$$\mathrm{Sr}^{2+} + 2\,\mathrm{Cl}^- \rightleftarrows \overline{\mathrm{SrCl}_2}, \quad \mathrm{K}_2'$$

It follows that

$$\mathrm{F} = \frac{\mathrm{K}_2}{\mathrm{K}_2'} \frac{1}{1 + \dfrac{\mathrm{x}}{\mathrm{K}}}$$

x is the ratio $Yb(III)/Yb^{2+}$, comprised between the limits 0 and 1. The constant K is obtained from measurement of F as a function of x. It is related to the potentials of the couples $Fm(III)/Fm^{2+}$ and $Yb(III)/Yb^{2+}$, i.e., $-RT\ 1nK = E^o_{Fm} - E^o_{Yb}$. Entities containing the M(III) species are not unequivocal because the precipitation of $SrCl_2$ requires addition of LiCl, which may complex the aqua ions M^{3+}.

(β) From a mixture at 1170 K of fused salts NdOCl, $NdCl_3$, and $NdCl_2$ containing Pu as a microcomponent, the solid phase NdOCl that appears on cooling extracts only $PuCl_3$ and $NdCl_3$. The Pu is reduced according to

$$\mathrm{PuCl}_3 + \mathrm{NdCl}_2 \rightleftarrows \mathrm{PuCl}_2 + \mathrm{NdCl}_3, \quad \mathrm{K}$$

The partition equilibria are

$$\mathrm{PuCl}_3 \rightleftarrows \overline{\mathrm{PuCl}_3}\ (\mathrm{NdOCl}), \quad \overline{\mathrm{P}}_{\mathrm{Pu}}$$

$$\mathrm{NdCl}_3 \rightleftarrows \overline{\mathrm{NdCl}_3}\ (\mathrm{NdOCl}), \quad \overline{\mathrm{P}}_{\mathrm{Nd}}$$

It follows that

$$\mathrm{F} = \frac{\overline{\mathrm{P}_{\mathrm{Pu}}}}{\overline{\mathrm{P}_{\mathrm{Nd}}}} \frac{1 - \mathrm{x}}{1 + \dfrac{\mathrm{x}}{\mathrm{K}}}$$

where x is the ratio $[NdCl_2]/[NdCl_3]$ in the fused salt, with $0 < x < 1$. For measurement of x, the mixture is dissolved in HCl in an inert atmosphere. NdOCl is insoluble and $NdCl_2$ is oxidized:

$$2\,NdCl_2 + 2\,HCl \rightarrow 2\,NdCl_3 + H_2$$

The amount of Nd(II) is calculated from the measured value of H_2. The difference of potentials of the redox couples is derived from K.

Formation of anomalous mixed crystals can be explained in this way if the various coefficients lead to constant F values.

In other cases, the partition equilibrium of M concerns only the solution and the *surface* of the crystals rather than the bulk of the solid. Although conditions for syncrystallization still hold, distribution of the tracer in the crystal will not be homogeneous. This situation can arise from various kinetic factors, e.g., when the crystals are formed from a supersaturated solution or by isothermal evaporation.

Setting x, the amount of M in the crystals, and y, the amount of M′ in the crystals; a, the total amount of M, and b, the total amount of M′, the Berthellot-Nernst law becomes

$$\frac{x}{y} = F\frac{a - x}{b - y}$$

When equilibrium is established not within the bulk of the solid but only between the liquid phase and the surface of the crystals,

$$\frac{dx}{dy} = \lambda\frac{a - x}{b - y}$$

Here λ is a kinetic parameter different from F, which is independent of the amount of the macrocomponent precipitated. Integration leads to the *Dorner-Hoskins* law:

$$\ln\frac{a}{a - x} = \lambda \ln\frac{b}{b - y}$$

which is equivalent to

$$\ln(1 + D_e) = \lambda \ln(1 + D'_e)$$

since $D_e = x/(a-x)$ and $D'_e = y/(b-y)$.

The law applies under given experimental conditions for which the kinetics of crystallization are well controlled. For instance, precipitation of Ba/Ra chlorides or nitrates followed by immediate filtration yields $\lambda = 4.3$ and $\lambda = 2$, respectively. The previous conclusions and remarks with respect to species of the microcomponent still hold, since it is a true syncrystallization, i.e., a coprecipitation via isomorphous replacement, with a continuous distribution of M in the solid.

The Berthellot-Nernst relation shows that M concentrates in the crystal when $F > 1$. The same holds for the Dorner-Hoskins distribution when $\lambda > 1$, the concentration of M being highest at the center of the crystals. Otherwise, the amount of M is higher near the surface of the crystals.

Despite the problems encountered, partition based on well-controlled co-precipitation with syncrystallization has particular advantages: (1) it can be applied to both aqueous solutions and organic solvents as well as to fused salts, a medium in which the reacting species are particularly simple; (2) the biphasic system can be chosen either as oxidizing or reducing, depending on the choice of the solid phase, and thus unstable oxidation states of tracer M can be preserved; and (3) coprecipitation provides the only means of estimating properties of solid compounds of a tracer M which cannot be obtained in weighable amounts (e.g., Fr, At, actinide elements at the end of the series, transactinides). Po forms a series of mixed crystals $(NH_4)_2\,MCl_6$, with M = Te, Pb, Sn, Pt; this behavior indicates that Po may exist as Po(IV) and gives a clue to the structure of $(NH_4)_2\,PoCl_6$. Predictions made on this basis were confirmed when weighable amounts of radioelements became available.

At this point, several comments on the thermodynamics of coprecipitation are appropriate. For a given set of experimental conditions, F is related to the free energies $\Delta G^\circ(M)$ and $\Delta G^\circ(M')$ associated to the process. Coprecipitation of aqua ions M^{n+} and M'^{n+} with formation of crystals ML_n and $M'L_n$ ($l = l' = 0$, $\bar{l} = n$, $y = y' = n$, $K_{0,n,n} = K$ and $K'_{0,n,n} = K'$) follows the relation

$$\begin{aligned} RT \log F &= \log K - \log K' = \Delta G^\circ(M') - \Delta G^\circ(M) \\ &= \Delta H^\circ(M') - \Delta H^\circ(M) - T[\Delta S^\circ(M') - \Delta S^\circ(M)] \end{aligned}$$

When M and M′ are identical, $F = 1$ and the reaction is a mere exchange process as mentioned earlier. For different M and M′, the entropy terms are negative because order increases, and the enthalpy terms depend on the hydration energy of the aqua ions and on the lattice energy of the pure crystals. Both energies increase with decreasing ionic radius r. Assuming that the entropy terms cancel out, high F values ($\log F > 0$) are expected when r_M is larger than $r_{M'}$. This is sometimes observed for coprecipitation of Ra with Ba compounds.

Predictions for F are much less reliable when the ions are complexed because of various effects, including those of the ligand field.

2. Adsorption

Adsorption of a tracer occurs at the surface of the precipitate $C^0_{\bar{l},n'}$ of M′. The surface may be that of a well-crystallized (ionic) or of a more or less amorphous, nonionic solid.

In the first instance, the surface of the precipitate is not in electrostatic equilibrium, but bears a charge which may have several origins: defects, impurities, or excess or lack of lattice ions in addition to the normal superficial charge. A surface of insoluble precipitates such as halides or sulfides immediately attracts species $E_i(M'_j)$ of the macrocomponents M'_j remaining in solution, or even those of other macrocomponents not participating in formation of the precipitate. In this way, electrostatic equilibrium is ensured at the surface of the solid and a double layer is formed.

For instance, a precipitate of AgI formed in a solution of KI, say by addition of $AgNO_3$, attracts I^- ions in excess; if the precipitate is introduced into a solution of KCl, Cl^- ions are attracted. In both cases the halide ions form with Ag^+ a superficial, insoluble, negatively charged precipitate of AgI or AgCl. To preserve electrical neutrality, K^+ ions are drained toward the precipitate to build up the double layer.

The positively charged complexes of M, $C^z_{l,y}$, can now exchange with the K^+ ions in the double layer and bind again to the surface of the crystal. The reverse would occur with a positively charged precipitate, such as AgI, formed by addition of KI to $AgNO_3$; in this case, the double layer would be Ag^+/NO_3^- and the complexes would exchange with NO_3^- ions.

This mechanism has similarities with exchange on cationic or anionic resins, but here the solid phase is ill-defined and poorly controlled since the composition of the double layer depends on the properties of the medium. Further exchange may occur between species $C^z_{l,y}$, with z positive or negative, and the lattice ions. Numerous factors (charge, radius and polarizability of the species, surface of the solid, aging of the precipitate) are involved in the adsorption process, and a satisfactory expression of D_e cannot be established. One can only propose qualitative rules which have little interest in the present context.

Most precipitates, including hydroxides, oxides, silicates, carbonates, phosphates, etc., being ionic or partially covalent, generally exhibit an ill-defined crystallinity. They are formed *in situ* or introduced into the system. In contact with aqueous solutions, these solids acquire superficial charges with pH-dependent signs, arising from acid-base reactions of their superficial OH groups. Charges on the solids, together with hydrolysis reactions of M influence the adsorption. Nonamphoteric elements M are best adsorbed as positive $C^z_{0,y}$ species, whereas amphoteric elements are fixed both in cationic and anionic form.

In these cases, partition follows one of the three classical isotherms defined in terms of the amounts q remaining in solution and $\overline{q}$ fixed on the solid phase:

(1) The Henry isotherm

$$\frac{\overline{q}}{q} = K$$

controlled by concentration,

(2) The Freundlich isotherm

$$\overline{q} = K\, q^n$$

with $0 < n < 1$, which assumes a heterogeneous distribution of the fixation sites without saturation limit,

(3) The Langmuir isotherm

$$\frac{\overline{q}}{\overline{q_m}} = \frac{q}{k + q}$$

valid for homogeneous distribution of the fixation sites and uniform coverage of the surface up to saturation. The amount $\overline{q}_m$ fixed on the surface corresponds to a monoatomic or monomolecular layer; k is a constant. The Langmuir isotherm can be established on the basis of thermodynamic or kinetic arguments. In the first instance, one obtains the expression

$$k = k_0 \exp \frac{\Delta H^\circ}{RT}$$

ΔH° is the enthalpy of bonding between the surface and the adsorbate. The Langmuir isotherm is the limit of the Freundlich isotherm when q tends toward 0 and $n = 1$.

A peculiar effect may be observed during adsorption. When the latter occurs at the surface of successive layers of a growing crystal (or in crystallites), the distribution of M in the bulk of the solid is discontinuous, with preferential localization at the edges or in planes of the lattice, or at lattice defects. This behavior, characteristic of *internal adsorption* is known for ^{212}Pb, ^{210}Po, and ^{224}Ra in M'_2SO_4 ($M' = K, Rb, NH_4$) and for ^{212}Pb in NaCl.

As in the case of ionic solids, no satisfactory expression for D_e can be formulated. Entrainment of a radionuclide by adsorption strongly depends on experimental conditions, contrary to the case of syncrystallization, and is poorly reproducible. Main factors are the charge and age of the precipitate, grain size, and the presence of electrolytes, all of which preclude quantitative information on the species of a microcomponent in solution.

As for speciation, it appears that the utility of coprecipitation is limited to syncrystallization. Before conclusions can be drawn, the mechanism of coprecipitation should be checked by performing a series of precipitations with different reagents, or using radioautography in order to verify that the radionuclide is incorporated homogeneously in the crystal, rather than adsorbed at the surface.

Formation of pseudocolloids is a particular case of adsorption of a tracer on microparticles of an ill-defined macroimpurity.

The partition of a radionuclide between an aqueous solution and a bulky solid such as a metal, silica, glass, or polymer, depends strongly on the nature and properties of the surface. Hence, expressions of D_e which would give information on the $E_i(m)$ can hardly be established. Adsorption isotherms of M on the solid are those already indicated.

Unwanted partitions always occur between a microcomponent in solution and the walls of containing vessels. Because the latter are indispensable in experiments with radionuclides, it must be assumed (and hoped) that species fixed on the walls do not participate in the radiochemical partition system investigated. Adsorption on walls can be minimized or prevented by using propylene, Teflon™, or other hydrophobic materials instead of glass. Alternatively, the glass surface can be coated with silicone.[43]

VI. DETERMINATION OF NON-EQUILIBRIUM DISTRIBUTION COEFFICIENTS

The measurement of equilibrium coefficients D_e may require a length of time incompatible with the half-life of a radionuclide. In other cases, the duration of the experiment may be limited because an intensive parameter $\mathbf{P}_i$ cannot be maintained constant until the partition equilibrium is attained. Such situations are often encountered in electrochemical partitions of a microcomponent between a metal and a solution.[44] The underlying chemistry consists of redox reactions in which the species of M are reduced to the elementary state. Expressions which relate D, D_e, and the time t and contain all accessible information on the $E_i(m)$ species may be rather complicated. Accordingly, it would appear preferable to search for simpler expressions that are independent of the time parameter.

With respect to other partition procedures, all electrochemical methods have a particular limitation in that the investigated species are those present in a diffusion layer close to the electrode, and are not necessarily identical with those present in the bulk of the solution. For instance, the oxidation state of some of the species may not be the same, or the concentration of these species may be higher in the diffusion layer. Strictly speaking, these species should be distinguished from those in the bulk. Such a procedure would render the notations more complicated, and for the sake of simplicity distinction will be omitted (see Appendix 8.5).

A. Spontaneous Deposition

There exists no metal $M' = C'^{0}_{0,0}$ that is strictly insoluble in an aqueous solution; however, quite often the dissolved species $E_i(M') = E_i(m') = C'^{z}_{l,y} = C^{(n'-y)}_{l,y}$ cannot be detected owing to their very low concentration, except when M′ has a very high specific activity.

With suitable kinetics, a potential can be established between M′ and the oxidized species $E_i(m')$ which equilibrates the potential imposed by the medium.

Under these conditions, when a metal M′ is introduced into a solution of a microcomponent M, species $E_i(m) = C_{l,y}^{(n-y)+}$ can be reduced to $M = C_{0,0}^{0}$. The result is a spontaneous deposition of M on M′ that can be easily detected by radioactivity. On the other hand, transfer of M′ into the solution is too small to be observed since, in general, M′ is not radioactive.

The distribution coefficient is

$$D = \frac{\overline{C_{0,0}^{0}}}{\sum_{l,y}[C_{l,y}^{(n-y)+}]}$$

The overall redox partition equilibrium

$$n\,\overline{M'} + n'\,C_{l,y}^{(n-y)+} \rightleftarrows n'\,\overline{M} + n\,C'^{(n'-y')+}_{l',y'} + (nl' - n'l)\,H_pL + (ny' - n'y)\,H^+, \quad K_e$$

can always be decomposed into two half-reactions involving the redox couples of M and M′

$$C_{l',y'}^{(n'-y')+} + n'\,e^- \rightleftarrows \overline{M'} + l'\,H_pL + (n' - y')\,H^+$$

$$C_{l,y}^{(n-y)+} + n\,e^- \rightleftarrows \overline{M} + l\,H_pL + (n - y)\,H^+$$

It can be shown that, at a given time t, the distribution coefficient D depends on the apparent standard potentials $E^o[M(N)/M]$ and $E^o[M'(N')/M']$, which are themselves a function of the standard potentials $E^o[M^{n+}/M]$ and $E^o[M'^{n'+}/M']$. It is recalled that N is the oxidation number and n the charge of the aqua ion; $n = N$ if $N \leq 4$. The coefficient D depends also on the constants $K_{l,y}$ and $K_{l',y'}$, on the dissolution energy of M in M′, on $[H_pL]$ and on the pH. In practice, the complicated expression for D cannot be used for deriving information on $E_i(m)$ because values of D depend on the way in which the support was prepared, as will be shown later. However, $E^o[M^{n+}/M]$ can be evaluated with respect to $E^o[M'^{n'+}/M']$ because the spontaneous deposit occurs only if

$$\Delta G^o = -RT \ln K_e = n\Delta G_a^o - n\Delta G_b^o < 0$$

The concentration $[H_pL]$ can be calculated since $C_{M'}$ remains very low, probably at the tracer scale. This is very different from partition by syncrystallization. Adsorption may perturb the spontaneous deposit.

Spontaneous deposition of Po(IV) on Ag from a 0.2 *M* HNO_3 solution is known since the early days of radioactivity as the simplest method for separation of the element. The potentials involved are E° Po(IV)/Po = 0.76 V; E° Ag^+/Ag = 0.80 V. This property has not been used for speciation of Po.

B. Electrolysis at Controlled Potential

If M does not deposit spontaneously and appreciably on M′, an external potential can be applied to the electrode M′ in order to reduce M(N). However, care must be taken to avoid discharge of the H^+ ions, which would modify the pH near the electrode and change the nature of the $E_i(m)$ species. At a given time t, the expression for D depends now on the equilibrium constant of the partition reaction

$$C_{l,y}^{(n-y)+} + N\,e^- \rightleftarrows \overline{M} + l\,H_pL + (n - N - y)\,H^+, \quad K_e$$

because it is related to D_e which depends on the applied potential E and on the apparent standard potential E°[M(N)/M] (in short, E°). At 25°C,

$$D_e = K_e = \exp\left[-0.06\,\frac{(E^\circ - E)}{N}\right]$$

The apparent standard potential of the couple M is

$$E^\circ[M(N)/M] = E^\circ[M^{n+}/M] + \frac{0.06}{N}\log\,[H_pL]^l\,[H^+]^{z-N}$$

and the measurement of E°[M(N)/M] as a function of $[H_pL]$ and $[H^+]$ leads to the standard potential E°[M^{n+}/M], to ⟨l⟩ and ⟨z⟩. This determination can be achieved in the following way.

Variations of D with time are always determined by the difference E − E° (Figure 8.11). The values of D are symmetric with respect to that for which E = E° (see Appendix 8.8). The plot of D as a function of the potential E for a constant time of electrolysis yields an S-shaped curve with an inflexion point for the value E°. Intersection of the tangent to the curve at this point with the potential axis defines the *critical deposition potential* E_c (lying close to E°), which is a useful, but not very precise parameter. The curve is analogous to the intensity-potential curve in classical electrochemistry which for a same medium would yield an E_c value equal to the apparent potential E° measured under static conditions. The same should hold for a radionuclide at the tracer scale, i.e., E_c = E°[M(N)/M]. From E_c values measured under different conditions, information can be obtained on the order and charge of the complexes of M, but this method is rather time-consuming.

All measured critical deposition potentials for Bi^{3+} in nitric acid solution on a polarized gold cathode in a concentration range from 10^{-16} to 10^{-5} *M* yield the recorded value for E°[Bi(III)/Bi].[44]

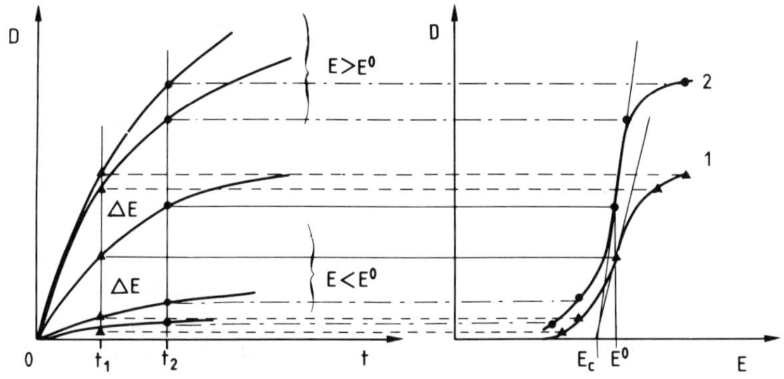

FIGURE 8.11. Critical deposition potential. The difference between E_c and E^o is exaggerated in order to illustrate that for any time t_1 and t_2 the curves 1 and 2 yield $E_c \approx E^o$, the apparent standard potential.

However, results obtained with a radioactive isotope of a stable element (Ag, Zn) at the tracer scale may show considerable deviations from the behavior of macroamounts of this element, depending on the nature and previous treatment of the electrode. This inconsistency arises from the heterogeneity of the surface of a metal, and is of paramount importance when one is dealing with radionuclides at the tracer level. In many cases, the amount is too low to cover the surface with a monolayer of atoms. For instance, it would require 2×10^8 Bq of a nuclide with half-life 10^4 s to cover with a monolayer a Pt wire with a diameter of 0.1 mm and 1 cm in length.

The deposition of M requires first its extraction from the complexes $C_{l,y}^{(n-y)+}$, followed by reduction of M^{n+} to M and fixation of the atom on the electrode. With ponderable amounts, the electrode is very rapidly covered by many monolayers of the metal M itself. The bonding energy to the electrode corresponds to the sublimation enthalpy of M.

At the tracer level, it may occur that the bonding energy of M on preferential superficial sites is equal to the (unknown) sublimation energy of M; in this case the behavior of M is "normal" and $E_c = E^o$. In other instances, the bonding energy may exceed the sublimation energy and deposition is observed at an *overpotential* $E_c > E^o$ or, conversely, at a lower potential, $E_c < E^o$. The actual behavior can be determined only from experiment.

Once the more active superficial centers are saturated, further deposition of M on less favorable sites requires more energy, i.e., a higher potential. The slope of the S-shaped curve has a lower value than in the case of electrolysis of ponderable amounts and the critical potential E_c may be altered. Accordingly, data from electrolysis on a metallic support at the tracer scale should be used with caution.

However, any change in the oxidation state of M would result in considerable changes in the values of D or E_c. Electrodeposition studies of M as a function of the redox properties of the medium provide a means of obtaining information on the stability of M in its various oxidation states. In the presence of the reducing agent hydrazine, the critical deposition potential of Po is less than that of Po(IV), which points to reduction to Po(II) by hydrazine.

C. Radiopolarography

Contrary to electrolysis on a solid support, deposits on a mercury electrode give satisfactory reproducibility, since many metals form an amalgam. In this case, E° and its variation with $[H_pL]$ and $[H^+]$ are again meaningful.

This fact is used to advantage in *radiopolarography*.[45-52] Here, the activity of M is determined in mercury drops maintained at a potential E during a constant time τ. The activity of each drop is next plotted as a function of potential. The distribution coefficient D_τ of M during the lifetime τ of the drop is given by

$$D_\tau = \frac{\overline{C_M}}{C_M}$$

and the activity of the drop is

$$\overline{A} = k\,\overline{C_M}$$

k being a proportionality constant.

For small drops, the transfer of activity is negligible with respect to the amount of radionuclide in solution and C_M is practically a constant. Hence,

$$\overline{A} = k'\,D_\tau$$

This is similar to electrolysis at constant potential and the plot of A vs. E is the S-shaped curve of a *radiopolarogram* (Figure 8.12A). However, the abcissa of the inflexion point does not coincide with E°, since the amalgamation energy must be taken into account. It is customary to denote this parameter as $E_{1/2}$. Formally, a radiopolarogram is similar to a classical i = f(E) polarogram. The variations of $E_{1/2}$ with parameters $\mathbf{P}_i$ convey all information with respect to the $E_i(m)$ species.

The theory of radiopolarography takes into account the kinetics of diffusion of $E_i(m)$ species in the diffusion layer of the solution in the vicinity of mercury drops and of the M atoms in the diffusion layer of mercury. The theory enables one: (1) to formulate the time-dependence of D and to express k as a function of the mercury flow and the diffusion coefficient of the reduced entity, and (2) to establish an expression for the radiopolarogram for a reversible or irreversible process, including the oxidation state of the entities of M which are reduced at the mercury.

For a sufficiently negative value of E, and when the diffusion limit is attained, the activity $\overline{A}$ tends toward a limiting value $\overline{A_d}$ related to E:

$$E = E_{1/2} + \frac{0.06}{N} \log \frac{\overline{A_d} - \overline{A}}{\overline{A}}$$

which can be readily used for determination of the oxidation state N of the tracer M.

Radiopolarography is practicable in relatively acidic media (pH < 2) and at very negative potentials (down to -2.3 V) since measurement is not perturbed by the discharge of H^+ ions. Kinetic data are also provided by this method. However, the nature of the $E_i(m)$ species near the mercury may be altered. Complications in interpretation of data arise when the redox reaction is not reversible, which is often the case.

Radiopolarograms of the actinides Cm to Md in 0.1 *M* LiCl at pH 2.2 have been recorded.[50,52]

D. Radiocoulometry

Radiopolarography is feasible if the activity of the radionuclide under study is sufficiently high.[49,51-53] Otherwise, the activity in solution decreases to the point that C_M can no longer be considered constant. If C_M^0 is the initial concentration of M, the partition coefficient of M between the solution and a mercury bath is

$$D = \frac{C_M^0 - C_M}{C_M}$$

and

$$\left[\frac{d\,C_M^0}{dt}\right]_{t=0} = -\,C_M^0\left[\frac{dD}{dt}\right]_{t=0}$$

The initial variation of the slope of the concentration of M, as a function of the potential E, yields the same information as the much more time-consuming determination of the equilibrium distribution coefficient D_e. Actually, the half-time τ of the reaction can be determined from the initial slope for each value of E; on the other hand, τ is related to D (see Appendixes 8.2 and 8.8). Accordingly, radiocoulograms can be obtained from $\tau = f(E)$. Radiocoulometric curves have the same apparent shape as radiopolarograms (see Figure 8.12B).

Radiocoulograms of the elements Cm to Md in 0.1 *M* lithium citrate at pH 6 have been published.[50]

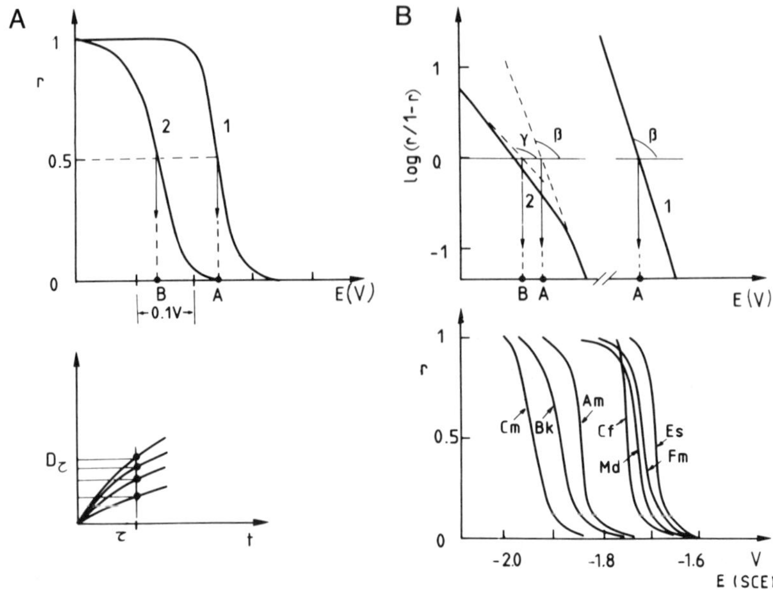

FIGURE 8.12. A. Radiopolarograms. The ordinate r is the ratio $A/A_d = D_\tau/A_d$; τ is the lifetime of the mercury drop and A_d the limiting activity for $E << 0$. Potentials are given with respect to SCE. Curve 1 refers, to a reversible or quasi-reversible amalgamation. (A) Half-wave potential $E_{1/2}$; $\beta = -0.06/N$ at 25°C. For Am in 0.15 *M* LiCl, pH = 2.2, $E_{1/2} = -1.846$ V, N = 3, $\alpha = 0.8$. Curve 2 refers to an irreversible amalgamation. (B) Half-wave potential $E_{1/2}$; $\gamma = -0.06/\alpha N$. For Cm in 0.15 *M* LiCl, pH = 2.2, $E_{1/2}$ (irrev.) = -1.950 V; $E_{1/2}$ (rev.) = -1.930 V, N = 3, $\alpha \approx 0.4$. Lower part, at right: radiopolarograms of actinides in 0.15 *M* LiCl, pH = 2.2. B. Radiocoulograms. The ordinate r is the ratio τ_1/τ; τ is the amalgamation half-time, and τ_1 is the limit of τ for $E << 0$. Curve 1 refers to a reversible or quasi-reversible amalgamation. (A) Half-wave potential $E_{1/2}$; $\beta = 0.06/N$ at 25°C. For Fm in 0.1 *M* Li citrate, pH = 6, $E_{1/2} = -1.79$ V, $\alpha < 1$, N = 2. Curve 2 refers to an irreversible amalgamation. (B) Half-wave potential $E_{1/2}$; $\gamma = -0.06/\alpha N$. For Cf in 0.1*M* Li citrate, pH = 6, $E_{1/2}$ (irrev.) = -2.15 V, $\alpha < 1$, N = 2. Radiocoulograms of actinides in 0.1 *M* Li citrate at pH 6. (Adapted from Samhoun, K. and David, F., *J. Electroanal. Chem.*, 106, 161, 1980.)

VII. RADIOCHROMATOGRAPHY

In chromatographic partition methods, as said before, information on the $E_i(m)$ species is inferred from the displacement x_M and from the velocity dx_M/dt of the radionuclide with respect to the static phase. The radionuclide is deposited at the origin of this phase and is allowed to migrate with the mobile phase. The common practice of percolating the solution through the static phase merely separates the entities but does not furnish additional information.

The nature of $E_i(m)$ species is determined by the *macroscopic* properties of the *mobile* phase. The latter may, or may not, have the same composition of the initial solution that contains the radionuclide.

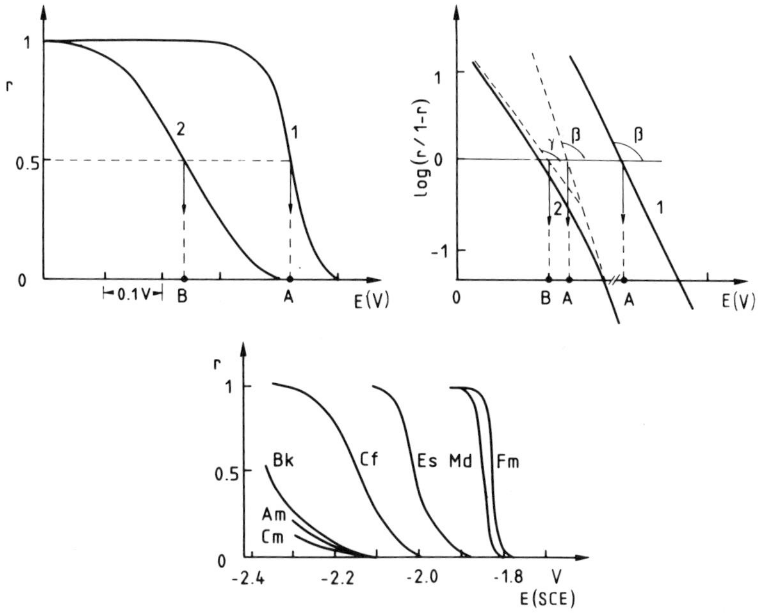

FIGURE 8.12 (continued).

A. Chromatographic Extraction

1. Paper Chromatography

In this method, the aqueous phase is fixed on a cellulose support and the mobile phase is an organic solvent. The static reference reaction is a partition by solvent extraction, usually of associated ion pairs. The mobile phase is a water-soluble mixture of a strong acid and the organic solvent. Displacements of the radioactive $E_i(m)$ or $E_i(m_j)$ entities and of the mobile phase (solvent front migration) are easily recorded and each $E_i(m_j)$ is characterized by the ratio

$$R_f = \frac{x_M}{d}$$

in which d is the displacement of the mobile phase.

2. Extraction Chromatography

In this method, the organic phase is adsorbed on an inert support and the aqueous phase moves with a constant velocity. The concentration of extraction reagent in the static reference partition is very high. The volume required for elution of the radionuclide depends on x_M and on dx_M/dt and in this way is characteristic of the $E_i(m)$ or $E_i(m_j)$ species formed in the eluate.

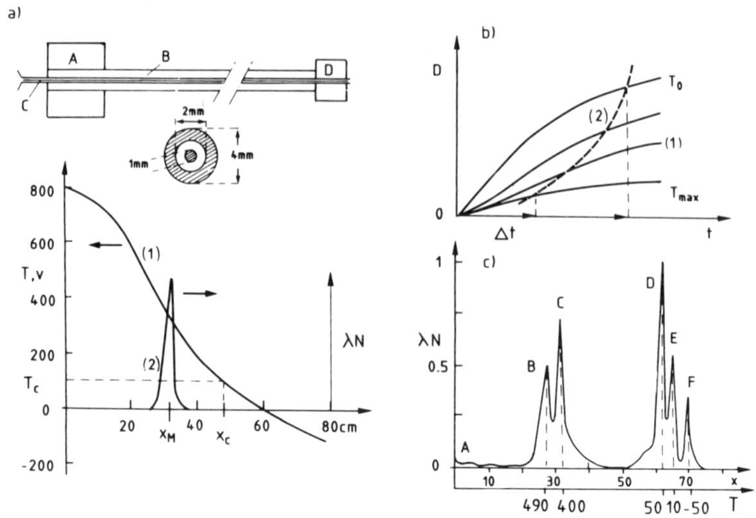

FIGURE 8.13. Thermochromatography. (a) Upper part: simplified scheme of a thermochromatograph. (A) Oven; (B) Copper mantle; (C) Ni migration tube for fluorine atmosphere; (D) Cryostat. A sectional view shows the removable Ni rod for α counting. Lower part: temperature (1) and migration velocity (2) of M along the tube. (b) Theoretical isothermal variation of D in the static reference system (1); and isothermal partition coefficient along the tube (2). (c) Thermochromatograms of Pu after fluoration of PuO_2 with $F_2 + BF_3$ (relative activities). (A) Deposition point; (B) $PuOF_3$; (C) PuF_5; (D) $PuOF_4$; (E) PuO_3F; (F) PuF_6. (Adapted from Fargeas, M., Fremont, R., Legoux, Y., and Merini, J., *J. Less Common Metals*, 121, 439, 1986.)

3. Chromatography on Resin

This is the most widely used chromatographic method, owing to easy control of the nature and composition of the mobile phase, in this case the aqueous one. The static reference partition is generally well established and can be used for planning separation procedures. As before, the volume for which M appears at the exit of the static phase is an indication of the nature of species present in the eluate. Selectivity may be remarkably high for elements M_j with nearly identical chemical properties, as well as for the $E_i(m)$ entities which incorporate M in different oxidation states. It may be improved by using one or several gradients.

B. Thermochromatography

The basic structure of a thermochromatographic system is a tube along which is established a temperature gradient dT/dx (Figure 8.13a).[54-58] The temperature is highest (T_{max}) at the entrance of the tube and decreases to a value T_o at the other extremity. An appropriate gas vector travels through the tube with a velocity $v_x(x)$ and carries a compound $C^0_{l,n}$ of M.

The partition occurs between the moving gas phase and the static solid phase constituted by the tube. Assuming there is no change in the oxidation state of M during partition, the static reference equilibrium is

$$C^0_{l,y}\ (\text{gas}) \rightleftarrows \overline{C^0_{l,\bar{y}}}\ (\text{solid}), \quad K_e$$

for which $D_e = K_e$ and

$$\ln K_e = -\frac{\Delta G^o}{RT}$$

At any given temperature T, the partition coefficient D_T of the reference equilibrium has a time evolution $D_T = f(t)$. The distribution coefficients D_T *decrease* with increasing temperature, i.e., $D_{To} > D_T$ (Figures 8.9c and 8.13b). For each static reference partition, the duration Δt of contact between the two phases depends on $v_x(x)$. The velocity decreases with the temperature; in turn, Δt and D_T increase, the latter toward its maximum value D_{To}. It follows that

$$\frac{dx_M}{dt} = v_x(x)\frac{1}{1 + D_T/A}$$

in which A is a characteristic of the experimental system. Hence, dx_M/dt always decreases and the migration of M should progressively cease, although dx_M/dt never reaches zero.

In practice, M stops at a distance x_c to which there corresponds a critical deposition temperature, T_c. The calculation of this distance is somewhat complicated, but it is easy to follow migration of M along the tube by radioactive counting.

Each species $C^0_{l,y}$ with its own oxidation state N and free energy ΔG^o has a specific migration behavior and a characteristic deposition temperature T_c.

Calculation of the migration time t_c until the species reaches its critical distance x_c can be performed on the basis of several hypotheses which lead to different expressions.

For instance, one could take $T = T_{max}\exp(-Cx)$ and $v_x(x) = BT$, in which B and C are constants, and assume that partition results from an adsorption isotherm such that

$$D = D_o \exp\frac{\Delta H^o}{RT}$$

D_o is a constant and ΔH^o the molar adsorption enthalpy, supposed to be constant in the temperature interval from T_{max} to T_o (Figure 8.13b). After integration,

$$t = \frac{1}{BC}\left(\frac{1}{T} + \frac{1}{T_{max}}\right) + \frac{\Delta_o}{AC\,\Delta H^o}\left(\exp\frac{\Delta H^o}{RT} - \exp\frac{\Delta H^o}{RT_{max}}\right)$$

For a high velocity of the carrier gas (i.e., high B value) and $T < T_{max}$, the migration time of a species with respect to the distance corresponding to its critical temperature deposit T_c is

$$t = \frac{D_o}{AC\,\Delta H^o}\exp\frac{\Delta H^o}{RT}$$

Alternative assumptions on the temperature gradient and on the adsorption isotherm would lead to different relationships.[57]

Thermochromatography is a very fast separation procedure, well adapted to investigation of short-lived radionuclides. Numerous investigations have been reported on chlorides, fluorides, or organic compounds of actinides and of the 6d elements 104 and 105.[57,58]

VIII. TRANSPORT METHODS

A. *Diffusion and Migration*

Transport methods are much less frequently used in radiochemistry than partition methods.[58-63] For one reason, they are more difficult to apply in the attempt to elucidate the nature of $E_i(m)$ species. Moreover, they require a very strict control of critical parameters and are slower.

In a solution maintained uniformly at a constant temperature T with a high degree of precision (± 0.1°C or better), entities $C^z_{l,y}$ can be transferred within the medium according to two modes: (1) *diffusion*, for which transport is determined by a gradient in chemical potential, and (2) *migration*, which is displacement of a charged species in the presence of an electric potential. Because of Brownian motion, diffusion always accompanies migration.

Both transport modes can be used for sorting out spatially the entities that contain a radionuclide. The corresponding techniques, although very simple in their principles, are rather cumbersome in practice.

The simplest case is that of linear transport in a given direction. An experimental parameter which can be measured within a reasonable time span t is the displacement x_i of a species $E_i(m)$ in the presence (or absence) of an electric field, i.e.,

$$x_i = v_i\,t$$

With x_i in centimeters v_i is the velocity in cm s^{-1}. The relation holds for electric fields not exceeding a few V cm^{-1}. It is also relatively easy to record the time variation of concentration $C_{E_i(m)}$ of a species at a given point, but concentration as a function of the distance x_i is less amenable to measurement.

In an electric field E derived from the applied potential, the transfer distance x_i or the velocity v_i is related to the mobility u_i ($cm^2\ V^{-1}\ s^{-1}$):

$$v_i = u_i$$

The mobility itself depends on the properties of the medium:

$$u_i = u_i^0\ (1 + H_i)$$

u_i^0 is the mobility at infinite dilution and H_i is a function expressing the nonideality of the solution. The first term is related to the diffusion coefficient D_i^0 ($cm^2\ s^{-1}$) at infinite dilution:

$$u_i^0 = \frac{z_i\ e\ D_i^0}{k\ T}$$

On the other hand,

$$D_i^0 = \frac{k\ T}{6\pi\ r_i \eta}$$

where η is the viscosity. Finally,

$$u_i^0 = \frac{z_i\ e}{6\pi\ r_i \eta}$$

It may be noted that mobility depends on both *charge* and *hydrodynamic radius* r_i (itself a function of ligands) of the species, whereas diffusion is only a function of *size* of the diffusing entity. With respect to a reference species,

$$\frac{u_i}{u_r} = \frac{z_i\ D_i^0(1 + H_i)}{z_r\ D_r^0(1 + H_r)} \quad \text{and} \quad \frac{D_i^0}{D_r^0} = \frac{r_r}{r_i}$$

For elements with closely similar mobilities, relative data can be obtained and the effects of secondary phenomena cancel out. The interpretation is simplified if it be assumed that the function H_i is the same for a series of complexes of M(N), which implies that the radii are also of the same order.

In the absence of an electric field, the diffusion coefficient D_i can extracted from combined measurements of x_i and $C_{E_i(m)}$; by changing the ionic strength of the solution, the limiting value D_i^0 can be obtained.

The method of the *integrated capillary* is well suited for measurements in aqueous solutions. A capillary tube with one open end, filled with a solution of the radionuclide M, is immerged in an *inactive* solution of the same composition. Diffusion of the E_i(m) species decreases the concentration

$C_{E_i(m)}$ in the capillary tube and is easily recorded. The activity A(t) measured after a diffusion time t is related to the diffusion coefficient D_i:

$$D_i = \frac{4\,L^2}{\pi^2\,t} \ln \frac{8\,A_0}{A(t)\,\pi^2 \exp(-\lambda t)}$$

in which L is the length of the capillary and A_0 is the initial activity. The coefficient D_i is obtained with a fair approximation when $D_i t/L^2 > 0.24$. This condition is fulfilled after 3 days for a tube of length 3 cm and $D_i = 10^{-5}$ $cm^2\,s^{-1}$. The error on D_i is minimal for a time $t = 0.4\,L^2/D$, i.e., $A(t)/A_0 = 0.3$, which also requires several days. For these measurements, the temperature of the device should be kept constant within 0.01°C.

This method has recently been used for determination of the diffusion coefficient of the actinides Am^{3+} to Es^{3+} aqua ions in a LiCl-HCl medium at pH 2.5.[62]

B. Electromigration and Electrophoresis

Electromigration refers to the free migration of M and electrophoresis to its migration on a support. In the latter case, it must be ensured that the support is inert, i.e., that it does not act as an ill-defined solid phase of a dynamic partition reaction between solution and solid. Electro-osmotic displacement must be taken into account. Furthermore, the systems should be stable in air.

The initial position of M on the support is recorded; the macroscopic composition of the solution must be kept constant during migration. Depending on the stability of the complexes, two situations may be encountered:

1. The complexes $C^z_{l,y}$ are inert and they move independently. One measures as many displacements as complexes present in the solution. The interpretation depends on assumptions introduced with respect to the relative radii and the H_i terms of the various species. Changes in the behavior and characteristics of the $E_i(m)$ species are recorded as a function of $[H_pL]$ and pH.
2. More frequently, the complexes are labile and are in equilibrium with dissociation products throughout their displacement. In this case, one measures an average value $\langle u \rangle$ of all individual displacements:

$$\langle u \rangle = \sum_{l,y} \alpha_{l,y}\, u_{l,y}$$

$\alpha_{l,y}$ is the fraction of the complex $C^z_{l,y}$ in the solution. It follows that

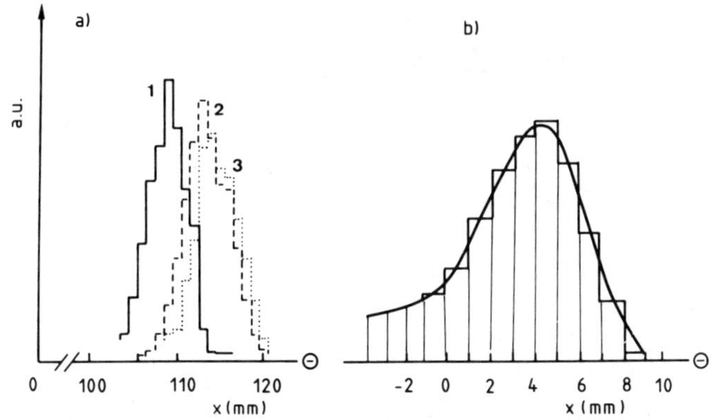

FIGURE 8.14. Electrophoregrams. (a) Aqua ions of (1) ^{170}Tb; (2) ^{153}Gd; (3) ^{152}Eu on Whatman cellulose 3 MM. Field, 12.5 V cm^{-1}; duration, 1 h; electrolyte: Na^+, H^+, ClO_4^-; pH 1; $\mu = 0.25$; T = 20 ± 0.5°C. Activity in arbitrary units. Precision on displacement x_M is 0.5%. (b) ^{230}U(VI) in 4.2 *M* H_3PO_4, 8.35 V cm^{-1}. Other conditions same as in (a). (Adapted from Fourest, B., Duplessis, J., and David, F., *Radiochim. Acta,* 46, 131, 1989.)

$$\langle u \rangle = \frac{\sum_{l,y} u_{l,y}\,[C^z_{l,y}]}{C_M} = \frac{\sum_{l,y} u_{l,y}\,K_{l,y}\,[H_pL]^l\,[H^+]^{-y}}{\sum_{l,y} K_{l,y}\,[H_pL]^l\,[H^+]^{-y}}$$

If a sufficient number of sets of the parameters $\langle u \rangle$, $[H_pL]$ and pH are collected, the $u_{l,y}$ and $K_{l,y}$ can be computed. Next, the $E_i(m)$ are identified more or less completely depending on the assumptions made with respect to the radii and H_i factors. In practice, this procedure is limited to solutions containing only a small number of species (Figure 8.14).

Electromigration and electrophoresis have been applied for a long time in chemistry, and recently, with renewed interest, to radiochemistry for speciation of d-series transition elements (Cu, Co, Zr, Tc, and Ru), several lanthanides, trivalent actinides, and Np(V) in various media.[64-79]

Autodiffusion experiments are excluded since they require ponderable amounts of the element corresponding to the radionuclide. In the heterodiffusion process, measurement of displacement leads to D_i rather than to D_i^0 and appropriate corrections must be introduced. Alternatively, measurements must be performed as a function of ionic strength.

The data obtained are similar to those from electrophoresis, but they depend only on radii and thus, are informative on the $E_i(m)$ species. In practice, only one entity of a single element can be characterized.

Coupled chromatographic and migration experiments in an electric field may provide further data for a more precise knowledge of the $E_i(m)$ species.

IX. PERTURBATION OF A SYSTEM BY RADIOCHEMICAL METHODS

Any radiochemical method applied to the speciation of radionuclides raises the problem of a possible perturbation of the system. Determination of whether or not the perturbation actually occurs and to what extent it can interfere may be a challenging problem.

The least perturbing procedures are certainly based on transport methods, and among these, the best ones rely on diffusion. However, their application is difficult, time-consuming, and limited to the most simple systems.

Perturbations due to the widely used partition methods are minimized under experimental conditions for which D values are *lowest*. This requirement is just the *opposite* to that which applies for most analytical separation methods. The determination of such low D values may lead to a loss in precision which, after all, is to be preferred to erroneous data.

Once the least perturbing macroscopic reagent is chosen, the perturbation can be inferred from a comparison of experimental results obtained on the one hand for a radionuclide, and on the other for the corresponding element in ponderable amount, providing the latter is available, and its behavior known. For this reference element, however, consumption of reagents must be taken into account and in addition to pH, the concentrations $[E_i(L_j)]$ must also be measured.

The best procedure, which is also the only one feasible for nonponderable amounts of the element, is to use several methods and compare the results. If they concur, one may reasonably conclude that the observed species are actually present in the system investigated.

X. HOW TO FORMULATE THE ENTITIES

Radiochemical methods provide knowledge only of the charge and order of the species of a radionuclide in solution. This information may be too restrictive for the chemist, who may also be interested in the ligands which contribute to the structure of the entities. Because the nature of the ligands in the coordination sphere of M cannot be determined experimentally, theoretical considerations are required. For this purpose, relatively simple predictions can be made from: (1) the coordination number x of the aqua ion $M(H_2O)_x^{n+}$; (2) the predominant anions $H_{p-x}L^{-x}$ present in the solution as a function of pH and the concentrations C_{H_pL}, C_{H_pQ} . . . ; and (3) the partial charges of the atoms or groups of atoms in the complexes.[80-84]

With regard to the second point, it is likely that the ligand in the first coordination sphere of M arises from the predominant anion present at a given pH and C_{H_pL}, and occupies one or several coordination sites in the complex $C_{l,y}^z$. Graphical representations, such as those discussed in Chapter 6, are helpful for deciding on the nature of the main anions present. If the charge z of the complexes does not match that calculated from the number l of attached ligands, several formulations can be proposed, such as a change in one or several of the ligands or an introduction of OH groups.

Hence, it must be decided whether or not the complex can contain such OH groups. A clue is provided by the sensitivity of M^{n+} to hydrolysis, which can be quantified with the partial charges δ_o^- and δ_H^+ carried by the O and H atoms in the aqua ion $M(H_2O)_x^{n+}$. Let $f_o = -(\delta_o - \delta_{H^+})$ be a factor which measures the attraction of O and H in a water molecule located in the first coordination sphere of M. Hydrolysis, i.e., loss of a proton from a water molecule, occurs if $f < f_o$. The value of f can be checked for the supposed structure of the complex from the partial charges carried by all atoms. If f is below its value in the aqua ion, there is a fair probability that OH groups are present in the complex.

This will be illustrated by the phosphato complex of Th, $C^+_{2,3} = Th(H_3PO_4)_2H^+_{-3}$ formed from $Th(H_2O)_8^{4+}$ in a pH range (pH > 1.32) for which the anion $H_2PO_4^-$ is dominant. At this pH, Th^{4+} would hydrolyze in the absence of complexing reagent. From the partial charges of O and H in the bidentate phosphato complex $(H_2O)_4Th(H_2PO_4)_2^{2+}$ it is found that $f_o(C^+_{2,3}) < f_o(C^{4+}_{0,0})$ (9.51 arbitrary units vs. 9.72). This result indicates that the attraction between O and H in water molecules bound to Th is not strong enough to oppose the departure of a proton in the presence of phosphato ligands. Hence, the formulation to be retained is $ThOH(H_2PO_4)_2^+$ rather than $ThHPO_4H_2PO_4^+$. Once the formulation has been determined, the formation constant of the complex can be computed.

The preceding systematization can be improved if the variation of f_o with pH is known, i.e., if the successive hydrolysis constants of the aqua ion are available.

XI. SEPARATION OF RADIONUCLIDES BASED ON RADIOCHEMICAL METHODS

Obviously, the radiochemical methods which have been described and discussed with the objective identifying the $E_i(m_j)$ species can also be used for separation of the latter if the working conditions ensure a high specificity.[85-99] But, more frequently, the elements M_j must be separated.

All methods can be used, even those which appear more qualitative than quantitative with respect to speciation. If the nature of the $E_i(m_j)$ species is known, the most appropriate systems can be selected, as well as the conditions which favor the transfer of a radionuclide in a single phase.

Many recipes are available for separation of radionuclides formed in fission or in nuclear reactions, or which are in filiation with a long-lived parent. These techniques apply for virtually all elements in the periodic table.

The rules described are based on the known behavior of matter at the tracer level. Quite often they have been established in a qualitative and empirical manner, but each problem of separation has its own complexity arising, for example, from radiolytic or chemical effects associated with the nuclear reaction (hot atom chemistry). Various methods can be combined which may permit both separation *and* identification of the species.

The choice of a radiochemical method is primarily governed by the half-life of the radionuclide in question and the amount available. Half-lives of more than a few minutes and activities in the kBq to the MBq range belong to "conventional" radiochemistry and do not require particular procedures.

Half-lives of less than a few minutes or seconds and radionuclides produced at a rate of several atoms during this time interval require special attention. Of all methods that can be applied, it is the fastest one that has to be chosen. The number and duration of steps in the selected experimental procedure must be decreased, which necessarily involves automation and rapidity of the operations. Separation of a static phase can be achieved by fast centrifugation or filtration under pressure; displacement of mobile phases is enhanced by pressurized injection. The radionuclides are transferred rapidly to the chemical laboratory (for $0.1 \text{ s} < T < 10 \text{ s}$) or processed online in the accelerator beam ($1 \text{ ms} < T < 0.1 \text{ s}$). In both cases a complicated setup is required. The experiment is limited to simple chemical identification, and may operate on a batchwise or continuous basis. The choice depends on the number of available atoms and the type of radiation emitted.

Justification of such unconventional radiochemistry may be questioned when one considers the meager chemical information that is obtained in spite of extreme complexity and high cost of the experiments. On the other hand, such experiments provide the only source of information on the heaviest actinides and transactinides, and from this point of view they are really rewarding (see Chapter 10).

When half-lives lie below 1 ms, the Z and A of radionuclides are determined by physical methods, but in the present state-of-the-art, all chemical tests are precluded.

In the investigation of very short-lived radionuclides, ultra-fast radiochemical separations and transport must be achieved. Available techniques for meeting these requirements are closely related to partition methods and deserve a few comments.

One procedure consists in the transport of the nuclides, in the atomic or molecular state, through a gaseous atmosphere inside a capillary. From the preceding considerations, it is evident that collection of the nuclides is hampered by inevitable partition of the entities between the gas flow and the walls of the capillary tube. In order to achieve a reasonable yield, a means must be found for avoiding establishment of thermodynamic equilibrium of the nuclides between the two phases.

The nuclides, which are produced with a high momentum in nuclear reactions, are slowed down and eventually thermalized in gaseous helium at atmospheric pressure and seeded with aerosols. The gas is sucked through an orifice of appropriate diameter such that a supersonic jet is produced. The gas jet may be stopped at the exit or transported along several meters through a capillary. In a typical experiment, the capillary is 10 m long and permits transport within 5 s with a 50% yield. The extraction yield of radionuclides depends to a large extent on the nature of aerosols, which determine the partition of the radioactive atoms between static, finely dispersed solid phase and the inert gas. Dust particles in helium may play the role of aerosols; alternatively, solid inorganic aerosols formed from alkali halides or liquid organic ones in the form of long-chain alcohols may be introduced into the gas stream. The ultimate choice is dictated by the further destiny of the aerosols rather than by the properties of this particular partition system.

Inorganic aerosols are better suited if dissolution in water is envisaged; organic aerosols are more convenient if thermal decomposition is preferred in view of vapor phase experiments.

It appears here that partition results from physical adsorption because all components of a mixture are collected without discrimination.

Another means for ensuring transport of a radionuclide is its incorporation into a gaseous molecule, followed by adsorption on aerosols. However, transport without the use of clusters can be achieved only for short distances and for short periods of time (0.1 s).

GENERAL REFERENCES

Benes, P. and Majer, V., *Trace Chemistry of Aqueous Solutions,* Elsevier, Amsterdam, 1980.

Bockris, O'M. J. and Reddy, A. K. N., *Modern Electrochemistry,* Vol. 1 and 2, Plenum Press, New York, 1973.

Duncan, J. F. and Cook, G. B., *Isotopes in Chemistry,* Oxford University Press, Oxford, 1968.

Evans, E. A. and Muramatsu, M., Eds., *Radiotracer Techniques and Applications,* Vol. 1, Marcel Dekker, New York, 1977.

MacKay, H. A. C., *Principles of Radiochemistry,* Butterworths, London, 1971.

Marcus, Y. and Kertes, A. S., *Ion Exchange and Solvent Extraction of Metal Complexes,* Wiley Interscience, New York, 1968.

Marinsky, J. A. and Marcus, Y., *Ion Exchange and Solvent Extraction, A Series of Advances,* Vol. 1 to 8, Marcel Dekker, New York, 1981.

Robinson, R. A. and Stokes, R. H., *Electrolyte Solutions,* 2nd ed., Butterworths, London, 1970.

Rossotti, F. C. J. and Rossotti, V. M., *The Determination of Stability Constants,* McGraw-Hill, New York, 1961.

Seaborg, G. T. and Loveland, W. D., *The Elements Beyond Uranium,* John Wiley & Sons, New York, 1990.

Stary, J., *The Solvent Extraction of Metal Chelates,* Pergamon Press, Oxford, 1964.

Wahl, A. C. and Bonner, N. A., Eds., *Radioactivity Applied to Chemistry,* 2nd print., John Wiley & Sons, New York, 1958.

REFERENCES

1. **Dunlap, B. D.,** Isomer shifts in the actinides, in *Mössbauer Isomer Shifts,* Shenoy, G. K. and Wagner, F. E., Eds., North-Holland, Amsterdam, 1978, 763.
2. **Harston, M. R. and Pyper, N. C.,** Chemical influences on nuclear beta decay rates, *J. Phys. B,* 17, L 839, 1984.
3. **Pyper, N. C. and Harston, M. R.,** Atomic effects on β-decay, *Proc. R. Soc. London,* A 420, 277, 1988.
4. **Fiser, M., Brabec, V., Dragoun, O., Kovalik, A., Rysavy, M., and Dragounova, N.,** ICES studies on ^{99m}Tc-halide complexes: formation, hydrolysis and ligand exchange, *Appl. Radiat. Isot.,* 39, 943, 1988.
5. **Burke, J. F., Archer, C. M., Wei Chu, K., Latham, I. A., and Egdell, R. G.,** A study of core electron binding energies in technetium-99m complexes by internal conversion spectroscopy, *Appl. Radiat. Isot.,* 42, 49, 1991.
6. **Sano, H. and Gütlich, P.,** Hot atom chemistry in relation to Mössbauer emission spectroscopy, in *Hot Atom Chemistry,* Matsuura, T., Ed., Kodansha, Tokyo, 1984, 265.
7. **Friedt, J. M.,** Mössbauer spectroscopy in actinide research, *Radiochim. Acta,* 32, 105, 1983.
8. **Ache, H. J., Ed.,** *Positronium and Muonium Chemistry,* Adv. Chem. Ser., 175, American Chemical Society, Washington, D.C., 1979.
9. **Schrader, D. M. and Jeans, Y. C., Eds.,** *Positron and Positronium Chemistry,* Stud. Phys. Theor. Chem. 57, Elsevier, Amsterdam, 1988.
10. **Duplâtre, G.,** Positronium chemistry in liquids: utility of combining independent techniques and of applying a magnetic field, in *Positron and Positronium Chemistry,* Jean, Y. C., Ed., World Scientific, Singapore, 1990, 329.
11. **Duplâtre, G., Abbé, J. C., Maddock, A. G., and Haessler, A.,** Inhibition and enhancement of positronium formation in aqueous solutions from the Doppler broadening of annihilation peaks, *J. Chem. Phys.,* 72, 89, 1980.
12. **Didierjean, F., Duplâtre, G., and Abbé, J. C.,** Consecutive captures of electrons and positrons in the positron spur by $HgCl_2$ and halide detachment in polar solvents, *Radiat. Phys. Chem.,* 38, 413, 1991.
13. **Hahn, H. P., Born, H. J., and Kim, J. I.,** Survey on the rate perturbation of nuclear decay, *Radiochim. Acta,* 23, 23, 1976.
14. **Vargas, J. I.,** The chemical applications of angular correlation and half-life measurements, in *Radiochemistry,* (MTP International Review of Sciences, Inorganic Chemistry, Series One), Vol. 8, Emeleus, H. J., Cons. Ed., and Maddock, A. G., Vol. Ed., Butterworths, London, 1972, 45.
15. **Yoshihara, K.,** Chemical nuclear probes using photon intensity ratios, in *Topics in Current Chemistry,* Vol. 157, Springer-Verlag, Berlin, 1990, 1.
16. **Adloff, J. P.,** Application to chemistry of electric quadrupole perturbation of gamma-gamma angular correlations, *Radiochim. Acta,* 25, 57, 1978.
17. **Boyer, P. and Baudry, A.,** Perturbed angular correlation of gamma rays, in *Hot Atom Chemistry,* Matsuura, T., Ed., Kodansha, Tokyo, 1984, 315.
18. **Rydberg, J.,** The formation of composite, mononuclear complexes: theoretical, *Ark. Kemi,* 8, 101, 1955.
19. **Sillen, L. G.,** Some graphical methods for determining equilibrium constants, *Acta Chem. Scand.,* 10, 803, 1956.
20. **Rossotti, F. C. J. and Rossotti, V. M.,** Mixed complexes, in *The Determination of Stability Constants,* McGraw-Hill, New York, 1961, 391.
21. **Guillaumont, R., Muxart, R., and Bouissières, G.,** Solvent extraction of elements by formation of chelates, *Bull. Soc. Chim.,* 1952, 1968 (in French).
22. **Hubert, S., Hussonnois, M., and Guillaumont, R.,** Measurement of complexing constants by radiochemical methods, in *Structure and Bonding,* Duntz, J. D., Ed., Springer-Verlag, Berlin, 34, 1978, 1.

23. **Stary, J.,** *The Solvent Extraction of Metal Chelates,* Pergamon, Oxford, 1964, 51.
24. **Marcus, Y. and Kertes, A. S.,** *Ion Exchange and Solvent Extraction of Metal Complexes,* Wiley Interscience, New York, 1968, 499.
25. **De, A. K., Khopar, S. M., and Chalmers, R.,** *Solvent Extraction of Metals,* Van Nostrand, London, 1970, 1.
26. **Musikas, C.,** Potentiality of monorganophosphorous extractants in chemical separation of actinides, *Separation Sci. Technol.,* 23, 1211, 1988.
27. **Cuillardier, C., Musikas, C., Hoel, P., Nigoud, L., and Vitart, X.,** Malonamides as new extractants for nuclear waste solutions, *Separation Sci. Technol.,* 26, 1229, 1991.
28. **Condamines, N. and Musikas, C.,** The extraction by N,N dialkylamides of actinide cations, *Solvent Extr. Ion Exch.,* 10, 69, 1992.
29. **Lemaire, M., Guy, A., Chomel, R., and Foos, J.,** Dicyclohexano-18-crown-6 ether, a new extractant for nuclear fuel processing, *J. Chem. Soc. D,* 1152, 1991.
30. **Elyahyaoui, A., Boulhassa, S., Brillard, L., and Guillaumont, R.,** Complexes of hexavalent uranium in industrial phosphoric acids, *J. Less Common Metals,* 135, 147, 1987 (in French).
31. **Elyahyaoui, A., Brillard, L., Boulhassa, S., Hussonnois, M., and Guillaumont, R.,** Complexes of thorium with phosphoric acid, *Radiochim. Acta,* 49, 39, 1990.
32. **Kolarik, Z.,** Extraction of actinides by tributylphosphate and other phosphonic and triesters, in *Handbook on the Physics and Chemistry of the Actinides,* Vol. 3, Freeman, A. J. and Keller, C., Eds., Elsevier, Amsterdam, 1985, 431.
33. **Zhin Young Jun.,** Extraction of trivalent actinides with phosphines oxydes, in *Handbook on the Physics and Chemistry of the Actinides,* Vol. 3, Freeman, A. J. and Keller, C., Eds., Elsevier, Amsterdam, 1985, 469.
34. **Guillaumont, R.,** Synergistic effect in the extraction of Pa(V) by TTA and TBP, *Bull. Soc. Chim.,* 132, 1965 (in French).
35. **Guillaumont, R. and Miranda, C.,** Solvent extraction of protactinium, *Solv. Extr. Rev.,* 1, 105, 1971.
36. **Korkisch, J.,** *Handbook of Ion Exchange Resins, their Application to Inorganic Analytical Chemistry,* Vol. I to VI, CRC Press, Boca Raton, FL, 1989.
37. **Clearfield, J.,** New ion exchange materials; their synthesis, structure and ion exchange properties, in *New Developments in Ion Exchange,* Proc. Intern. Conf. on Ion Exchange, Tokyo, 1991, 131; also Inorganic Ion Exchange Material, CRC Press, Boca Raton, FL, 1982.
38. **Bonner, N. A. and Kahn, M.,** Behaviour of carrier free tracers, in *Radioactivity Applied to Chemistry,* Wahl, A. C. and Bonner, N. A., Eds., John Wiley & Sons, 2nd print., 1958, 102.
39. **Haissinsky, M.,** The distribution of a microcomponent between two phases, in *Nuclear Chemistry and Its Applications,* Addison-Wesley, Reading, 1964, 597.
40. **McKay, H. A. C.,** Physicochemical applications of radiotracers, in *Principles of Radiochemistry,* Butterworths, London, 1971, 327.
41. **Mikeev, N. B.,** Lower oxydation states of actinides, *Inorg. Chim. Acta,* 94, 241, 1984.
42. **Mikeev, N. B. and Myassoedov, B. J.,** Lower and higher oxidation states of transplutonium elements in solutions and melts, in *Handbook on the Physics and Chemistry of the Actinides,* Vol. 3, Freeman, A. J. and Keller, C., Eds., Elsevier, Amsterdam, 1985, 347.
43. **Caceci, M. and Choppin, G. R.,** An improved technique to minimize cation adsorption in neutral solutions, *Radiochim. Acta,* 33, 113, 183.
44. **Haissinsky, M.,** Electrochemical applications, in *Nuclear Chemistry and its Applications,* Addison-Wesley, Reading, 1964, 556.
45. **Rogelet, P. and David, F.,** Application of radiopolarography to the investigation of several elements at the tracer level, *J. Electroanal. Chem. Interfacial Electrochem.,* 61, 131, 1975 (in French).

46. **Samhoun, K. and David, F.,** Radiocoulometry: its applications and mechanism studies, *J. Electroanal. Chem.*, 106, 161, 1980.
47. **Fourest, B., David, F., Guillaumont, R., Rogelet, P., and Samhoun, K.,** Radiopolarography of Ba and Ra, *J. Inorg. Nucl. Chem.*, 43, 1331, 1981.
48. **Samhoun, K., David, F., Hahn, R. L., O'Kelly, G. D., Tarrant, J. R., and Hobart, D. E.,** Electrochemical study of Md in aqueous solution: no evidence for monovalent ion, *J. Inorg. Nucl. Chem.*, 41, 1749, 1979.
49. **David, F.,** Radiochemical studies of transplutonium elements, *Pure Appl. Chem.*, 53, 997, 1981.
50. **David, F.,** Oxidation, reduction and thermodynamic properties of curium and heavier actinide elements, in *Handbook on the Physics and Chemistry of the Actinides,* Vol. 4, Freeman, A. J. and Keller, C., Eds., Elsevier, Amsterdam, 1986, 97.
51. **David, F., Samhoun, K., and Guillaumont, R.,** Reducibility of transplutonium elements in aqueous solution, *Rev. Chimie Min.*, 14, 1977, 199 (in French).
52. **David, F., Peretrukine, V. F., Masennikov, A. G., and Fourest, B.,** Separation of californium from actinides and lanthanides in aqueous solutions by electrochemical formation of amalgams, *Radiochim. Acta,* 50, 151, 1990.
53. **Samhoun, K. and David, F.,** The radiocoulometry: its application to kinetics and mechanism studies of amalgamation reaction of Ba^{2+}, Ca^{2+}, Eu^{3+}, Sm^{3+}, and Cf^{3+} in aqueous media, *J. Electroanal. Chem.*, 106, 161, 1980.
54. **Zvara, I.,** Thermochromatographic method of separation of chemical elements in nuclear and radiochemistry, *Isotopenpraxis,* 26, 251, 1990.
55. **Fargeas, M., Fremont, R., Legoux, Y., and Merini, J.,** Thermochromatographic investigation of fluorides and oxyfluorides of actinides in high valence states, *J. Less Common Metals,* 121, 439, 1986 (in French).
56. **Merinis, J. and Bouissières, G.,** Migration of radioelements in a tube with temperature gradient, *Radiochim. Acta,* 12, 140, 1969 (in French).
57. **Fremont-Lamouranne, R., Legoux, Y., Merini, J., Zhuikov, B. L., and Eichler, B.,** Thermochromatography of the actinides and their compounds, in *Handbook on the Physics and Chemistry of the Actinides,* Vol. 3, Freeman, A. J. and Keller, C., Eds., Elsevier, Amsterdam, 1985, 331.
58. **Zvara, I.,** One atom at a time chemical studies of transactinide elements, in *Transplutonium Elements,* Muller, W. and Lindner, R., Eds., North-Holland, Amsterdam, 1976, 11.
59. **Rossotti, F. C. J. and Rossotti, V. M.,** Miscellaneous methods in the determination of stability constants, in *The Determination of Stability Constants,* McGraw-Hill, New York, 1961, 309.
60. **Turq, P., Orcil, L., Simonin, J. P., and Barthel, J.,** Transport properties and their coupling with ion aggregate formation, *Pure Appl. Chem.*, 59, 1083, 1987.
61. **Magdalenat, H., Turq, P., Tirant, P., Chemla, M., Menez, R., and Drifford, M.,** Apparent ionic charge in electrolyte and polyelectroyte solution, *J. Chem. Ed.,* 55, 12, 1978.
62. **Doberenz, W., Fischer, S., Pliquet, H., Dreyer, R., Süb, R., and Bruchertseifer, H.,** Electromigration investigation into hydrolysis and citrate complex formation of $^{152/154}Eu$ and ^{87m}Sr ions as model for experiments with ^{256}Md(III) and (II), *Isotopenpraxis,* 24, 379, 1988.
63. **Orcil, L., Fatouros, N., Turq, P., and Chemla, M.,** Electrophoretic transport and chemical reactions in electrolyte solutions, *Phys. Chemie, Neue Folg.*, 138, 129, 1983.
64. **Orcil, L., Turq, P., Musikas, C., Pruliere, Y., and Chemla, M.,** Electrophoresis and ligand exchange for nitrato complexes of nitrosylruthenium, *J. Phys. Chem.*, 88, 5265, 1984.
65. **Jolk, V.,** Studies on complex compounds in solution with paper electrophoresis, electrophoretic mobility and composition of complexes, *J. Chromatography,* 13, 451, 1964.

66. **Milanov, M., Dobrerenz, W., Chalkin, V. A., and Marinov, V.,** Determination of ion mobilities of radionuclides in free electrolyte, method and experimental organization, *J. Radioanal. Nucl. Chem. Articles,* 82, 291, 1984.
67. **Rösch, F., Reimann, T., Buglanov, V., Chalkin, V. A., and Dreyer, R.,** Electromigration of carrier free radionuclides. Ion mobilities and hydrolysis of ^{241}Am, *J. Radioanal. Nucl. Chem. Articles,* 134, 109, 1989.
68. **Rösch, F., Reimann, T., Buglanov, V., Chalkin, V. A., and Dreyer, R.,** Complexation of ^{241}Am with oxalate and sulphate in aqueous solutions, *J. Radioanal. Nucl. Chem. Articles,* 140, 159, 1990.
69. **Ludwig, R., Dreyer, R., and Fischer, S.,** First investigation of complex formation of At(I) with phosphorous compounds, *Radiochim. Acta,* 47, 129, 1989.
70. **Kulikov, E. V., Novogorodov, A. F., and Schumann, D.,** Hydrolysis of ^{225}Ac trace quantities, *J. Radioanal. Nucl. Chem. Letters,* 164, 103, 1992.
71. **Doberenz, W., Fischer, S., Hussein, H., and Dreyer, R.,** The determination of the physico-chemical state of ^{211}At and ^{131}I in aqueous solutions, *Isotopenpraxis,* 23, 269, 1987.
72. **Dreyer, R., Fischer, S., Ludwig, R., and Doberenz, W.,** The application of electromigration in no carrier electrolyte solution for characterization of carrier free radiohalogenides in non aqueous solution, *Isotopenpraxis,* 24, 389, 1988.
73. **Fourest, B., Duplessis, J., and David, F.,** Electromigration of trivalent lanthanide ions in aqueous solutions, *Radiochim. Acta,* 46, 131, 1989.
74. **Chatt, A., Bidoglio, G., and De Plano, A.,** An electromigration method for studying Tc in ground water under oxic and anoxic conditions, *Anal. Chim. Acta,* 151, 203, 1983.
75. **Bidoglio, G., Chatt, A., De Plano, A., and Zorn, F.,** Investigations on Tc species in ground water and migration through clay columns, *J. Radioanal. Chem.,* 79, 153, 1983.
76. **Rösch, F. and Khalkin, V. A.,** Ion mobilities of trivalent f elements in aqueous electrolytes, *Radiochim. Acta,* 51, 101, 1990.
77. **Rösch, F., Dittrich, S., Buklanov, G. V., Milanov, M., Khalkin, V. A., and Dreyer, R.,** Electromigration of carrier free radionuclides. II. Complex formation of ^{239}Np(V) with oxalate, tartrate and sulphate in neutral inert electrolytes, *Radiochim. Acta,* 48, 205, 1989.
78. **Bovington, C. H. and Dacre, B.,** Solution properties, in *Radiotracers Techniques and Application,* Vol. 1, Evans, E. A. and Muramatsu, M., Eds., Marcel Dekker, New York, 1977, 457.
79. **Fourest, B., David, F., and Haltier, E.,** Electromigration of Am and Cf in aqueous solution, *Lanthanide Actinide Res.,* 2, 393, 1988.
80. **Brandel, V., Genet, M., and Guillaumont, R.,** Phosphate complexes and inorganic gels of thorium, *N. J. Chem.,* 13, 473, 1989 (in French).
81. **Livage, J., Henry, M., and Sanchez, C.,** Sol gel chemistry of transition metal oxides, *Progr. Solid State Chem.,* 18, 259, 1988.
82. **Henry, M., Jolivet, J. P., and Livage, J.,** Aqueous chemistry of metal cations: hydrolysis, condensation and complexation in *Chemistry, Spectroscopy and Application of Sol Gel Glasses, Structure and Bonding,* Springer-Verlag, Berlin, Vol. 77, 1992, 155.
83. **Joergensen, C. K.,** Characteristics of transition group ions, in *Oxidation Numbers and Oxidation States,* Springer-Verlag, Berlin, 1969, 47.
84. **Joergensen, C. K.,** Optical electronegativities, in *Modern Aspects of Ligand Field Theory,* North-Holland, Amsterdam, 1971, 363.
85. **Herrmann, G. and Trautmann, N.,** Rapid chemical methods for identification and study of short lived nuclides, *Annu. Rev. Nucl. Part. Sci.,* 32, 117, 1982.
86. **Armbruster, P.,** On the production of heavy element by cold fusion, the elements 106 to 109, *Annu. Rev. Nucl. Part. Sci.,* 5, 135, 1985.
87. **Schädel, M.,** Radiochemical studies of transactinides elements, in School-Seminar on Heavy Ion Physics, Dubna, 1989 and Preprint GSI 89-72, D-6100, Darmstadt, Post 110552.

88. **Schädel, M., Brüchle, W., Jäeger, M., Schimpf, E., Kratz, J. V., Scheerer, U. W., and Zimmermann, H. P.,** Arca II, a new apparatus for fast rapid repetitive HPLC separation, *Radiochim. Acta,* 48, 177, 1989.
89. **Person, H., Skarnemark, G., Skälberg, M., Alstad, J., Liljenzin, J. O., Bauer, G., Halberger, F., Kaffrell, N., Rogowski, J., and Trautmann, N.,** Sisak 3, An improved system for rapid radiochemical separation by solvent extraction, *Radiochim. Acta,* 48, 177, 1989.
90. **Szeglowski, Z., Bruchertseifer, H., Domanov, V. P., Gleisberg, B., Guseva, L. J., Hussonnois, M., Tikhomirova, G. S., Zvara, I., and Oganessian, Y. T.,** Study of the solution chemistry of element 104, kurtchatovium, *Radiochim. Acta,* 51, 71, 1990.
91. **Oganessian, Yu. Ts., Hussonnois, M., Demin, A. G., Khaitonov, Yu. P., Bruchertseifer, H., Constantinescu, O., Korotkin, Yu. S., Tretyakova, S. P., Utyonkov, V. K., Shirokovsky, I. V., and Estevez, J.,** Experimental studies of the formation and radioactive decay of isotopes with Z = 104–109, *Radiochim. Acta,* 37, 113, 1984.
92. **Hulet, E. K.,** Chemistry of the elements einsteinium through 105, *Radiochim. Acta,* 32, 7, 1983.
93. **Kratz, J. V., Zimmermann, H. P., Scherer, U. W., Schädel, M., Brüchle, W., Gregorich, K. E., Gannett, C. M., Hall, H. L., Henderson, R. A., Lee, D. M., Leyba, J. D., Nurmia, M. J., Hoffman, D. C., Gäggeler, H., Jost, D., Baltensperger, U., Ya Nai-Qi, Türler, A., and Lienert, Ch.,** Chemical properties of element 105 in aqueous solution: halide complex formation and anion exchange into triisooctyl amine, *Radiochim. Acta,* 48, 121, 1989.
94. **Gregorich, K. E., Henderson, R. A., Lee, D. M., Nurmia, M. J., Chasteler, R. M., Hall, H. L., Bennett, D. A., Gannett, C. M., Chadwick, R. B., Hoffman, D. C., and Herrmann, G.,** Aqueous chemistry of element 105, *Radiochim. Acta,* 43, 223, 1988.
95. **Shinagawa, E. M.,** Electrophoretically focusing chromatography for the separation of short lived nuclides and some other techniques, in *Artificial Radioactivity,* Narayama, R. K. and Arnikar, H. J., Eds., Tata McGraw-Hill, New Delhi, 1985, 579.
96. **Trautmann, N.,** Study of short lived fission products by rapid separation methods, in *Artificial Radioactivity,* Narayama, R. K. and Arnikar, H. J., Eds., Tata McGraw-Hill, New Delhi, 1985, 12.
97. **Renagan, K. and Griffin, H. C.,** Gas jet system for the radiochemical study of fission products, in *Artificial Radioactivity,* Narayama, R. K. and Arnikar, H. J., Eds., Tata McGraw-Hill, New Delhi, 1985, 61.
98. **Bruchertseifer, H., Eichler, B., Estevez, J., and Zvara, I.,** Fast continuous radiochemical isolation of the short lived isotopes of Hf, Ta, and W produced by heavy ion induced reactions, *Radiochim. Acta,* 47, 41, 1989.
99. **Ghiorso, A., Hulet, E. K., Nitsche, J. M., Alonso, J. R., Lougheed, R. W., Alonso, G. T., Nurmia, N., and Seaborg, G. T.,** Identification of element 106, in *Transplutonium Elements,* Muller, W. and Lindner, R., Eds., North-Holland, Amsterdam, 1976, 323.

APPENDIX 8.1

HYPERFINE INTERACTIONS

Hyperfine interactions result from the interaction of the nucleus with the electronic environment. In each of the three types of hyperfine interactions, a nuclear parameter couples with a parameter characteristic of the distribution of electrons at the site of the nucleus and in the atomic shells.

1. Electric Monopole Interaction E0

This results from the electrostatic interaction of the electrons in the nuclear volume with the electric charge of the nucleus. The interaction raises the energy of the bare nucleus by an amount which is proportional:

(a) to the density of the electrons at the nuclear site

$$\rho_0 = |\Psi(0)|^2$$

(b) to the size of the nucleus

In a simple model, the nucleus is taken as a sphere of radius R with the nuclear charge Ze distributed uniformly throughout the volume.

The intensity of the interaction is then

$$W_0 = \frac{2\pi}{5} Ze^2R^2\rho_0$$

and is typically of the order of $\approx 10^{-8}$ eV. It is determined by the oxidation state of the atom, the ionicity of its bonds and the population of electron orbitals. The E0 interaction has the properties of a scalar; it is independent of the nuclear spin and does not lift the degeneracy of nuclear levels.

2. Electric Quadrupolar Interaction E2

The nuclear parameter is the nuclear quadrupole moment Q, which is a measure of the distortion of the nucleus from a spherical shape. Nuclei with a spin $I = 0$ or $I = 1/2$ possess no quadrupole moment. In an inhomogeneous electric field the nonspherical nucleus can acquire several orientations which are fixed by quantum mechanical rules. To each orientation corresponds an interaction energy

$$W_Q = eQ\, V_{zz} \left[\frac{3M^2\, I(I + 1)}{4I\,(2I - 1)}\right]$$

I is the spin of the nucleus and M the associated quantum number, with $-I < M < I$, and V_{zz} is the main component of the electric field gradient tensor. It is assumed here that all other terms of the tensor are nil.

The electric field at the nucleus has its origin in the ions of a crystal lattice, in the charge of ligands bond to the atom, and in the distribution of valence electrons. If all these species are distributed symmetrically around the nucleus, the derivative of the electric field is nil and W_Q vanishes.

The E2 interaction partially lifts the degeneracy of the nuclear levels with $I \geq 1$. Transitions between the separate levels follow the selection rule $\Delta M = \pm 1, 0$. The intensity of the interaction is of the order of 10^{-8} eV.

3. Magnetic Dipolar Interaction M1

Nuclei with a spin $I > 0$ have a magnetic moment μ. In a magnetic field the nucleus can assume various orientations fixed by quantum mechanical rules, each corresponding to an energy

$$W_M = -\mu \frac{M}{I} H$$

The magnetic field at the site of the nucleus originates in the surrounding electrons (hyperfine field); alternatively, it may be an externally applied field or the internal field in magnetically ordered materials.

The M1 interaction completely lifts the degeneracy of nuclear levels. Its intensity is in the range 10^{-6} to 10^{-8} eV.

REFERENCE

Greenwood, N. N. and Gibb, T. C., *Mössbauer Spectroscopy,* Chapman and Hall, London, 1971, chap. 3.

APPENDIX 8.2

TRANSFER KINETICS AT THE TRACER LEVEL

The most simple but rather general transfer mechanism of M from a phase A into a phase B is the rapid formation, in phase A, of an entity E_2; the latter transfers M into the phase B in which it exists in a form $\overline{E}_3$. The initial entity E_1 may be the aqua ion M^{n+} from which derives E_2 and subsequently $\overline{E}_3$. Formally,

$$E_1 \rightleftarrows E_2 \rightleftarrows \overline{E}_3$$

and at this point it is not necessary to deal explicitly with the nature of the entities E_2 and $\overline{E}_3$.

The start of the transfer is taken as the initial time $t = 0$; the initial equilibrium concentrations of E_1 and E_2 immediately before the beginning of the transfer are $[E_1]_0$ and $[E_2]_0$. The concentrations $[E_1]$ and $[E_2]$ will be calculated as a function of time, of the initial concentrations and of the concentrations $[E_1]_e$ and $[E_2]_e$ when the partition equilibrium is established.

If the transfer of M is governed by two *slow* steps $E_1 \rightarrow E_2$ and $E_2 \rightarrow \overline{E}_3$,

$$\frac{d\,[E_1]}{dt} = \lambda_{21}\,[E_2] - \lambda_{12}\,[E_1]$$

$$\frac{d\,[E_2]}{dt} = \lambda_{32}\,[\overline{E}_3] - \lambda_{23}\,[E_2]$$

Taking $[E_1] = x$, $C_0 = x_0 + y_0$ for $t = 0$, $C_e = x_e + y_e$ for $t = \infty$,

$$\frac{dx}{dt} = \lambda_{21}\,y - \lambda_{12}\,x = ax + b \qquad (1)$$

$$\frac{dy}{dt} = \lambda_{32}\,[C_0 - (x + y)] + \lambda_{12}\,x - (\lambda_{23} + \lambda_{21})\,y = c + dx + fy \qquad (2)$$

The simple case treated in the text corresponds to $x_0 = x_e = x = 0$ and $\lambda_{12} = \lambda_{21} = 0$.

For solution of the system of differential equations, Equation 1 is derived and dy/dt (Equation 2) is introduced into $d^2x/dt.^2$ This yields

$$\frac{d^2x}{dt^2} - (a + f)\,\frac{dx}{dt} - (bd - fa)\,x - bc = 0$$

or with a simpler notation,

$$\frac{d^2x}{dt^2} + \alpha \frac{dx}{dt} + \beta x - bc = 0$$

Setting now

$$u = \beta x - bc \qquad u' = \beta \frac{dx}{dt} \qquad u'' = \beta \frac{d^2x}{dt^2}$$

one obtains

$$u'' + \alpha u' + \beta u = 0 \tag{3}$$

for which the characteristic equation $r^2 + \alpha r + \beta = 0$ has real, positive roots r_1 and r_2 only for $\Delta = \alpha^2 - 4\beta > 0$. The solution of Equation 3 is now

$$u = K_1 e^{r_1 t} + K_2 e^{r_2 t}$$

Returning to x and y,

$$C = x + y = \frac{K_1 (a - b) - r_1}{\beta b} e^{r_1 t} + \frac{K_2 (a - b) - r_2}{\beta b} e^{r_2 t} - \frac{bc + ac}{\beta}$$

The determination of constants K_1 and K_2 with limits $C = C_e$ for $t = \infty$ and $C = C_0$ for $t = 0$, would lead to rather intricate calculations. Therefore it is more convenient to write first

$$C = A e^{r_1 t} + B e^{r_2 t} \frac{bc + ab}{\beta}$$

For $t = \infty$,

$$C_e = -\frac{bc + ab}{\beta}$$

and for $t = 0$,

$$C_0 - C_e = A + B \tag{4}$$

Introducing the derivative

$$p_0 = \left(\frac{\partial C}{\partial t}\right)_{t=0}$$

one obtains

$$p_0 = r_1 A + r_2 B \tag{5}$$

With Equations 4 and 5, A and B can be calculated and the final expression for the time evolution of C becomes

$$C = C_e + (C_0 - C_e)\frac{r_2 e^{r_1 t} - r_1 e^{r_2 t}}{r_2 - r_1} + p_0 \frac{e^{r_1 t} - e^{r_2 t}}{r_1 - r_2}$$

in which $r_1 = -a - \sqrt{\Delta}$ and $r_2 = -a + \sqrt{\Delta}$.

In case $\Delta = 0$, the solutions are of the form

$$u = e^{rt}(K_1 t + K_2)$$

and

$$C = C_e + (C_0 - C_e)\,e^{-\alpha t/2} + \left[p_0 + \frac{\alpha}{2}(C_0 - C_e)\right] t\, e^{\alpha t/2}$$

Finally, when $\Delta < 0$ and posing $-\Delta = |\Delta|$, the solutions are of the form

$$u = e^{-\alpha t}(K_1 \cos \sqrt{|\Delta|}t + K_2 \cos \sqrt{|\Delta|}t)$$

and

$$C = C_e + (C_0 - C_e)\left(\cos \sqrt{|\Delta|}t + \frac{\alpha}{\sqrt{|\Delta|}} \sin \sqrt{|\Delta|}t\right) e^{-\alpha t}$$
$$+ \frac{p_0}{\sqrt{|\Delta|}} \sin \sqrt{|\Delta|}t\, e^{-\alpha t}$$

Expressions for the partition coefficient D as a function of D_e and the equilibrium half-time τ can be derived. They are rather complicated and will not be dealt with here.

The various expressions of C show that in all cases the values of C, p_0, and C_e are related as well as D, D_e, and

$$\left(\frac{\partial C}{\partial t}\right)_{t=0} = -\frac{1}{C_0} p_0$$

APPENDIX 8.3

THERMODYNAMIC ACTIVITY COEFFICIENTS

It is generally considered that the thermodynamic activity coefficients of the entities of a radionuclide at the tracer scale in both phases of a biphasic system are equal to unity, or at least constant. The validity of this assumption may be questioned, in particular for aqueous solutions.

If the activities of the *microcomponent* entities are neglected, any effect of thermodynamic activity must be attributed to the *macrocomponents*, especially at high concentration.

Activity coefficients γ of complexing reagents in the organic phase are generally close to one because these compounds are in the molecular state. Accordingly, no major corrections are required. For instance, for 2-thenoyl trifluoroacetone in benzene,

$$\gamma_{HA} = 1 - 0.24\ [\overline{HA}]^{0.48}$$

and $\gamma \approx 1$ when $[\overline{HA}] \approx \overline{C}_{HA}$ is less than 0.2 *M*.[1] The activity coefficients depend strongly on the diluent.

On the other hand, the activity coefficients of the solvating reagents S increase rapidly with the concentration $\overline{C}_S$. For low values of $\overline{C}_S$, $\gamma \approx 1$; for high concentrations, log γ is approximately proportional to $\overline{C}_S$.[2,3] The true values of the slopes $\partial \log D_e / \partial \log [\overline{S}] = \langle \overline{s} \rangle$ are the limiting ones when C_S tends toward 0. The same holds for amine salts SHL, but in this case log γ is proportional to log $[\overline{S}]$.[4]

For neutral and charged species, as is well known, activity coefficients depend mainly on the ionic strength of aqueous solutions. In certain cases, the activities can be calculated with the aid of various models, among them the more recent *Specific Interaction Theory*.[5-8] If these calculations cannot be performed, or their validity remains dubious, it is customary to carry out experiments at *constant* ionic strength. The value of μ is fixed, e.g., by the highest concentration which must be used for one of the reagents in a given experiment, and all activity coefficients are assumed to remain constant. In reality, the situation is more complicated.

A commonly considered reference case is a mixture of two electrolytes A and B with a common ion at constant molality $m = m_A + m_B$. The activity coefficient can be expressed for example as

$$\log \gamma_{\pm}(A) = \log \gamma_0(A) + \alpha_A m_A$$

This relation is valid at constant molality m and is based on the Harned rule derived by Guggenheim.[9] In the first term, $\gamma_0(A)$ is the limiting mean activity coefficient of A in B (i.e., when $m_A \rightarrow 0$, $m_b \rightarrow m$). The Harned coefficient

α is independent of m, such that $\alpha_A = \alpha_B$. This is illustrated by the hydrochloric acid complexes $C^z_{l,y}$ of a microcomponent M in a solution containing Cl^- ions (e.g., HCl + LiCl or HCl + $HClO_4$). To introduce the mean thermodynamic activity coefficients, it is assumed that the complex $C^z_{l,y}$ derives from an imaginary electrolyte C_f defined as follows: for $z > 0$, $M(HCl)\ H_{-y}\ Cl_{(n-y)} = M(HCl)_{(l-y)}\ Cl_n$; for $z < 0$, $M(HCl)_l\ H_{-y}\ H_{(y-n)} = M(HCl)_l H_n$.

In hydrochloric acid media at constant ionic strength, the composition changes from HCl to $HClO_4$ or from HCl to LiCl. Setting A ≡ HCl and B ≡ $HClO_4$ or LiCl, and since $m_{Cf} = 0$, $\log \gamma_\pm(C_f)$ varies in a generally *unknown* manner with m_{HCl} from $[\log \gamma_0(C_f)]_A$ to $[\log \gamma_0(C_f)]_B$. These values are the limiting, not necessarily equal, values for given μ. Accordingly, it is not certain that the factor which unites all $\gamma_\pm(C_f)$ coefficients in the expression of the law of mass action is really constant at constant ionic strength.

On the other hand, the terms $\gamma_\pm(A)$ and $\gamma_\pm(B)$ are not changed by the presence of complexes $C^z_{l,y}$ and the values $\gamma_\pm(HCl)$ in the mixtures HCl + LiCl or HCl + $HClO_4$ are given by the preceding relation with appropriate values for α. They are known for these media as a function of m_{HCl} or C_{HCl}.[10]

The influence of activity coefficients on partition data is illustrated by the extraction of M as a chelate MA_3 from a HCl + $HClO_4$ medium with ionic strength μ, in which the dominant entity is MCl_4^- for $C_{HCl} \approx \mu/2$.

The four equations

$$M^{3+} + 4\ HCl \rightleftarrows M(HCl)_4\ H_{-4} + 3\ H^+$$

$$M^{3+} + \bar{3}\ HA \rightleftarrows M(HA)_3^-\ H_{-3} + 3\ H^+$$

$$M(HA)_3^-\ H_{-3} \rightleftarrows \overline{M(HA)_3^-\ H_{-3}}$$

$$\overline{HA} \rightleftarrows HA$$

are combined in a single expression

$$M(HCl)_4\ H_{-4} + \bar{3}\ \overline{HA} \rightleftarrows \overline{M(HA)_3^-\ H_{-3}} + 3\ HCl + Cl^-$$

In order to verify the influence of $[Cl^-]$ (since $[H^+] = \mu$), this equation is reduced to the following one

$$MCl_4^- + \bar{3}\ HA \rightleftarrows \overline{MA_3^-} + 3\ H^+ + 4\ Cl^-, \quad K^\circ$$

with

$$K^\circ = \frac{\gamma(\overline{MA_3^-})\ [\overline{MA_3^-}]\ \gamma^3(H^+)\ [H^+]^3\ \gamma^4(Cl^-)\ [Cl^-]^4}{\gamma(MCl_4^-)\ [MCl_4^-]\ \gamma^3(\overline{HA})\ [\overline{HA}]^3}$$

The $\gamma_\pm$ are developed by multiplying numerator and denominator by $\gamma(H^+)[H^+]$, which leads to

$$K^o = \frac{\gamma(\overline{MA_3^-})\,[\overline{MA_3^-})\,\gamma_\pm^8(HCl)\,[H^+]^4\,[Cl^-]^4}{\gamma_\pm(HMCl_4)\,[HMCl_4]\,\gamma^3(\overline{HA})\,[\overline{HA}]^3}$$

Here, $HMCl_4$ is the imaginary electrolyte which has a common ion with $HClO_4$ or HCl, and

$$\log \gamma_\pm(HMCl_4) = [\log \gamma_0(HMCl_4)]_{HClO_4,HCl} = f([HCl])$$

The expression for the equilibrium partition coefficients is now

$$D_e = K^o \frac{\gamma^3(\overline{HA})\,[\overline{HA}]^3\,\gamma_\pm(HMCl_4)}{\gamma(\overline{MA_3^-})\,[\overline{MA_3^-}]\,\gamma_\pm^8(HCl)\,[H^+]^4\,[Cl^-]^4}$$

For $[Cl^-] \approx \mu/2$,

$$\frac{\partial \log D_e}{\partial \log [Cl^-]} = -4 + \frac{\partial \log \gamma_\pm(HMCl_4)}{\partial \log [Cl^-]} - 8\,\frac{\partial \log \gamma_\pm(HCl)}{\partial \log [Cl^-]}$$
$$= -4 + \beta - 8\,\delta$$

In the media considered here, $[Cl^-] = C_{HCl}$ and $[H^+] = \mu$:

$$\beta = \frac{\partial\, f(HCl)}{\partial\, C_{HCl}} \qquad \delta = \frac{\partial \log \gamma_\pm(HCl)}{\partial \log C_{HCl}} = \alpha_{HCl}\quad C_{HCl} = \frac{\alpha_{HCl}\,\mu}{2}$$

The term β is expected to be small and term δ becomes appreciable only at high C_{HCl}, since α_{HCl} is not large. Hence, the significance of slopes of the log curves is questionable at high ionic strength. In the present case, the slopes would indicate the existence of a complex of order smaller than 4. The correction to be applied is proportional to C_{HCl}.

Similar considerations hold for more complex systems involving several complexes, such as MCl_2^+, MCl_3, or MCl_4^-. Each case should be considered individually, since rather complicated situations may arise. The hydrochloric acid medium is a privileged one since the α_{HCl} are known or may be evaluated.[11,12] In other media, corrections with respect to variations of the activity coefficient cannot be performed and this probably accounts for the noninteger values of slopes in the log curves which are frequently observed.

REFERENCES

1. **King, E. L. and Reas, W. H.,** The activity coefficient of thenoyltrifluoroacetone in benzene solution, *J. Am. Chem. Soc.*, 73, 1804, 1951.
2. **Alcock, K., Grimley, S. S., Healy, T. V., Kennedy, J., and McKay, H. A. C.,** *Trans. Faraday Soc.*, 52, 39, 1956.
3. **Siekiersky, S.,** The influence of diluent on extraction of europium and thorium nitrates by tri-*n*-butylphosphate, *J. Inorg. Nucl. Chem.*, 24, 205, 1962.
4. **Smulek, W. and Siekiersky, S.,** The influence of diluent on the extraction of iron(III) from hydrochloric acid solutions by tertiary amines, *J. Inorg. Nucl. Chem.*, 24, 1651, 1962.
5. **Grenthe, I. and Wanner, H.,** Ionic strength corrections, in *Chemical Thermodynamics of Uranium,* (Chemical Thermodynamics Series), Vol. 1, North-Holland, Amsterdam, 1992, 683.
6. **Riglet, Ch., Vitorge, O., and Grenthe, I.,** Standard potentials of the (MO_2^{2+}/MO_2^+) system for U and other actinides, *Inorg. Chim. Acta,* 133, 323, 1987.
7. **Wolery, T. J. and Jackson, K. J.,** Activity coefficients in aqueous salt solutions, in *Chemical Modelling of Aqueous Systems,* Melchior, D. C. and Bassett, R. L., Eds., American Chemical Society, Washington D.C., 16, 1990.
8. **Parkhurst, D. L.,** Ion association models and mean activity coefficients of various salts, in *Chemical Modelling of Aqueous Systems,* Melchior, D. C. and Bassett, R. L., Eds., American Chemical Society, Washington D.C., 30, 1990.
9. **Robinson, R. A. and Stokes, R. H.,** The thermodynamics of mixed electrolytes, in *Electrolyte Solutions,* 2nd ed., Butterworths, London, 1959, 432.
10. **Duplessis, J. and Guillaumont, R.,** Solvent extraction and thermodynamic activity, *Radiochem. Radioanal. Lett.*, 31, 283, 1977 (in French).
11. **Vaezi Nasr, F., Duplessis, J., and Guillaumont, R.,** Solvent extraction of Nb(V) with dichloro-5-7-hydroxy-8-quinoline, *Analusis,* 7, 302, 1979 (in French).
12. **Duplessis, J. and Guillaumont, R.,** Extraction of Pa(V) by TTA from hydrochloric media with high ionic strength, *Radiochem. Radioanal. Lett.*, 37, 159, 1979 (in French).

APPENDIX 8.4

CONCENTRATION-DEPENDENT PARTITION OF A MICROCOMPONENT

The effect of ion-pair dissociation in an organic phase is illustrated by the extraction of U(IV) from a HCl solution with an amine S (e.g., TNOA).[1-3] It is assumed that the species extracted from the aqueous phase are UCl_6^{2-} and HCl.

The equilibria involved are

$$2\,H^+ + UCl_6^{2-} + 2\,\overline{S} \rightleftarrows \overline{UCl_6(SH)_2}, \qquad K_1$$

$$H^+ + Cl^- + \overline{S} \rightleftarrows \overline{SHCl}, \qquad K_2$$

$$\overline{UCl_6(SH)_2} \rightleftarrows \overline{UCl_6^{2-}} + 2\,\overline{SH^+}, \qquad K_3$$

$$\overline{SHCl} \rightleftarrows \overline{SH^+} + \overline{Cl^-}, \qquad K_4$$

Further,

$$[\overline{SH^+}] = 2\,[\overline{UCl_6^{2-}}] + [\overline{Cl^-}]$$

Omitting charges, the distribution coefficient is written

$$D_e = \frac{[\overline{UCl_6}] + [\overline{UCl_6(SH)_2}]}{C_M}$$

Under conditions for which activity coefficients are constant

$$D_e = K_1\,[H]^2\,[\overline{S}]^2\left(1 + \frac{K_3}{[\overline{SH}]^2}\right)$$

From the relation

$$[\overline{SH}] = 2\,K_3\frac{[\overline{UCl_6(SH)_2}]}{[\overline{SH}]^2} + K_4\,\frac{[\overline{SHCl}]}{[\overline{SH}]}$$

one can write

$$[\overline{SH}]^3 - K_2K_4\,[H]\,[Cl]\,[\overline{S}]\,[\overline{SH}] - 2\,K_1K_3\,C_M\,[H]^2\,[\overline{S}]^2 = 0$$

or

$$[\overline{SH}]^3 - p\,[\overline{SH}] - q = 0$$

This equation has one positive root which is a complicated expression of C_M and $[\overline{S}]$ and shows that D is dependent on C_M. For very low values of this concentration, it can be assumed that $q \approx 0$, from which results

$$[\overline{SH}] \approx (K_2K_4\ [H]\ [Cl]\ [\overline{S}])^{1/2}$$

D_e is now a function of $[\overline{S}]^2$ *and* $[\overline{S}]$ instead of $[\overline{S}]^2$ if the dissociation of the ion pair is neglected.[1-2]

There are two other cases for which partition of a microcomponent M in a biphasic system may depend on the concentration C_M.

The first one occurs when M is partially oxidized or reduced in the partition system, for instance by impurities present in concentration of the same order as that of M. The yield of the redox reaction depends on the relative concentrations of M and the impurity (see Appendix 7.1). Because D_e values are not the same for M in two different oxidation states, the overall D_e values depend on C_M.

A similar situation may be encountered in a triphasic partition system including a gas phase, when the gaseous component forms a redox couple with a species in solution.[4-7] The relative proportions of the oxidation states of M may depend on the partial pressure of the gas and on C_M. An example is provided by the behavior of an oxidable trivalent element in the partition system: fused alcaline nitrates/polyphenyl eutectic $-$ TBP/NO_2, O_2.[3,4] The oxidation of M^{3+} occurs according to:

$$2\ M^{3+} + 6\ NO_3^- \rightleftarrows 2\ MO_2 + 6\ NO_2 + O_2, \quad K$$

In a closed volume initially free of O_2 and NO_2 the equilibrium concentration of $M^{3+} = n/V$ (n is the number of mole and V the volume of fused salts) is related to $C_M = n_t/V$. From the balance

$$n_t = n + \bar{n} \qquad n_{O_2} = \frac{1}{2}\bar{n} \qquad n_{NO_2} = 3\ \bar{n}$$

in which $\bar{n}$ is the number of mole of MO_2, it results

$$K' = \frac{1/2\ (n_t - n)\ (3\ (n_t - n))^6}{n^2\ (C_{NO_3^-})^2}$$

The constant K′ incorporates the gaseous and fused phases. The latter is expressed equivalently as

$$(n_t - n)^7 - n^2\ K'' = 0 \quad \text{or} \quad C_M = \alpha[M^{3+}]^{2/7} + [M^{3+}]$$

Because MO_2 is not extracted, the partition of M or the solvate of M^{3+} depends on C_M and K′. The case is more complicated if O_2 is initially present.

The second case refers to formation of polymeric species of M at concentrations for which M is still a microcomponent. With the notation used in Chapter 6, the formation equilibrium of polymers of order m is

$$\mathrm{m\,M^{n+} + \mathit{l}\,H_pL + q\,H_{p'}Q \rightleftarrows C^{z}_{m,\mathit{l},q,y} + (z - mn)\,H^{+}}$$

The concentrations of all macrocomponents are constant, including that of H^+, (which may require an acidic medium). Hence, the effect of C_M on the distribution coefficient D_e can be inferred from the formal equilibria

$$\mathrm{m\,M^{n+} \rightleftarrows C_m,} \qquad K_m \quad m = 1, 2, 3, \ldots \qquad K_1 = 1$$

Neglecting charges,

$$C_M = \sum_1^m m\,K_m\,[M]^m$$

Setting

$$P_{\overline{m}} = \frac{[\overline{C}_m]}{[C_m]}$$

$$\overline{C}_M = \sum_1^{\overline{m}} m\,P_{\overline{m}}K_{\overline{m}}\,[M]^{\overline{m}}$$

$$\frac{\partial \ln D_e}{\partial \ln C_M} = \frac{\partial \ln \overline{C}_M}{\partial \ln C_M} - 1 = \frac{\partial \ln \overline{C}_M/\partial \ln [M]}{\partial \ln C_M/\partial \ln [M]} - 1$$

$$\frac{\partial \ln C_M}{\partial \ln [M]} = \frac{\partial \ln C_M}{\partial C_M}\,\frac{\partial C_M}{\partial [M]}\,\frac{\partial [M]}{\partial \ln [M]} = \frac{1}{C_M}\left\{\sum_1^{m-1} m^2\,K_m\,[M]^{m-1}\right\}[M]$$

$$= \frac{\sum_1^m m(m\,K_m\,[M]^m)}{C_M}$$

The fraction of M in the polymer of order m is defined as

$$f_m = \frac{m\,K_m\,[M]^m}{C_M} = \frac{m\,[C_m]}{C_M}$$

from which

$$\frac{\partial \ln C_M}{\partial \ln [M]} = \frac{1}{C_M}\sum_1^m m\,f_m$$

In the same way,

$$\frac{\partial \ln \overline{C}_M}{\partial \ln [M]} = \sum_1^{\overline{m}} \overline{m}\, \overline{f}_{\overline{m}}$$

and finally,

$$\frac{\partial \ln D_e}{\partial \ln C_M} = \frac{\sum_1^{\overline{m}} \overline{m}\, \overline{f}_{\overline{m}}}{\sum_1^{m} m\, f_m} - 1$$

This expression does not consider the mean values of m,

$$\langle m \rangle = \frac{C_M - [M]}{C_M} = \frac{\sum_1^{m} m\, K_m\, [M]^m}{C_M} = \sum_1^{m} f_m$$

$$\langle \overline{m} \rangle = \sum_1^{\overline{m}} \overline{f}_{\overline{m}}$$

However, it can be noted

(1) in the absence of polymers in the system, $m = \overline{m} = f_m = \overline{f}_{\overline{m}} = 1$ and

$$\frac{\partial \ln D_e}{\partial \ln C_M} = 0$$

(2) in the absence of polymers in the organic phase, which is often the case,

$$\frac{\partial \ln D_e}{\partial \ln C_M} = \frac{1}{\sum_1^{m} m\, f_m} - 1$$

In this instance, $\ln D_e$ vs. $\ln C_M$ tends more or less rapidly toward -1 as the polymerization reaction proceeds.

In noncomplexing aqueous phases, real colloids or pseudocolloids of the microcomponent may be present. These forms cannot be extracted; furthermore, part of the microcomponent may be adsorbed on the walls of the container. The fraction of nonextractable forms can be estimated in the following way.[8]

It is assumed that the partition equilibrium of the microcomponent between two phases of equal volume V is established. The partition coefficient is D_1.

Now a volume V′ is taken from the organic phase and is replaced by an identical volume of a fresh organic solution. A new equilibrium is instaured which is characterized by a partition coefficient D_2. Setting for the two successive extractions the concentrations and activities of the extractable forms $\overline{C_1}$, $\overline{C_2}$, $\overline{A_1}$, and $\overline{A_2}$ in the organic phase, C_1, C_2, A_1, and A_2 in the aqueous phase and C_i the concentration of the nonextractable species, it follows that

$$D_1 = \frac{\overline{C_1}}{C_1 + C_i} = \frac{\overline{A_1}}{A_1} \quad D_2 = \frac{\overline{C_2}}{C_2 + C_i} = \frac{\overline{A_2}}{A_2} \quad D = \frac{\overline{C_1}}{C_1} = \frac{\overline{C_2}}{C_2}$$

D being the true distribution coefficient. The difference d in the amount of the microcomponent adsorbed on the walls along in two extractions satisfies the balance

$$V\,C_1 + (V - V')\,\overline{C_1} + V\,C_i = d = V\,C_2 + \overline{C_2} + V\,C_i$$

Setting $r = V - V'/V$ and $a = d/V$, it follows

$$C_1 + r\,\overline{C_1} + a = C_2 + \overline{C_2}$$

The parameter *a* is positive or negative, depending if the adsorption of M on the walls decreases or increases between the two extractions. It can be calculated from the measured activities if the ratio k between activities and concentrations is known:

$$A_1 + r\,\overline{A_1} + \frac{a}{k} = A_2 + \overline{A_2}$$

Combining the previous relations and setting $q = a/\overline{C_1}$, the distribution coefficient can be calculated as

$$D = \frac{(1 - r - q)\,D_1\,D_2 + D_1 - D_2}{D_2 - D_1\,(p + q)}$$

The percent of nonextractable forms is

$$p = 100\,\frac{D - D_1}{D\,(1 + D_1)}$$

When the organic phase is completely separated at the first extraction and the adsorption of the microcomponent is identical in the two successive steps ($r = 0$ and $q = 0$),

$$D = D_1 + \frac{D_1}{D_2} - 1$$

An alternative procedure can be applied in the absence of adsorption. The values of D and p are calculated from the activities which accompany a change in the ratio of the volumes of the two phases:

$$\frac{\overline{A}}{\overline{V}} = D\frac{A}{V} - \frac{D\,p}{100}C$$

The term C is $(\overline{A} + A)/V$ or $(\overline{A} + A)/\overline{V}$ according to the phase, aqueous or organic, which contains initially the microcomponent. The ratio $\overline{A}/\overline{V}$ is plotted against A/V; the straight line yields D and p.

REFERENCES

1. **Sadlick, J.,** Concentration effect on solvent extraction coefficients of some trivalent metal halides, *J. Phys. Chem.*, 60, 500, 1956.
2. **Maddock, A. G., Smulek, W., and Tench, A. J.,** Solvent extraction from halide solutions, *Trans. Faraday Soc.*, 58, 923, 1962.
3. **Erickson, R. L. and MacDonald, R. L.,** Solvatation of extracted complex metal acids, the $HFeCl_4$-$HFeBr_4$ and $HInBr_4$ nitrobenzene systems, *J. Am. Chem. Soc.*, 88, 2099, 1966.
4. **Foos, J. and Guillaumont, R.,** Solvent extraction of 4f and 5f elements from fused nitrates, *Bull. Soc. Chim. France,* 3129, 1971 (in French).
5. **Foos, J. and Guillaumont, R.,** Solvent extraction of lanthanides as thenoyltrifluoroacetonates from fused alkali nitrates, *Radiochem. Radioanal. Lett.*, 16, 315, 1974 (in French).
6. **Foos, J. and Guillaumont, R.,** Solvent extraction of cis- and trans-uranium elements from fused alkali nitrates, *J. Inorg. Nucl. Chem.*, 36, 3833, 1974 (in French).
7. **Tanipanichskul, P., Foos, J., Iuchenko, A., and Guillaumont, R.,** Solvent extraction of lanthanides as thenoyltrifluoroacetonates from a KNO_3-$LiNO_3$ eutectic, *J. Inorg. Nucl. Chem.*, 38, 315, 1976 (in French).
8. **Guillaumont, R. and Désiré, B.,** Characterization of forms nonextractable by solvents, *Radiochem. Radioanal. Lett.*, 8, 183, 1971; C. R. Acad. Science, Paris, 269, 448, 1969 (in French).

APPENDIX 8.5

TRANSFER OF A MICROCOMPONENT INTO A METAL

The partition of an element between a solution and a metal at a negative potential with respect to the first phase requires special attention. A classical example is the biphasic system of an aqueous solution and Hg. The equilibrium

$$E_2 \rightleftarrows \overline{E_3}$$

described in the text is considered in which E_2 is an entity in the solution and $\overline{E_3}$ the amalgamated metal M. In a general case, the transfer kinetics involves three steps:

1. Diffusion of E_2 toward the surface of Hg. E_2 may reach the surface unchanged or as a different entity E_2'.
2. Electron exchange between E_2' and Hg by which the *atom* E_3' is formed.
3. Diffusion of E_3' into the bulk of Hg and conversion to $\overline{E_3}$:

$$E_2 \rightarrow E_2' \rightleftarrows \overline{E_3'} \rightarrow \overline{E_3}$$

The concentration gradients inside the diffusion layers of thickness δ_i are assumed to be linear

$$\frac{dC_{Ei}}{dx} = \frac{C_{Ei} - [E_i']}{\delta_i}$$

The partition equilibrium involves only the entities E_2' and E_3':

$$E_2' \,(C'^{(n'-y)+}_{l',y'}) + N\,e^- \rightleftarrows \overline{E_3'}\,(M) + l'\,H_pL + (n' - N - y)\,H^+, \quad K_e$$

In the expression of the rate constants

$$\lambda_{23}' = C\,k^o \exp\left[-\alpha \frac{NF}{RT}(E - E^o)\right]$$

$$\lambda_{32}' = C\,k^o \exp\left[(1 - \alpha)\frac{NF}{RT}(E - E^o)\right]$$

α is the transfer coefficient (equal to unity if the discharge is reversible), C is a constant which includes the concentrations of the macroscopic reagents, k^o is the standard heterogeneous kinetic constant and E^o the apparent standard potential.

Case a. The rate-determining step is the transfer $E_2' \rightarrow E_3'$. This is a situation described in the text for which the equilibrium half-time is

$$\tau = -\frac{0.7}{f}$$

with

$$f = -(\lambda'_{23} + \lambda'_{32}) = g(E - E^{o}, \alpha)$$

and

$$D = \frac{1 - e^{-0.7t/\tau}}{\frac{1}{D_e} + e^{-0.7t/\tau}}$$

For a given time t, D is a function of $E - E_0$. A variation ΔE of E with respect to E^{o} results in changes $\Delta\tau$ and ΔD given by

$$\Delta D = \frac{\partial D}{\partial \tau}\Delta\tau = -\frac{(0.7)^2 e^{-0.7/\tau}\left(1 + \frac{1}{D_e}\right)}{\tau^2\left(\frac{1}{D_e} + e^{-0.7/\tau}\right)^2}\Delta\tau$$

The variations of D are symmetric with respect to the D value for E^{o}.

Case b. The transfer $E'_2 \rightarrow E'_3$ is fast ($\alpha = 1$) and the rate-determining step is the diffusion.

Setting

$$[E_2] = Y \quad [E'_2] = y \quad [\overline{E'_3}] = x \quad [\overline{E_3}] = X$$

under stationary conditions with conservation of the transfer flux,

$$-\frac{dY}{dt} = d_2(Y - y) \qquad \frac{dx}{dt} = d_3(x - X)$$

where d_2 and d_3 are diffusion coefficients.

Further, with $X/Y = K$ and $-RT \ln K = Ne\,[E - E^{o}]$, it results

$$\frac{X + \frac{1}{d_3}\frac{dX}{dt}}{Y + \frac{1}{d_2}\frac{dY}{dt}} = K$$

This implies that X and Y must satisfy the conditions that

$$\frac{dX}{dt} = 1 - d_3 X \quad \text{and} \quad \frac{dY}{dt} = \frac{d_3}{d_2} K - d_2 Y$$

These expressions are of type $dz/dt = c + fz$ ($z = X$ or Y) already encountered. The variation of Y can be established in a similar way,

$$d_2 Y = \frac{d_3}{d_2} K e^{-d_2 t} e^{-d_2 L}$$

where L is the integration constant. The equivalent expressions of C_M and D can be derived as before. Here the term $f = - d_2 < 0$ and τ depends only on the diffusion coefficients. The term K can be included in $Y_e = (d_3/d_2) K$ and finally in D_e.

Case c. This is the general situation, with combination of the two preceding cases. It would require a lengthy treatment. The first step will always appear in the solution. For a highly negative potential E, $[E_2'] \approx 0$ and the amalgamation rate depends only on C_M.

CHAPTER 9

CHEMISTRY WITH A FEW ATOMS

PART ONE — RADIOACTIVE PROPERTIES

When the number of radionuclide-containing entities in a given system decreases to about 100 or less, some of the points previously discussed must be reconsidered.[1] This is particularly true for time-dependent radioactive properties and conclusions drawn from the law of mass action as applied to the reactions of the $E_i(m_j)$ entities incorporating the radionuclides. These reactions may involve microcomponents in mutual interaction or reactions of the $E_i(m_j)$ with $E_i(M_j)$ and $E_i(L_j)$ macrocomponents.

For such low populations, the properties and behavior of the entities cannot rely on large numbers being present, but depend on peculiar statistics that may exhibit large fluctuations. Other reasons for the special treatments will appear progressively throughout the present chapter.

This topic is meaningful only because individual entities can be observed from the energy associated with their radioactive decay. Actual situations involving sub-trace amounts of matter have already been invoked, e.g., elements with $Z > 100$, radon and cosmogenic radionuclides in the atmosphere, radionuclides dispersed in large natural water reservoirs, and possibly matter in space. Indeed, the chemical behavior of small numbers of species in interstellar space is rather peculiar when compared to ''ordinary'' chemistry. Encounters of free atoms or entities are extremely rare and occur in a virtually matter-free environment. These conditions in which no ''third body'' is available may lead to unusual situations. In a diffuse interstellar cloud in which the density is 10^2 to 10^3 entities per cubic centimeter and T between 10 K and 150 K, the formation of a molecule in an exothermic process is immediately followed by the decomposition because the energy cannot be transferred. For similar reasons, slightly endothermic reactions are precluded. In the following, it is assumed that the systems are diluted solutions.

Finally, any population of radioactive entities must inevitably traverse a sub-trace level, in which there exists a strictly denumerable ensemble of entities, before complete extinction.

The ultimate conceivable dilution of matter is a single atom. Hence, it appears of interest to consider the behavior of this atom as a limiting theoretical case. Its description requires the introduction of specific notions that will be useful for the case of several atoms.[2,5]

I. DISINTEGRATION OF A SINGLE RADIOACTIVE ATOM

The instant at which a given radionuclide will undergo disintegration cannot be predicted, but the probability $\mathcal{P}(1)$ of the decay during a fixed observation time can be calculated. This time may be, for instance, an interval Δt, or an infinitesimal time dt at instant t, or the interval from t_1 to t_2. Alternatively, the decay probability at a given time t can be estimated. The choice of observation time depends on experimental conditions and on a preliminary knowledge of the radioactive properties of the atom, which may or may not be available. This point will be further considered later.

Quite generally, a radioactive atom may decay in several ways, e.g., α emission and spontaneous fission. Consider that two decay channels are open, with decay constants λ_1 and λ_2. The total decay probability is $\lambda_1 + \lambda_2$, but the probability for observing the disintegration depends on the detection system. Here, where a single atom is considered, λ is the decay constant with respect to a specific detector.

The decay probability of the atom during a time interval Δt at any given time t is

$$P(1)_{t,t+\Delta t} = \lambda \Delta t$$

and the probability of survival is

$$P(0)_{t,t+\Delta t} = (1 - \lambda \Delta t)$$

In these expressions, Δt can be replaced by the differential dt, which is the limit when Δt becomes vanishingly small with respect to t. The differential probabilities are

$$dP(1)_{t,t+dt} = \lambda dt$$

and

$$dP(0)_{t,t+dt} = (1 - \lambda dt)$$

Two cases will be considered.

A. Δt and t are Random

The time t is counted from an initial time $t = 0$, which is the birth of the atom. A dimensionless variable n is introduced in order to set time intervals $t = n\,\Delta t$, $t_1 = n_1\,\Delta t$, $t_2 = n_2\,\Delta t$ . . . It follows that the probability that the atom will decay in the interval between t and $t + \Delta t$ at a given time t is

$$\mathcal{P}(1)_{t,t+\Delta t} = P(0)_{0,t}\, P(1)_{t,t+\Delta t}$$

with

$$P(0)_{0,t} = P(0)_{0,\Delta t}\, P(0)_{\Delta t,2\Delta t}\, P(0)_{2\Delta t,3\Delta t} \ldots = (1 - \lambda\Delta t)^n$$

and

$$P(1)_{t,t+\Delta t} = \lambda\Delta t$$

from which one obtains

$$\mathcal{P}(1)_{t,t+\Delta t} = \lambda\,(1 - \lambda\Delta t)^n\Delta t$$

Similarly, for a finite time interval between t_1 and t_2

$$\mathcal{P}(1)_{t_1,t_2} = \lambda(1 - \lambda\Delta t)^{n_1}\,\Delta t + \lambda(1 - \lambda\Delta t)^{n_1+1}\,\Delta t + \ldots + \lambda(1 - \lambda\Delta t)^{n_2-1}\Delta t$$

or

$$\mathcal{P}(1)_{t_1,t_2} = \sum_{n'=n_1}^{n'=n_2-1} \lambda(1 - \lambda\Delta t)^{n'}\Delta t$$

with

$$\mathcal{P}(1)_{0,\infty} = \sum_{n'=0}^{n'=\infty} \lambda(1 - \lambda\Delta t)^{n'}\,\Delta t = 1$$

B. Δt is Replaced by dt

The differential probability that the atom will decay at time t in the interval between t and $t + dt$, is

$$d\mathcal{P}(1)_{t,t+dt} = \lambda(1 - \lambda\,dt)^n\,dt = \lambda\left(1 - \lambda\,\frac{t}{n}\right)^n dt$$

It can be shown (see Appendix 9.1) that for $n > 100$, a condition which in fact corresponds to $dt << t$, the term $(1 - \lambda\, t/n)^n$ is equivalent to $e^{-\lambda t}$. Accordingly,

$$d\mathcal{P}(1)_{t,t+dt} = \lambda e^{-\lambda t}\, dt$$

The probability that the atom will decay between t_1 and t_2 is

$$\mathcal{P}(1)_{t_1,t_2} = \int_{t_1}^{t_2} \lambda e^{-\lambda t}\, dt = [-e^{-\lambda t}]_{t_1}^{t_2} = e^{-\lambda t_1} - e^{-\lambda t_2}$$

The decay probability between $t_1 = 0$ and $t_2 = t$, is

$$\mathcal{P}(1)_{0,t} = 1 - e^{-\lambda t}$$

and between $t_1 = t$ and $t_2 = \infty$,

$$\mathcal{P}(1)_{t,\infty} = e^{-\lambda t}$$

The probability that the atom will decay precisely at time t is the product of the probabilities that it will not decay between $t = 0$ and t, nor between t and $t = \infty$:

$$\mathcal{P}(1)_t = \mathcal{P}(0)_{0,t}\, \mathcal{P}(0)_{t,\infty} = [1 - (1 - e^{-\lambda t})] \quad [1 - e^{-\lambda t}]$$
$$= (1 - e^{-\lambda t})\, e^{-\lambda t}$$

This probability is maximal when

$$\frac{d\mathcal{P}(1)_t}{dt} = \lambda e^{-\lambda t}\,(2e^{-\lambda t} - 1) = 0$$

which occurs at a time

$$t_{max} = \frac{\ln 2}{\lambda} = T$$

The probability that the atom will decay according to a given mode has a maximum value

$$\mathcal{P}(1)_t = (1 - e^{-0.7})\, e^{-0.7} = 0.25$$

at a time equal to the half-life of this mode for the statistical ensemble to which it would pertain.

II. DISINTEGRATION OF SEVERAL RADIOACTIVE ATOMS

It is necessary to distinguish the number N_i of disintegrating atoms observed with a given detector from the total number N of decaying atoms when two (or more) decay modes are available. N_i depends on, say, λ_1, and N on $\lambda_1 + \lambda_2$. In the following the discussion is simplified by assuming that only one decay mode is open.

In Chapter 3, it was shown that among an ensemble of N atoms, the number N_i which will disintegrate during the interval Δt at time t is given by the binomial law

$$P(N_i) = \frac{N!}{N_i!(N - N_i)!} (\lambda \Delta t)^{N_i} (1 - \lambda \Delta t)^{N - N_i}$$

The mean value of the decaying atoms is

$$\overline{N_i} = N \lambda \Delta t$$

and the variance is

$$\sigma^2 = \overline{N_i} (1 - \lambda \Delta t)$$

In the present context, this is the only rigorous probability law which can be used, since the conditions for which the Poisson distribution applies, i.e., $N > 100$ and $\overline{N_i} << N$ are not necessarily fulfilled.

The zero-time may not coincide with the instant at which the atoms are created, but may be arbitrary, corresponding to the beginning of the observation. The N_i are integral and discontinuous values and the time interval Δt can and will be replaced in the following by the differential dt.

A theoretical histogram of the decay of an initial population of N_0 atoms can be obtained for any arbitrary value of the product λ dt. With the binomial law, P(Ni) is meaningful only if $0 < \lambda\ dt < 1$. At the limits, i.e., $\lambda\ dt = 0$ and $\lambda\ dt = 1$, the probability $P(N_i)$ is zero, whatever the value of N_i between 0 and N. The probability is maximal for a value $N_i = N_i^{max}$ such that

$$\frac{P(N_i^{max} + 1)}{P(N_i^{max})} \leq 1$$

or

$$N_i^{max} \geq N\lambda dt - 1 = \overline{N_i} - 1$$

The maximum value of N_i is close to N for λ dt $\approx$ 1 and close to zero for small values of λ dt. The mean number of atoms which will decay during dt is N for λ dt = 1 and nil for λ dt = 0.

The probabilistic decay histogram which expresses the theoretical number of surviving atoms at time t after n_i time intervals dt, $N_{th}(n_i)$, can be established in several ways. The theoretical number $[N_i]_{th}$ of atoms which decays during successive time intervals equal to dt may be taken as: (1) the next integer above or below $\overline{N}i = N \lambda\ dt$, or as, (2) the value N_i^{max} which can be calculated from the $P(N_i)$ for each value of $N = N_0, N_0 - 1, N_0 - 2$ . . .

The shapes of the histograms

$$N_{th}(n_i) = N_0 - \sum_{i=0}^{i=t/dt} [N_i]_{th}$$

as a function of the parameter

$$t = \sum_{i=0}^{i=t/dt} n_i dt$$

are strongly dependent on λ dt. An arbitrary choice of dt may lead to odd theoretical disintegration events. On the other hand, when choosing $[N_i]_{th} = N_i^{max}$, predictions may be ambiguous since two values of this parameter yield the same $P(N_i^{max})$.

The shape of the histogram that is closest to an exponential decay for given λ and N_0 corresponds to a specific choice of the parameters dt and $[N_i]_{th}$. Theoretical histograms for $N_0 = 20$ are represented in Figure 9.1. For $N_0 > 20$, they fit well with the nearest integral values given by the classical exponential decay law.

Experimental measurement of the decay of a few radioactive atoms can be achieved in two different ways, depending on whether or not the half-life T is known.

A. The Value of the Half-Life is Known

Here, an observation time $\Delta t < T/5$ may be chosen for establishing the histogram of the number of decays during a time interval Δt at time t,

$$A(n_i) = \lambda N(n_i)\Delta t = f\left(t = \sum_{i=0}^{i=t/\Delta t} n_i \Delta t\right)$$

At this level of activity, it is necessary to take the background into account, for which a Poisson distribution usually applies, depending on the type of disintegration which is recorded. A histogram constructed from a single measurement during each consecutive time interval Δt may differ completely from the theoretical one for low values of N_i, owing to large fluctuations of $A(n_i)$. Depending on the values of A_0, the histogram may become more or less rapidly indistinguishable in relation to the background and, therefore, meaningless.

Next, the uncertainty in the measurement of a number N_i of decays will be examined. It will be recalled from Chapter 5 that $G(\sqrt{2})$ is the probability that for a unique measurement N_i lies between the limits $\overline{N_i} \pm \sigma\sqrt{2}$. In other terms, this is the probability that the mean value $\overline{N_i}$ which is *not measured*, but represents the best value which *could be measured*, lies between these limits. For a binomial distribution,

$$G(\sqrt{2}) = \sum_{N_i = \overline{N_i} - \sigma\sqrt{2}}^{N_i = \overline{N_i} + \sigma\sqrt{2}} \frac{N!}{N_i!(N - N_i)!} (\lambda\Delta t)^{N_i} (1 - \lambda\Delta t)^{N - N_i}$$

This function must be calculated step by step for the different values of N and σ. As an example, for $N = 20$ and $\lambda = 0.25\ s^{-1}$, 5 decays should be registered during a time $\Delta t = 1$ s (see Figure 9.1). $N_i = 5$ corresponds to $\sigma = 2.2$, $\sigma\sqrt{2} \approx 2$ and $G(\sqrt{2}) = 0.934$. The chance is 93% that the experimental N_i^{exp} lies between 2 and 8. The corresponding figures for $N = 3$ and $N_i = 1$ are $\sigma = 1.5$ and $G(\sqrt{2}) = 0.911$. The probability that $0 < N_i^{exp} < 3$ is about the same as in the previous example.

The error on the background is calculated from the Poisson distribution,

$$G(\sqrt{2}) = \sum_{N_i = \sigma^2 - \sigma\sqrt{2}}^{N_i = \sigma^2 + \sigma\sqrt{2}} \frac{\sigma^{2N_i}}{N_i!} e^{-\sigma^2}$$

The experimental observation of extinction of a population of several atoms implies that the half-life is relatively short, or conversely, that λ is relatively high. In this case, the true mean life θ of *all* atoms can be deduced from calculation of the approximate experimental mean life θ_{exp} of *several* atoms which is

$$\theta_{exp} = \frac{\Delta t}{2} \frac{\sum_{i=1}^{n} (2n_i - 1)\Delta N_i^c}{\sum_{i=1}^{n} \Delta N_i^c}$$

ΔN_i^c is the number of disintegrations during the time interval Δt corrected for background. For $i = 1$, ΔN_1 atoms have lived for a time $\Delta N_1(\Delta t/2)$; for $i = 2$, ΔN_2 atoms have lived a time $\Delta N_2(3\Delta t/2)$, and so on. The experimental mean life is set equal to the theoretical one

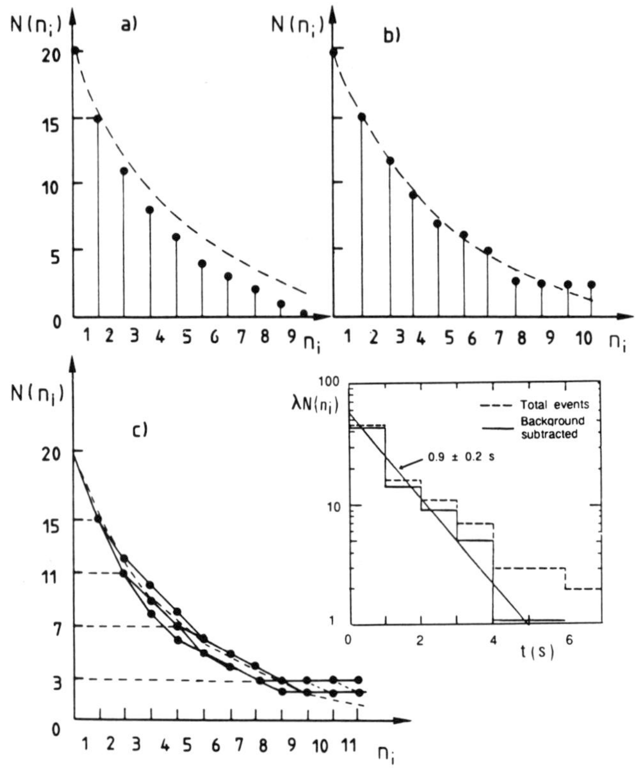

FIGURE 9.1. Extinction of an initial population of $N_0 = 20$ radioactive atoms with half-life $T = 2.8$ s ($\lambda = 0.25\ s^{-1}$). Histograms are based on the theoretical number $[N_i]_{th}$ of atoms which disappear at each n_i step during an observation time dt = 1 s. The dashed line represents the exponential decay law. (a) $[N_i]_{th}$ is the integer above $\overline{N}_i = \lambda N(n_i)dt$ and $N(n_i)$ is the value of N after a time idt. The limit of $N(n_i)$ is zero. (b) $[N_i]_{th}$ is the integer below $\overline{N}_i$; the limit of $N(n_i)$ is 3. (c) $[N_i]_{th} = N_i^{max}$, maximal value of $N(n_i)$. The limit of $N(n_i)$ lies between 3 and 2. $N(n_i)$ values to which correspond two N_i^{max} are indicated, e.g., for N = 15, P(3) = P(4) = 0.225 and for N = 11, P(2) = P(3) = 0.258. Insert: Alpha decay curve of ^{263}Unh (E_α from 9.02 to 9.229 MeV). (Adapted from Ghiorso, A., et al., *Transplutonium 1975*, Müller, W. and Lindner, R., Eds., North-Holland, Amsterdam, 1976, 323.) (d) $P(N_i)$ values for $0 \le N(n_i) \le N$ and $N \le 20$. Values of N are indicated on the curve.

$$\theta = \frac{\int_0^{n\Delta t} t\,dN}{\int_0^{n\Delta t} dN}$$

Together with $N = N_0\,e^{-\lambda t}$, this becomes

$$\theta_{exp} = \theta\left[1 - \frac{n\Delta t/\theta}{(e^{n\Delta t/\theta} - 1)}\right]$$

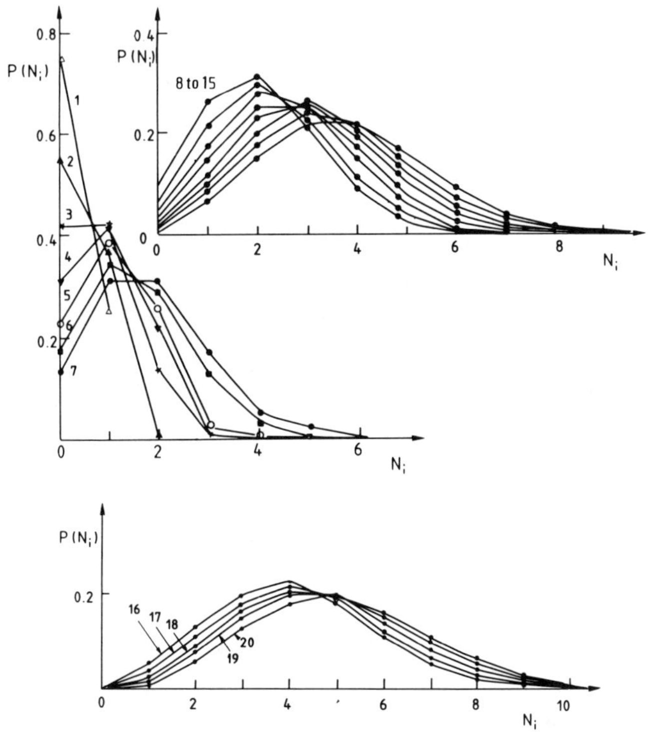

FIGURE 9.1c (continued).

The error on the mean life θ depends on the observation time and on the background.

B. The Half-Life is Unknown

If the half-life T is not known, the disintegrations must be recorded one by one and the time elapsed between two successive events should be considered carefully. This is a special aspect of radioactivity that becomes important if the succession of decays is not too fast and the disintegrations not too numerous; under these conditions the decays can be registered chronologically.

The differential probability $dP(t_s = dt)$ that the time interval t_s between two successive disintegrations is equal to dt is

$$P(t_s = dt) = P(0)_{0,t}\, N\, dP(1)_{t,t+dt}$$

The first term represents the probability that no atom will disintegrate in the time interval between 0 and t; the second term corresponds to the probability that one of the N atoms present at time t will decay between t and t + dt. In this expression dt is taken small enough with respect to the counting time Δt considered before, in order that time t will be meaningful.

The first term is

$$P(0)_{0,t} = \sum_{n=1}^{n} P(0)_{t,t+dt}$$

and from the binomial law,

$$P(0)_{t,t+dt} = (1 - \lambda dt)^n$$

In Appendix 9.1 it is shown that

$$P(0)_{0,t} = (1 - \lambda dt)^{Nn} = \left[\left(1 - \lambda \frac{t}{n}\right)^n\right]^N = [e^{-\lambda t}]^N = e^{-\lambda Nt}$$

On the other hand,

$$dP(1)_{t,t+dt} = \lambda dt$$

and finally,

$$dP(t_s = dt) = \lambda N e^{-\lambda Nt} dt$$

This result is similar to that which would result if the distribution of the disintegrations were to follow a Poisson law.

Two different situations may be considered:

(α) *In the first instance*, the observation time Δt is such that

$$\overline{N_i} = \lambda N \Delta t \ll N$$

so that λ N remains constant. One expresses the probability that the time interval between two successive disintegrations, among an average of λ N Δt decays occurring during Δt, is larger than t ($0 < t < \Delta t$). This probability is the distribution law of t_s:

$$P(t_s > t) = \int_t^{\infty} dP(t_s = dt) = e^{-\lambda Nt}$$

Similarly,

$$P(t_s < t) = \int_t^{\infty} dP(t_s = dt) = 1 - e^{-\lambda Nt}$$

The probability that t_s is precisely equal to t is the product of the probability $1 - P(t_s > t)$ that t_s is smaller than t and of the probability $1 - P(t_s < t)$ that t_s is greater than t. Hence,

$$P(t_s = t) = (1 - e^{-\lambda N t})\, e^{-\lambda N t}$$

This probability is maximal for

$$\frac{dP\,(t_s = t)}{dt} = \lambda N e^{-\lambda N t}\,(2e^{-\lambda N t} - 1) = 0$$

which leads to

$$t = t_{max} = \frac{\ln 2}{\lambda N} = \frac{T}{N}$$

The value of the half-life T can be estimated from the frequency of the time intervals between two successive disintegrations during a chosen time Δt or multiples thereof, such that the activity remains practically constant. For a single interval, statistics are poor and all disintegrations must be measured. Statistics improve with the number of intervals, and it is no longer necessary to measure the totality of decays. Each case requires separate examination. An example is given in Figure 9.2A.

The mean value of t_s is, as expected,

$$\langle t_s \rangle = \int_0^\infty t\, dP(t_s = dt) = \frac{1}{\lambda N}$$

When λN is no longer constant during the time interval Δt, the establishment of $P(t_s = t)$ becomes rather complicated, since N is time-dependent in the expression $dP\,(t_s = dt)$. In general, the relation $N = N_0 e^{-\lambda t}$ may be used.

This is a very interesting case in the present context, but the treatment is difficult because the integration of $dP\,(t_s = dt)$ requires the series expansion of the exponential term (see Appendix 9.2). It can be shown that when the number of atoms is very small, a limiting situation appears from which the value of the half-life can be derived. This case is similar to that which is discussed below.

(β) *The extreme situation* is that for a unique atom, $N = 1$, with Δt infinite. This implies that the atoms are produced one by one and that the time interval between the appearance of two atoms is much greater than the recording time of the disintegrations. It is assumed here that the time elapsed between the birth and death of each atom is known. This condition is satisfied if the nuclear event which produces the atom can, in fact, be recorded. In a radioactive filiation, appearance of the daughter atom is signaled by the instant of decay of the parent; in a nuclear reaction, the birth of the atom is recorded from observation of the particle or photon emitted by the compound nucleus.

The differential probability that the time t_s between two disintegrations is equal to dt is now

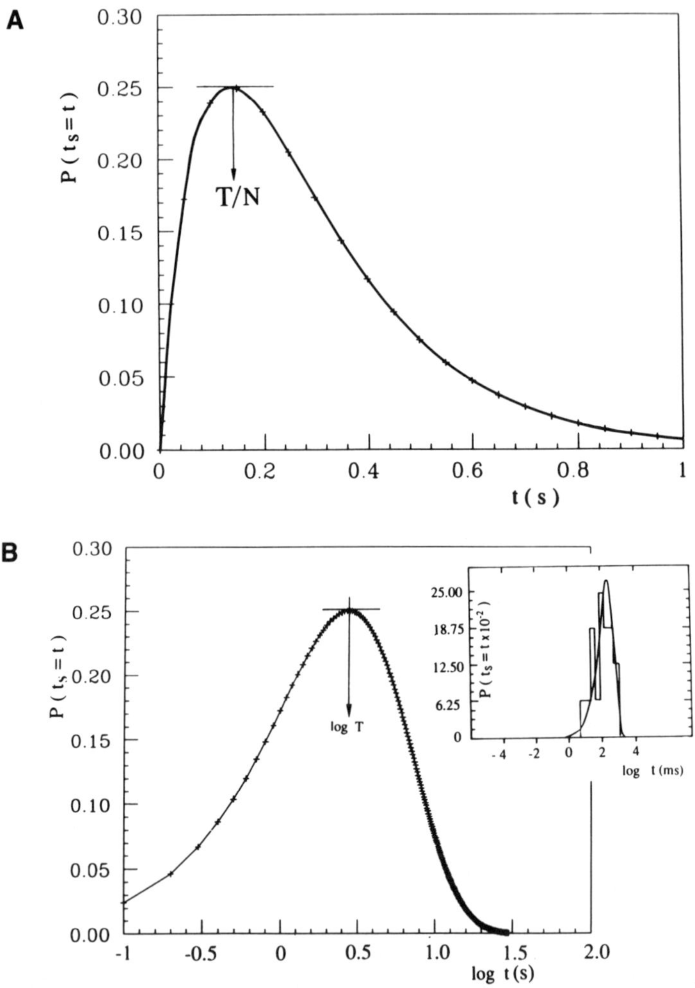

FIGURE 9.2. Disintegration of several atoms and half-life unknown. (A) Probability curve for the condition that the time interval between two successive disintegrations be equal to t ($0 \leq t \leq 1$ s). The curve is drawn for $N_0 = 20$ atoms, $\lambda = 0.25$ s,$^{-1}$ observation time $\Delta t = 1$ s. The maximum is at T/N = 0.14 s. This case corresponds to the limit of applicability of $P(t_s = t) = (1 - e^{-\lambda N t})e$ since N decreases from 20 to 15 within Δt = 1 s. (B) Theoretical distribution of the frequency of time intervals between the disintegration of a parent nuclide A and its daughter B as a function of log t. ($T_A << T_B = 2.8$ s). Insert: Frequency of time intervals in the disintegration of ^{243}Fm (T = 0.18s) for 16 registered α decays. (Adapted from Armbruster, P., *Annu. Rev. Nucl. Part. Sci.*, 35, 113, 1985.) (C) Variation of dN/dln t as a function of ln t for the total disintegration of N_0 = 20 atoms, $\lambda = 0.25$ s^{-1}. Insert: Corresponding data for the decay of ^{246}Cf (T = 35.7 h). Time in abscissa is normalized vs. the half-life T; hence, the maximum corresponds to log 1.44–0.16. (Adapted from Oganessian, Y. T. et al., *Radiochim. Acta*, 37, 113, 1984.)

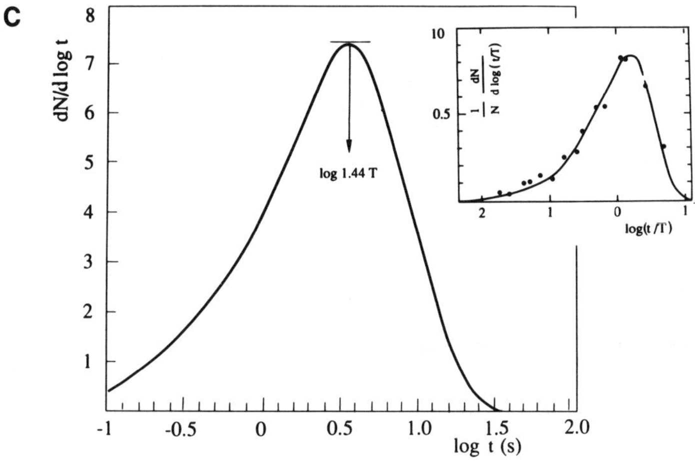

FIGURE 9.2 (continued).

$$dP(t_s = dt) = \lambda e^{-\lambda t}$$

from which

$$P(t_s = t) = (1 - e^{-\lambda t})\, e^{-\lambda t}$$

In practice, data acquisition and treatment over long periods of time are easier to handle if ln t rather than t itself is taken as a variable. Obviously, this case is very different from the previous one in which decays are registered over a short time Δt.

Setting $\ln t = u$,

$$P(t_s = t) = (1 - e^{-\lambda e^u})\, e^{-\lambda e^u}$$

The maximum of this distribution corresponds to the value $u = u_{max}$, for which

$$\frac{dP(t_s = t)}{du} = \lambda e^{-\lambda e^u} [2e^{-\lambda e^u} - 1] = 0$$

i.e.,

$$u_{max} = \ln T$$

Figure 9.2B shows the theoretical case of the disappearance of several atoms of ^{243}Fm (0.18 s) produced in the reaction $^{206}Pb(^{40}Ar, 3n)$. The implantation of the recoil atoms in a surface barrier detector which records the α particle of ^{243}Fm is an indication of the birth of the Fm nuclides.[3]

In the same line, replacement of t by ln t = u transforms

$$\frac{dN}{dt} = N_0 \lambda e^{-\lambda t}$$

to

$$\frac{dN}{du} = N_0 \lambda e^{ue^{-\lambda e^u}}$$

The maximum value u = u_{max} of this distribution is obtained for

$$\frac{d}{du}\left(\frac{dN}{du}\right) = e^u e^{-\lambda e^u} (1 - \lambda e^u) = 0$$

i.e.,

$$u_{max} = \ln \theta$$

The calculated distribution dN/dln t as a function of log t is represented in Figure 9.2C. It can be compared to the experimental sets of data corresponding to the extinction of several ^{246}Cf atoms (T = 35.7 h) formed rapidly from decay of ^{258}Unp (4.2 s) and ^{262}Uns (10^{-3} s to 0.1 s).[4] The half-lives of the nuclides involved in the filiation are convenient for accumulation of ^{246}Cf. This case is quite different from the previous one, since the knowledge of the birth and decay times of each atom is no longer required.

III. ACTIVITY MEASUREMENTS

Depending on the case, the activity can be measured in different ways. When the half-life and decay modes of a set of radionuclides are known, the observation time Δt or dt can be chosen conveniently.

The number N of atoms required for observing an activity of one disintegration per minute (1.66×10^{-2} Bq) is related to the half-life T (s) by

$$T = 42\ N$$

from which the detection limit of populations of several radioactive atoms can be estimated.

When half-life and decay modes are not known, successive events must be recorded as a function of time using different types of detectors. This is particularly necessary when new nuclides are produced in nuclear reactions. All accessible information is contained in these measurements. The plot of the frequency of time intervals between two successive disintegrations as a function of ln t provides a convenient means of determining the half-life.

For very small values of N and λ, radioactivity measurements are not feasible. Even the sensitivity of powerful methods such as cyclotron mass spectrometry or resonance ion spectroscopy followed by time-of-flight mass spectrometry is too low by many orders of magnitude.[6,7]

PART TWO — CHEMICAL BEHAVIOR OF A HUNDRED ATOMS OR LESS

I. PROBLEMS ENCOUNTERED

When the number of atoms becomes very small, the usual fundamental macroscopic and microscopic thermodynamic descriptions no longer hold.[1]

The phase concept disappears when the medium contains only one or several radioactive atoms, since the distribution of $E_i(m_j)$ entities becomes less and less homogeneous with the decrease in the number N_{m_j} and consequently in the number $N_{E_i(m_j)}$.

The instantaneous distribution coefficient of a radionuclide M between two phases also loses its significance: at the extreme limit of one atom, the latter can exist only in one of the two phases at the moment of separation. In this case, average distribution coefficients can be determined by performing a large number of identical experiments. The number of decay events observed in each phase is recorded. The decay statistics are replaced by the statistical distribution of experimental results.

In the microscopic description of the system, the number of $N_{E_i(m_j)}$ entities can no longer be treated as a continuous variable. This restriction has two far-reaching consequences: (1) the use of the Stirling expression is excluded for calculation of factorials in the partition function $Z[T,V,N_{E_i(m_j)}]$ of the $E_i(m_j)$ entities and (2) the free energy (Gibbs function) of the $E_i(m_j)$ species can no longer be obtained from the differentiation of the logarithm of the partition function with respect to $N_{E_i(m_j)}$. This is a very important point; the equilibrium state of $E_i(m_j)$ with the other entities in the medium cannot be described in the usual way by using the chemical potential of these $E_i(m_j)$ entities.

Two fundamental questions arise: (1) how should the behavior of matter be described in the state of extreme dilution? and (2) are the predictions deduced from application of the classical law of mass action to the sub-trace level still valid?

An answer is not straightforward and requires more far-reaching considerations. First, mutual reactions between microcomponent entities $E_i(m_j)$ and reactions between micro- and macrocomponent entities $E_i(M_j)$ must be distinguished. The $E_i(m_j)$ entities are complexes C_l^z of M_j present at trace and sub-trace levels. The $E_i(M_j)$ entities are C_l^z complexes of the M_j present in ponderable amounts or, alternatively, $E_i(L_j)$ entities. The first category includes reactions in which all stoichiometric coefficients of a microcomponent are equal to unity:

$$\sum_{i=0}^{i} E_i(m) + \sum_{i=0,j=0}^{i,j} \nu_i(M_j)\, E_i(M_j) = 0$$

The second category encompasses all other types of reaction:

$$\sum_{i=0,j=0}^{i,j} v_i(m_j)E_i(m_j) + \sum_{i=0,j=0}^{i,j} \nu_i(M_j)\, E_i(M_j) = 0$$

The equilibrium constants of these reactions are denoted K_0; the coefficients ν_i are positive for the reagents and negative for the products. The reactions belong to the classes 2A, 2B, 3A, and 3B defined in Chapter 6. For each reaction, the classical law of mass action reads

$$\prod [c_i]^{\nu_i} = K_0$$

Terms in $[c_i]$ are equilibrium concentrations of the entities of micro- and macrocomponents.

In the following, only thermodynamic aspects of the reactions will be treated. Kinetics at the trace scale have been considered in Chapter 7. It is recalled that the reaction rate depends on the volume accessible to the entities, and that it increases when the volume decreases. In order to circumvent the kinetic aspects, it is assumed that the reaction volume accessible to the $E_i(m_j)$ entities is sufficiently small, and the time allotted for the reaction is long enough to permit establishment of equilibrium.

II. STRATEGY FOR A SOLUTION

Extensive and intensive parameters used to describe the composition of a phase of volume V include the concentration $C_{E(M)}$ of an entity E(M) given by

$$C_{E(M)} = \frac{\overline{N_{E(M)}}}{V} = \frac{\Sigma\, PN_{E(M)}}{V}$$

$\overline{N}_{E(M)}$ is the average value of the E(M) entities per unit volume. P is the canonic probability for realizing the states of the closed system represented by the phase. This probability is defined in terms of the partition functions of all entities contained in the phase. The summation is performed over all possible values of $N_{E(M)}$. Under ordinary conditions, $N_{E(M)}$ is usually of the order of magnitude of the Avogadro number; accordingly, the mean value is always equal to the exact one and fluctuations around the mean value vanish. From

$$\overline{N_{E(M)}} = N_{E(M)}$$

the concentration can be written in the usual form

$$C_{E(M)} = \frac{N_{E(M)}}{V}$$

In consequence, the law of mass action is *independent* of the number of reaction entities.

On the other hand, with finite values of $N_{E(m)}$ characteristic of the subtracer level, the concentration *must* be expressed in terms of the average value of the number of entities. It is calculated from the expression of P for the allowed integral values of $N_{E(m)}$ which, of course, cannot exceed the number N_m of radioactive atoms. A more rigorous demonstration of this statement is given in Appendix 9.3.

To answer the second question, the expression of the law of mass action computed with microcomponents as a function of the average concentrations, is compared to that obtained in the classical way for macrocomponents, i.e., one compares

$$\Pi\ \overline{[c_i]}^{\nu_i} = \overline{K_0} \qquad \text{and} \qquad \Pi\ [c_i]^{\nu_i} = K_0$$

For this purpose, it is convenient to introduce the parameter

$$\rho = \frac{\overline{K_0}}{K_0}$$

Obviously, the limiting value of ρ when the number of all $N_{Ei(mj)}$ entities increases well beyond a few hundred is equal to unity.

Hence, the problem is treated in two steps: (1) calculation of the canonic probability for realizing all states of the system and of the mean values of the $N_{E_i(m_j)}$ entities; and (2) check for possible discrepancies between the requirements of the classical and the newly derived law of mass action.

It may be noted that throughout the reasoning the chemical potential is not used because it is no longer a necessary concept.

In the following, only closed systems with a limited number of entities will be considered.

III. WORKING OUT THE CALCULATION

The canonic probability for preparing the states of the system is given by the expression (see Appendix 9.4)

$$P_N = \frac{Z_0}{Z}\,Z(TVN)$$

The terms involved are defined as follows: Z is the partition function of all entities of the micro- and macrocomponents and represents a normalization factor. Z_0 is the partition function of the $E_i(M_j)$ and $E_i(m_j)$ entities whose number greatly exceeds a few hundred. This term is not pertinent to sub-tracer scale behavior. Z(TVN) is the partition function of the entities whose numbers $N_{E_i(m_j)}$ are below a few hundred. Its explicit form is

$$Z(TVN) = \prod_{ij} \left[\frac{z_{E_i(m_j)}V}{\Gamma_{E_i(m_j)}}\right]^{N_{E_i(m_j)}} \frac{1}{N_{E_i(m_j)}!}$$

In this last expression, $z_{E(m)}\,V = Z\,(T\ V\ 1)$ is the partition function of an entity E(m) in an ideal system of volume V. For an ideal gaseous system, it takes the familiar form

$$Z_{E(m)} = \frac{(2\pi kT)^{3/2}}{h^3} Z_{rot.}\, Z_{vib.}\, Z_{el.}$$

$\Gamma_{E(m)}$ is a constant characteristic of E(m) in the system considered which depends only on the macrocomponents.

Depending on values i and j and on the initial numbers of the entities $E_{i(mj)}$, the expression of Z (TVN) may be more or less complicated. Once the P_N have been obtained, the mean values $\overline{N_{E_i(m_j)}}$ are derived for each case and the value of ρ is calculated as a function of the initial values $N_{E_i(m_j)}(0)$ and K_0.

A further assumption is made with respect to the half-lives of the radionuclides. It is supposed that they are long enough to permit any nuclide to occupy all quantum states corresponding to the different degrees of freedom of each $E_i(m)$. Since the residence time in these states is very short, the condition is usually satisfied for the radionuclides suitable for radiochemical investigations. Further, it is considered that the $E_i(m_j)$ entities occupy all quantum states of the system, so that a dynamic equilibrium is established.

Several cases must be distinguished, depending on the stoichiometry of the reactions involving micro- and macrocomponents. Formulation of the reactions can be simplified by noticing that only the $E_i(m_j)$ entities determine specific behavior at the sub-tracer level. Hence, the reactions are formally written

$$\sum_{i=0}^{i} E_i(m_j) = 0$$

and

$$\sum_{i=0,\ j=0}^{i,j} v_i(m_j)E_i(m_j) = 0$$

The equilibrium constants K of the formal reactions are related to the constants K_0 and to the total concentration of all other entities which are obviously equal to their equilibrium concentrations. It follows that

$$\rho = \frac{\overline{K_0}}{K_0} = \frac{\overline{K}}{K}$$

IV. APPLICATION TO SPECIFIC CASES

The very general conclusions derived from the previous treatment are summarized (see Appendix 9.5):

1. For reactions with stoichiometry 1/1 denoted

$$E_1 + E_2 = 0$$

   and involving only one microcomponent (j = 1), the ratio ρ is always equal to unity for any number N_{E_1} of entities E_1 and for any value of K. The law of mass action applies without restriction.
2. For reactions with stoichiometry ν_1/ν_2 of type

$$v_1E_1 + v_2E_2 = 0$$

   the ratio ρ depends on the initial number $N_{E_1}(0)$ of the entities E_1 and on the equilibrium constant K. No general solution exists for the variation of ρ with $N_{E_1}(0)$ and each case requires particular treatment. The same conclusion holds for more complex cases.

The most common examples are those in which several atoms of a unique radionuclide M react with macrocomponents. Since the stoichiometric coefficients of the M entities are all equal to one, the thermodynamic behavior at the sub-tracer level is the same (or should be) as at the tracer level. Strictly speaking, however, comparative data are lacking, because experiments are performed here with a few atoms or at concentrations that correspond to the tracer level.

At present, examples in which the behavior of radionuclides depends on the number of atoms are still theoretical. They will be illustrated by two types of reactions:

1. The dimerization of an M(N) entity

$$2M(N) \rightleftarrows M(N)_2$$

   formally written

$$E_1 + E_1 \rightleftarrows E_2 \qquad (1)$$

To begin, the increase of the number N_{E_i} of entities E_i (i = 1,2) will be investigated in order to show how equilibrium would be attained as a function of initial conditions, either in the forward or backward direction. For the sake of simplicity, the radioactive decay of M during the reaction will be ignored. Following this case, the effect of the diminution of the number of the M(N) atoms will be examined, starting from the reaction at equilibrium.

2. The simple disproportionation of M(N)

$$2M(N) \rightleftarrows M(N+1) + M(N-1)$$

represented by

$$2E_1 \rightleftarrows E_2 + E_3 \quad (2)$$

In this case it will also be considered in what manner an increase in the number N_{E_i} of E_i entities (i = 1,2,3) influences the equilibrium. Similarly, it will be examined how the equilibrium, once established, would vary with a decrease in the number of atoms M.

In both examples it is assumed that the values of N_{E_i} change unit by unit and may take all possible values. Restrictions imposed by stoichiometry are evidently taken into account.

A. The Number of Atoms Increases

A dimerization reaction can be easily observed provided log K is comprised between the limits $-1.60 < \log K < 2.25$, which correspond, respectively, to 5 and 95% of M present as a monomer at equilibrium.

Figure 9.3A represents the variation of $\rho = \overline{K}/K$ in the case of Reaction 1 as a function of the number of entities E_1, for the two values $K = 10^2$ and $K = 10^{-1}$; the initial number of E_2 entities is $N_{E_2}(0) = 0$. Curves for intermediate values of K are located between those drawn.

It is seen that for $K < 1$, dimerization of a few M entities would be less advanced than under ordinary conditions. The variation is more complicated for $K > 1$: the extent of dimerization of a few M entities may be *more* pronounced or *less* pronounced with respect to the behavior of a large number (> over a few hundred) of entities. The progress of the reaction depends on the values $N_{E_i} = N_M$, which may be even (2 modulo 2) or odd (2 modulo 3). In a case where N_M and consequently N_{E_1} fluctuate within the limits of a few and 100 units, the behavior appears chaotic. With increasing K, the values of N_{E_1} for which $\rho = 1$ become higher, e.g., $N \approx 800$ for $K = 10.^2$

Variations of ρ for Reaction 1 are shown in Figure 9.3B as a function of $N_{E_2} = 2\,N_M$ for K values 10 and 10^{-2}, and an initial number $N_{E_1}(0) = 0$.

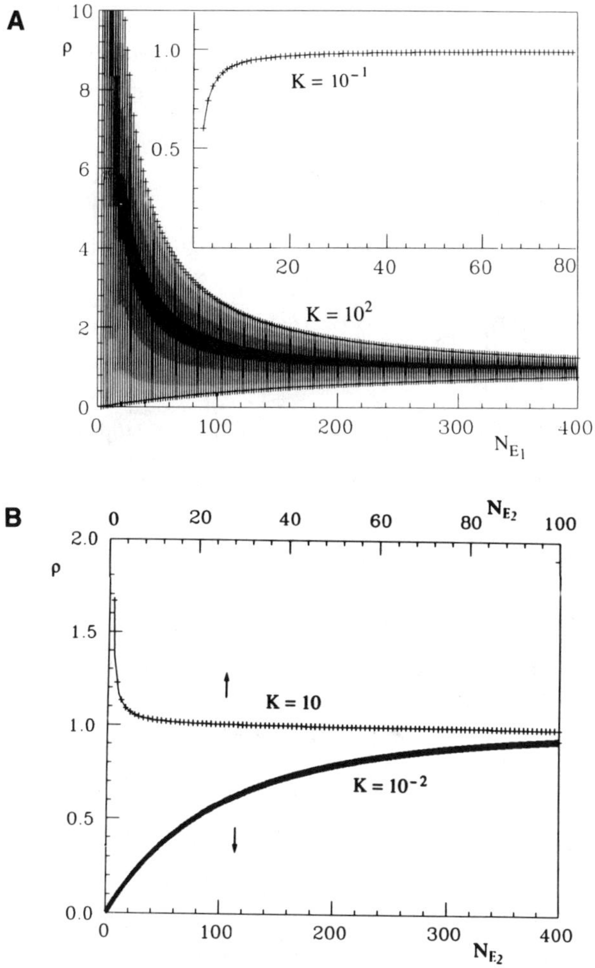

FIGURE 9.3. (A) Variation of ρ for a reaction of type $2E_1 \leftrightarrows E_2$ as a function of $N = N_{E_1} \equiv N_M = 2, 3, 4, \ldots$ with the initial condition $N_{E_2}(0) = 0$. K values are indicated on the curves. Values of ρ are computed for N = 2 modulo 2 (upper branch) and N = 3 modulo 2 (lower branch). For $K = 10^2$, the first values of the couples N/ρ are: 2/100, 3/0.009, 4/50, 5/0.019, 6/33, 7/0.029, 8/25, 9/0.038, 10/20, 11/0048, 12/17, 13/0.06, 14/15, 15/0.066, 16/13, 17/0.075, 18/11, 19/0.08, 20/10. (B) Variation of ρ for the reverse reaction $E_2 \leftrightarrows 2E_1$. K values are indicated on the curve.

It is clearly apparent that the behavior of dimer populations with less than a few tens of units differs from the classical one. The value ρ = 1 is not yet attained for $N_{E_2} = 400$, when $K = 10^{-2}$.

A phenomenon which is close to the present case is the adsorption, assumed reversible, of an entity E_1 on some colloidal form E_2 of an element M*, denoted M*(Col) resulting in a pseudocolloid M*M(Col) which is the entity E_3. The reaction is

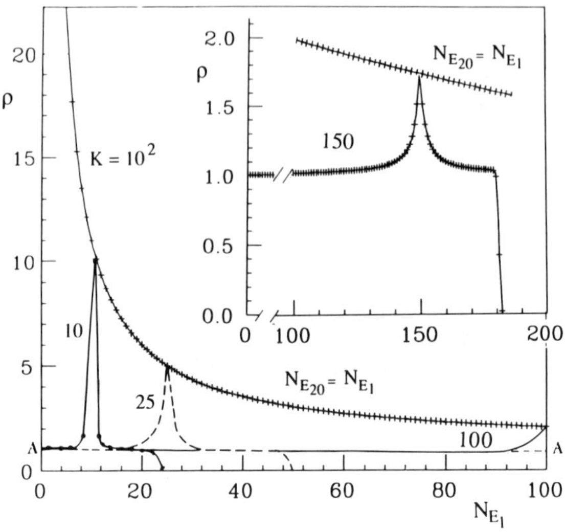

FIGURE 9.4. Variation of ρ for a reaction of type $E_1 + E_2 \leftrightarrows E_3$, $K = 10^2$, as a function of $N = N_{E_1}$ for different initial values of $N_{E_2}(0)$ indicated on the curves and $N_{E_3}(0) = 0$ $N_{E_2}(0) = N_{E_1}$ (reaction stoichiometry) and $N_{E_2}(0) = 10, 25, 100, 150$. The first values of the couples N/ρ for $N_{E_2}(0) = N_{E_1}$ and $K = 10^2$ are 1/75, 2/51, 3/35, 4/26, 5/21. For $K = 10^{-2}$, $\rho \approx 1$ whatever N_{E_1} and $N_{E_2}(0)$. All values lie on line A.

$$M + M^*(Col) \rightleftarrows MM^*(Col)$$

or equivalently,

$$E_1 + E_2 \rightleftarrows E_3$$

which is formally the same as Reaction 1.

To describe the behavior of the microcomponent, situations must be considered in which the initial number $N_{E_2}(0)$ of colloids is larger than the initial number of M entities. Figure 9.4 shows that with the increase of $N_{E_2}(0)$, ρ rapidly reaches the limit of one and takes the stoichiometric value for $N_{E_1} = N_{E_2}(0)$.

It is interesting to look at the sensitivity of the behavior of M with respect to the initial number of reacting entities.

A simple disproportionation reaction can be measured when log K lies within the limits $-3.16 < \log K < 1.95$, corresponding to the two extreme measurable situations: $[M(N)] = 5 \times 10^{-2}\ C_M$ and $[M(N + 1)] = [M(N - 1)] = 0.45\ C_M$.

Figure 9.5A shows the variation of ρ for the forward Reaction 2 as a function of the number of N_{E_1} entities, for selected values of K and under simple initial conditions. When $K << 1$, and N_{E_1} is less than about 100 entities, the disproportionation of M(N) is expected to be *less* advanced than

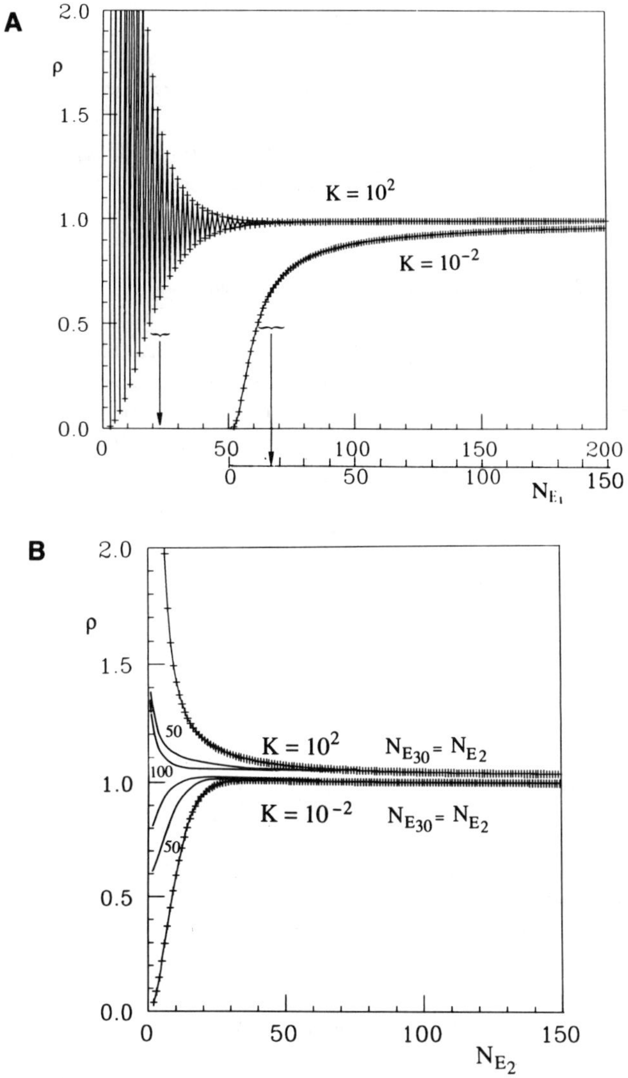

FIGURE 9.5. (A) Variation of ρ for a reaction of type $2E_1 \leftrightarrows E_2 + E_3$ as a function of $N = N_{E_1}$ for $N_{E_2}(0) = N_{E_3}(0) = 0$. Values of K are indicated on the curves. Ratios ρ are given for N = 2 modulo 2 (upper branch) and N = 3 modulo 2 (lower branch). For $K = 10^2$, the first values of the couples N/ρ are 2/100, 4/25, 6/11, 8/6.8, 10/4.5, 12/3.3, 14/2.67, 16/2.2. (B) Variation of ρ for the reverse reaction $E_2 + E_3 \leftrightarrows 2E_1$ as a function of $N = N_{E_2}$ for $N_{E_1}(0) = 0$ and different initial values of N_{E_3}: $N_{E_3}(0) = N_{E_2}$ (reaction stoichiometry), $N_{E_3}(0) = 50$ and 100. Values of K and $N_{E_3}(0)$ are indicated in the figure.

at the tracer scale or at higher concentrations. As before, for $K = 1$ and $K >> 1$, the problem is more complicated, since the reaction of a population below 100 could just as well be *more* advanced than *less* pronounced at higher concentration.

Figure 9.5B corresponds to the backward Reaction 2 for typical K values and selected initial conditions. The preceding comments apply also to this case.

B. The Number of Atoms Decreases

It is assumed that Reactions 1 and 2 are at equilibrium for a large number N_M of atoms M. With the progressive decay of M, the sub-trace situation is attained when N has decreased to a few hundred. Since the decay is slow, it is assumed that the equilibrium is restored after each individual disintegration. Accordingly, the initial numbers of reacting entities, at each step, are equal to those prevailing in the preceding equilibrium. They are related to the successive values of K and N. Thus, for Reaction 1,

$$N_{E_1} = \frac{\sqrt{1 + 8N_M} - 1}{4K} \qquad N_{E_2} = \frac{N_M - N_{E_1}}{2}$$

and for Reaction 2,

$$N_{E_1} = \frac{N_M}{1 + 2\sqrt{K}} \qquad N_{E_2} = N_{E_3} = N_{E_1}\sqrt{K}$$

The variations of ρ as a function of the stoichiometrically allowed and denumerable N_M values are shown in Figures 9.6 and 9.7. Computation of ρ is carried out according to the following scheme: starting with a given set of K and N values ($N >> 100$), a first ρ value is calculated, which yields $\overline{K_1}$. The transient equilibrium values $\overline{N_{1,E_i}}$ are calculated with the previous expressions, setting $K = \overline{K_1}$. This yields a new average value $\overline{\rho_1}$, and consequently a new value for K, i.e., $\overline{K_2}$. Next, a new set $\overline{K_2}$ and N_2 is chosen, N_2 being the integer value of the sum of the $\overline{N_{2,E_i}}$ values calculated with $\overline{K_2}$, which is immediately below N_M, and so on. It is seen that the equilibrium constant decreases from its initial value to that allowed by the smallest value of N_M compatible with the stoichiometry of the reaction.

Figure 9.6 shows that dimerization is remarkably sensitive to the decrease of N_M when $K > 1$. Sudden changes are observed in the ρ values with decreasing N_M, as a consequence of the appearance of odd or even values of N_{E_1} (see Figure 9.3A). In this range of N_M values, the system behaves chaotically.

In Figure 9.7, ρ decreases very rapidly with decreasing N_M for $K < 1$. This signifies that the extent of disproportionation of M(N) also decreases

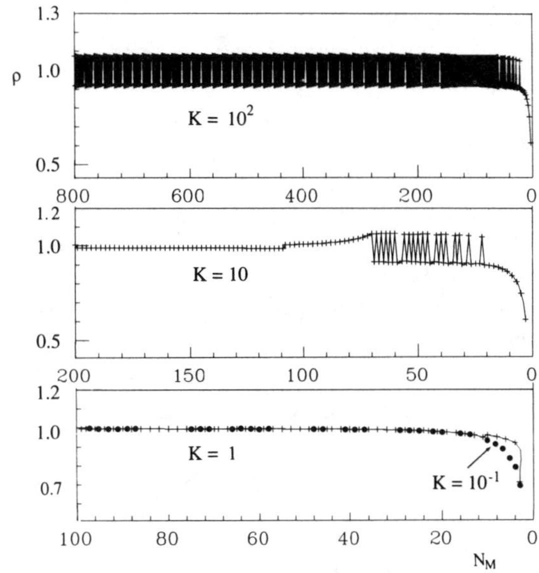

FIGURE 9.6. Variation of ρ for a reaction of type $2E_1 \leftrightarrows E_2$ as a function of N_M, starting from an equilibrium state with ρ = 1. Values of K are indicated on the curves. For K = 10^2, initial values of N_M are even (800 is chosen since for this number, ρ = 1) for the upper branch and odd (801) for the lower branch. For other K values, distinction between even and odd values disappear.

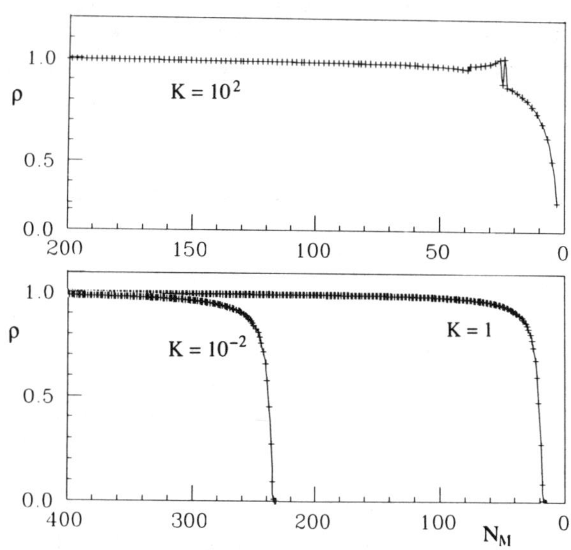

FIGURE 9.7. Variation of ρ for a reaction of type $2E_1 \leftrightarrows E_2 + E_3$ as a function of N_M, starting from an equlibrium state. K values are indicated on the curves. Initial N_M values are even or odd.

with diminution in the number of atoms M. The same situation prevails for $K > 1$, but discontinuities and chaos are again observed when N_{E_1} changes from odd to even. This should be compared with the results in Figure 9.5.

C. Fluctuations

Fluctuations of the calculated ρ values have their origin in those of the average $N_{E_i(m_j)}$ values. For a given E(m) entity they are due to two causes: (1) fluctuations ΔP in the probability law P_N, and (2) fluctuations ΔN in the decay curve of $N_{E(m)}$, which can be calculated using the binomial or Poisson distribution law. Thus,

$$\Delta\overline{(N_{E_i}(m_j)} = \Delta P + \Delta N$$

As long as ρ is close to unity, the fluctuations are practically negligible, as in the classical situation. When ρ depends on the number of reacting entities, the fluctuations may be very important, but their magnitude remains unpredictable and erratic, as is also the case for the behavior of extremely diluted matter.

V. EXPERIMENTAL APPROACH

So far, the chemical behavior of a small number of atoms predicted by theory has not been effectively confirmed by experiments, which in this instance are very difficult to perform, or even to conceive.[7]

The experimental approach is, by nature, severely limited. In the case of confined systems which are normally encountered in the laboratory, only two types of experiment are feasible, namely, partition between two phases or transport methods, as described in Chapter 8. These concern, of course, only observable reactions.

Partition experiments rely on the measurement of average concentrations of the entities present in the two phases. In the static procedure, experiments are sequential and the accuracy on the average concentrations increases with the number of trials. The main problem is the preparation of successive samples containing the same number of atoms of the microcomponent.

This difficulty is overcome in dynamic experiments such as chromatography, which essentially constitutes a practically infinite succession of static distributions. Information on the average concentration of the moving or eluted entities is obtained directly. The drawback of required reproducibility of the initial amount of the atoms is obviated.

Both types of experiment have been performed on isotopes of the heaviest transfermium elements. These isotopes are produced in nuclear reactions which satisfy the criteria of observability of the reaction (which is of type 3B). However, the number of entities cannot be increased up to a tracer level concentration and comparison of the behavior of sub-tracer and tracer amounts is precluded at the present. Conversely, no experiment has yet been designed in which the behavior of a tracer can be followed down to a few atoms.

The theoretical arguments developed only confirm the validity of the conclusions about the main chemical characteristics of the elements considered in these experiments. Transport methods have not yet been used in experiments on atoms produced one at a time, probably because they are too time consuming.

Difficulties for performing experiments with reactions governed by the law of mass action are of merely technical nature and may be solved in the future. This is no longer the case when a deviation from the classical law of mass action is sought, because one must essentially investigate processes which have been developing over a protracted period of time. In addition, the following conditions should be fulfilled:

1. Selection of an element with a suitable half-life to allow detection of a few atoms, and whose oxidation state can be identified without perturbation of the initial state.
2. Selection of a system and of reaction conditions which exhibit the largest possible deviation from the law of mass action for the largest accessible number of atoms.
3. Choice of a system in which entities are in true equilibrium, without risk of formation of colloids or other dispersed particles.
4. Finally, it is necessary to check whether the behavior of atoms at the sub-tracer level, at which no solid phase can be formed, differs from that of a macrocomponent; the best way to detect any discrepancy is to use dynamic partition methods, which avoid the necessity of maintaining the initial numbers of atoms constant. Any kind of test will probably be qualitative rather than quantitative.

Systems which fulfill all requirements are not easy to find and experimental proof of deviation from the law of mass action at extreme dilution may remain a challenge for a long time to come. Speculative sub-tracer chemistry of Pu in environmental conditions is discussed in Reference 8.

The solution may lie outside the field of radiochemistry. In space, extended systems, such as interstellar clouds, exist in which gaseous chemistry has been going on since remote times between ions, radicals, and molecules at concentrations of 100 or less entities per cubic centimeter. Some of the characteristics of such reactions are known, and it may be rewarding to compare them with those of reactions of the same species at much higher concentration.

VI. BEYOND RADIOCHEMISTRY

The manifold problems encountered by radiochemists have provided a stimulus for the theoretical treatment of questions illustrated by the examples given. However, similar situations of extremely diluted matter occur in other areas of chemistry and are governed by the particular statistics of the Z (TVN) partition function.

For instance, in the case of forbidden chemical reactions between macro-amounts of reagents, the number of product entities may not be nil, but extremely small. The chemical equilibrium required by the law of mass action could only be attained if this number increases beyond a limit which depends on the type of reaction and on the equilibrium constant K. It can be shown (see Appendix 9.6) that the law of mass action always applies to a reaction of type

$$E_1(M_1) + E_2(M_2) \rightleftarrows E_3(M_1M_2)$$

for any number N_3 of entities E_3. On the other hand, this conclusion does not hold for the reaction

$$E_1(M_1) + E_2(M_2) \rightleftarrows \nu_3E_3(M_1M_2)$$

These considerations can be extended to more complicated situations.

REFERENCES

1. **Guillaumont, R., Adloff, J. P., Peneloux, A., and Delamoye, P.,** Sub-tracer scale behaviour of radionuclides. Application to actinide chemistry, *Radiochim. Acta,* 54, 1, 1991, and references therein.
2. **Evans, R. D.,** *The Atomic Nucleus,* McGraw-Hill, New York, 1955, chap. 26.
3. **Armbruster, P.,** On the production of heavy elements by cold fusion; the elements 106 to 109, *Annu. Rev. Nucl. Part. Sci.,* 35, 113, 1985.
4. **Oganessian, Y. T., Hussonnois, M., Demin, A. G., Kharitonov, Y. P., Bruchertseifer, H., Constantinescu, O., Korotkin, Y. S., Tretyakova, S. P., Utyonkov, V. K., Shirokovsky, I. V., and Estevez, J.,** Experimental studies of the formation and radioactive decay of isotopes with Z = 104–109, *Radiochim. Acta,* 37, 113, 1984.
5. **Schmitt, K. H., Sahm, C. C., Pielenz, K., and Clerc, H. G.,** Some remarks on the error analysis in the case of poor statistics, *Z. Phys. A. Atoms Nuclei,* 316, 19, 1984.
6. **Peuser, P., Herrmann, G., Rimke, H., Sattelberger, P., Trautmann, N., Ruster, W., Ames, F., Bonn, J., Kluge, H. J., Krönert, U., and Otten, E. W.,** Trace detection of plutonium by three steps photoionization with a laser system pumped by a copper vapor laser, *Appl. Phys. B,* 38, 249, 1985.
7. **Trautmann, N., Peuser, P., Rimke, H., Sattelberger, P., Herrmann, G., Ames, F., Krönert, U., Ruster, W., Bonn, J., Kluge, H. J., and Otten, E. W.,** Laser resonant-ionization mass spectrometry of actinides, *J. Less Common Met.,* 122, 533, 1986.
8. **Guillaumont, R. and Adloff, J. P.,** Behaviour of environmental plutonium at very low concentration, *Radiochim. Acta,* 58, 53, 1992.

APPENDIX 9.1

LIMITING VALUE OF $(1 - \lambda t/n)^n$

$$\left(1 - \frac{X}{n}\right)^n = 1 - X + \frac{n(n-1)}{2!n^2}X^2 - \frac{n(n-1)(n-2)}{3!n^3}X^3 + \ldots$$

$$e^{-X} = 1 - X + \frac{X^2}{2!} - \frac{X^3}{3!}$$

The expansion coincides with the function to within 1% if

$$\frac{n(n-1)}{n^2} = 0.99 \qquad \text{and} \qquad \frac{n(n-1)(n-2)}{n^3} = 0.99$$

These conditions are satisfied for $n > 100$.

APPENDIX 9.2

INTEGRATION OF P (t_s = t) WHEN λN IS NOT CONSTANT DURING THE OBSERVATION TIME

Starting from

$$\frac{dP}{dt} = \lambda N e^{-\lambda N t} = \lambda N_0 e^{-\lambda t} e^{-\lambda N t}$$

and expanding the second exponential term,

$$\frac{dP}{dt} = \lambda N_0 e^{-\lambda t} - \lambda^2 N_0^2\, t e^{-2\lambda t} + \frac{\lambda^3 N_0^3 t^2 e^{-3\lambda t}}{2!} - \frac{\lambda^4 N_0^4 t^3 e^{-4\lambda t}}{3!} + \ldots$$

$$dP(t_s = dt) = \sum_{n=1}^{n} \frac{N_0^n \lambda^n t^{n-1} e^{-n\lambda t}}{(n-1)!}\, dt$$

$$P(t_s < t) = \int_0^t dP = \sum_{n=1}^{n} \int_0^t \frac{N_0^n}{n^n} \frac{(\lambda n)^n}{(n-1)!} t^{n-1} e^{-n\lambda t}\, dt$$

After setting λn = a, each integral in the summation is calculated according to

$$I_n = \frac{N_0^n}{n^n} \int_0^t \frac{a^n}{(n-1)!} t^{n-1} e^{-at}\, dt$$

$$= \frac{N_0^n}{n^n} \left[1 - e^{-at} - \frac{at}{1!} e^{-at} - \frac{(at)^2}{2!} e^{-at} - \ldots - \frac{(at)^{n-1}}{(n-1)!} e^{-at} \right]$$

$$I_n = \frac{N_0^n}{n^n} [1 - (P_0 + P_1 + P_2 + \ldots P_{n-1}]$$

In the latter expression, P_i is the Poisson probability of 0, 1, 2, . . . i events during time t. Hence,

$$P(t_s < t) = \sum_{n=1}^{n} \frac{N_0^n}{n^n} [1 - (P_0 + P_1 + P_2 + \ldots + P_{n-1}]$$

The normalization factor is obtained from

$$P(0 < t_s < \infty) = \sum_{n=1}^{n} \frac{N_0^n}{n^n} = 1$$

It follows that

$$P(t_s < t) = \sum_{n=1}^{n} [1 - (P_0 + P_1 + P_2 + \ldots P_{n-1})]$$

and

$$P(t_s > t) = 1 - P(t_s < t)$$

Keeping n = 1,

$$P(t_s < t) = 1 - P_0 = 1 - e^{-\lambda t} \qquad P(t_s > t) = e^{-\lambda t}$$

$$P(t_s = t) = (1 - e^{-\lambda t})e^{-\lambda t}$$

The maximum of the distribution is given for $t = T$, which is equivalent to $e^{-\lambda Nt} \approx 1$, i.e., $\lambda Nt = \lambda t N_0 e^{-\lambda t} < \epsilon$ or

$$N_0 < \frac{e^{0.7t/T}}{0.7t/T}\epsilon$$

If the decays are recorded during a time $t \approx 10$ T (a situation which is close to the total number of disintegrations), N_0 should be less than 160 ϵ. Taking $\epsilon \approx 0.1$, it appears that for a population of several atoms, the maximum of the distribution yields an approximate value of T.

REFERENCE

Evans, R. D., *The Atomic Nucleus,* McGraw-Hill, New York, 1955, 812.

APPENDIX 9.3

MEAN CONCENTRATION FOR A VERY SMALL NUMBER OF ATOMS

The concept of *mean concentration* for very small values of the number $N_{E(M)}$ of entities E(M) is very important in the present context. It can be derived on a rigorous basis from the grand canonic ensemble of statistical thermodynamics.

The grand canonic ensemble of a system S(T,V) which contains only one type of molecule E(M) includes this system and a reservoir of an infinite number of molecules E(M) at temperature T. The thermodynamic properties of S at equilibrium, such as the number of molecules $N_{E(M)}$ contained in the system, are mean values $\overline{N_{E(M)}}$. They are calculated from the grand partition function, which includes the mean partition function $\overline{Z}$ of S:

$$\overline{Z} = \frac{Z(TV1)^{\overline{N_{E(M)}}}}{\overline{N_{E(M)}}^{\overline{N_{E(M)}}}} \exp(\overline{N_{E(M)}})$$

The term Z (TV1) is the molecular partition function. The mean values are not necessarily integers. However, for very large $N_{E(M)}$, the latter can be approximately represented by an integer and it follows that

$$\frac{\exp(N_{E(M)})}{N_{E(M)}^{N_{E(M)}}} = \frac{1}{N_{E(M)}!}$$

In this equality, one accepts the Stirling approximation, which applies to integer values of $N_{E(M)}$. In this case, the partition function of S becomes that of a canonical ensemble

$$Z = \frac{Z(TV1)^{N_{E(M)}}}{N_{E(M)}!}$$

This expression no longer includes mean values. Conversely, starting from the canonic ensemble, the mean values $\overline{N_{E(M)}}$ must be used in order to describe a system which comprises only a limited number of entities.

APPENDIX 9.4

CANONIC PROBABILITY FOR REALIZING THE STATES OF A SYSTEM

A closed system of a phase of volume V at temperature T is not disturbed by the introduction of one or several E(m) species of a microcomponent. Accordingly, the properties of the system are determined only by the macrocomponents which it contains. The energy levels of a single species E(m) are given by a sum of two terms:

$$\epsilon_n = \epsilon_n^* + \epsilon_n^\infty$$

where ϵ_n^* are the energy levels of E(m) in an ideal system and ϵ_n^∞ an additional term depending only on the overall properties of the phase which, as said, remain constant. The partition function of E(m) is

$$Z(TV1) = \sum_{n=0}^{\infty} \exp(-\epsilon_n/kT) = \sum_{n=0}^{\infty} \exp(-\epsilon_n^*/kT) \sum_{n=0}^{\infty} \exp(-\epsilon_n^\infty/kT)$$

Setting

$$\sum_{n=0}^{\infty} \exp(-\epsilon_n^*/kT) = z_{E(M)}\,(T)\,V$$

and

$$\Gamma_{E(m)} = \sum_{n=0}^{\infty} \exp(\epsilon_n^\infty/kT)$$

one obtains

$$Z(TV1) = Z_{E(m)}(T)\,V/\Gamma_{E(m)} \equiv Z_E\,V/\Gamma_E$$

in which it appears explicitly that the molecular partition function depends on the volume.

The last part of this expression is used as an abbreviated form. The numerator is the partition function of a single species in an ideal system which depends on the temperature T and on the volume V. The denominator is a characteristic coefficient of E(m), constant for a given real system, and equal to unity in an ideal gaseous phase.

The partition function of the $N_{E(m)}$ species E(m) is

$$Z(TVN_E) = \left[\frac{z_E V}{\Gamma_E}\right]^{N_E} \frac{1}{N_E!}$$

and for several species i of j elements

$$Z(TVN) = \prod_{ij} \left[\frac{z_{E_i(m_j)} V}{\Gamma_{E_i(m_j)}}\right]^{N_{E_i(m_j)}} \frac{1}{N_{E_i(m_j)}!}$$

The expression of the probability of realizing the microstates can be separated into two parts. The first part is the partition function Z_0 of all macrocomponent species $E_i(M_j)$ and of the microcomponent species $E_i(m_j)$ whose number greatly exceeds 100. The number of these species is irrevelant with respect to sub-tracer chemistry. The second part refers to the partition function Z (TVN) of the microcomponents, whose number is less than 100.

The probability of realizing the microstates reduces to

$$P_N = \left(\frac{Z_0}{Z}\right) Z(TVN)$$

This probability is normalized by dividing by Z, which is the total partition function of all species present in the system.

APPENDIX 9.5

CALCULATION OF ρ RATIOS

1. Case of a Single Microcomponent and Reaction Stoichiometry $\nu_1 = \nu_2 = 1$

In the simple case of several atoms (<< 100 atoms) of a single element, the reactions are of type

$$E_1 + E_2 = 0$$

The extent of the reaction is expressed by the discrete number i of species which have reacted. To simplify the expressions, following notation is used:

$$z_{E_1} = z_1;\ z_{E_2} = z_2;\ \Gamma_{E_1} = \Gamma_1;\ \Gamma_{E_2} = \Gamma_2;\ N_{E_1} = N_1;\ N_{E_2} = N_2;$$
$$N = N_1 + N_2 = N_1(0) + N_2(0)$$

where $N_1(0)$ and $N_2(0)$ are the initial numbers (i.e., for $i = 0$) of E_1 and E_2. When i species have reacted, the number of product species E_2 is $N_2 = N_2(0) + i$, which coexist with $N_1 = N_1(0) - i$ species of E_1. The canonic probability for realizing the states is obtained from the general expression for P_N, setting $P_i = P_N$:

$$P_i = \frac{Z_0}{Z}\left(\frac{z_1 V}{\Gamma_1}\right)^{N_1(0) - i}\left(\frac{z_2 V}{\Gamma_2}\right)^{N_2(0) + i}\frac{1}{[N_1(0) - i)]!\,[N_2(0) + i]!}$$

For simplification, it is assumed that the initial number of E_2 species is zero, i.e., $N_2(0) = 0$ and $N_1(0) = N$. It follows that

$$P_i = \frac{Z_0}{Z}\left(\frac{z_1 V}{\Gamma_1}\right)^{N - i}\left(\frac{z_2 V}{\Gamma_2}\right)^{N + i}\frac{1}{(N - i)!\,i!}$$

which can be written in the alternative form

$$P_i = \frac{Z_0}{Z}\left(\frac{z_1 V}{\Gamma_1}\right)^{N}\left(\frac{z_2 \Gamma_1}{z_1 \Gamma_2}\right)^{i}\frac{1}{(N - i)!\,i!}$$

Setting

$$k = \frac{Z_0}{Z}\left(\frac{z_1 V}{\Gamma_1}\right)^{N} \qquad \chi = \frac{z_2}{z_1}\frac{\Gamma_1}{\Gamma_2} \qquad \theta_i = \frac{\chi^i}{(N - i)!\,i!}$$

the expression of P_i reduces to

$$P_i = k\theta_i$$

The parameter χ is characteristic of the system. The value of k is obtained from the normalization

$$\sum_{i=0}^{N} P_i = 1$$

from which it follows that

$$k = \frac{1}{\sum_{i=0}^{N} \theta_i} = \frac{1}{S}$$

The average populations N_1 and N_2 are given by

$$S\,\overline{N_1} = \sum_{i=0}^{N} (N - i)\,\theta_i = \frac{1}{(N-1)!} + \frac{\chi}{(N-2)!} + \frac{\chi^2}{(N-3)!2!} + \ldots + \frac{\chi^{N-1}}{(N-1)!}$$

$$S\,\overline{N_2} = \sum_{i=0}^{N} i\,\theta_i = \frac{\chi}{(N-1} + \frac{\chi^2}{(N-2)!} + \frac{\chi^3}{(N-3)!2!} + \ldots + \frac{\chi^N}{(N-1)!}$$

from which

$$\frac{\overline{N_2}}{\overline{N_1}} = \chi$$

With increasing N, the numbers N_1 and N_2 also increase; classical equilibrium is progressively reached and the most probable canonical state is that for which P_i is highest. Successive values P_i and P_{i-1} become closer until the limit

$$\frac{P_i}{P_{i-1}} = \chi\,\frac{N - i + 1}{i}$$

is equal to unity, which means that

$$\chi = \frac{i}{N - i} = \frac{N_2}{N_1}$$

or

$$\rho = \frac{\overline{N_1}/\overline{N_2}}{N_2/N_1} = \frac{\chi}{\chi} = 1$$

This is the only case for which the expression of the *classical* law of mass action holds, independently of the number of entities present, provided care is taken to express the law in terms of *average* concentrations.

When the system contains only a single atom which may form two distinct species, the probability of finding the atom in one species is

$$P_0 = \frac{1}{1 + \chi}$$

and in the other species

$$P_1 = \frac{\chi}{1 + \chi}$$

The law of mass action can still be used to describe the behavior of a single atom if average concentrations are used which are proportional to P_0 and P_1.

2. Case of a Single Microcomponent and Stoichiometry ν_1/ν_2
The formal expression of the reaction is

$$\nu E_1 + \nu E_2 = 0 \qquad \nu_1 > 0 \qquad \nu_2 < 0$$

With the notation of the preceding section,

$$N_1 = N_1(0) - i\nu_1 \qquad N_2 = N_2(0) + i\nu_2$$

the probability P_i becomes

$$P_i = \frac{Z_0}{Z}\left(\frac{z_1 V}{\Gamma_1}\right)^{N_1(0) - i\nu_1}\left(\frac{z_2 V}{\Gamma_2}\right)^{N_2(0) + i\nu_2}\frac{1}{[N_1(0) - i\nu_1]!\,[N_2(0) + i\nu_2]!}$$

Assuming initial conditions $N_2(0) = 0$, and hence, $N = N_1(0)$,

$$P_i = \frac{Z_0}{Z}\left(\frac{z_1 V}{\Gamma_1}\right)^{N}\left[\left(\frac{z_2 V}{\Gamma_2}\right)^{\nu_2}\left(\frac{\Gamma_2}{z_1 V}\right)^{\nu_1}\right]^{i}\frac{1}{(N - i\nu_1)!\,i\nu_2!}$$

Setting

$$k = \frac{Z_0}{Z}\left(\frac{z_1 V}{\Gamma_1}\right)^N \qquad \chi = \left(\frac{z_2 V}{\Gamma_2}\right)^{\nu_2}\left(\frac{\Gamma_2}{z_1 V}\right)^{\nu_1} \qquad \theta_i = \frac{\chi^i}{(N - i\nu_1)!\,i\nu_2!}$$

the probability P_i again takes the simple form

$$P_i = k\theta_i = \frac{\theta_i}{S} \qquad S = \sum_i \theta_i \qquad 0 \leq i \leq N$$

The summation is extended over all integral values of i allowed by the stoichiometry of the reaction. The average values $\overline{N_1}$ and $\overline{N_2}$ are obtained from

$$S\overline{N_1} = \sum_i (N - i\nu_1)\theta_i \qquad S\overline{N_2} = \sum_i i\nu_2\theta_i$$

as a function of N and χ, but contrary to the previous case, it is no longer possible to find a relation between $\overline{N_1}$ and $\overline{N_2}$ which is independent of N. This is a consequence of the strictly limited number of entities E_1 and E_2, whose behavior is governed by the special statistics included in the law of probability. If the classical law of mass action were respected, the ratio

$$\overline{K} = \frac{\overline{N_2}^{\nu_2}}{\overline{N_1}^{\nu_1}}$$

would be independent of N, which is not the case. A general solution cannot be found when the stoichiometric coefficients differ from unity. Specific examples must be considered for which these coefficients have fixed values.

It should be noted that, for all possible values of ν_1 and ν_2, the coefficients χ are required for calculation of $\overline{N_1}$ and $\overline{N_2}$. It will be shown, in fact, that the values of χ are equal to those of K.

A large number N (N >> 100) of the available entities also implies high values of N_1, N_2, and i. This situation is that of a classical equilibrium in which the most probable canonical state is that for which P_i is the highest. For the value of i which maximizes the probability P_i, the ratio

$$\frac{P_i}{P_{i-1}}$$

is equal to unity. Expanding this ratio, one obtains

$$\frac{P_i}{P_{i-1}} = \chi\frac{(N - i\nu_1 + \nu_1)\ldots(N - i\nu_2 + 2)(N - i\nu_1 + 1)}{i\nu_2(i\nu_2 - 1)\ldots(i\nu_2 - \nu_2 + 1)}$$

Because the number N of entities obviously exceeds the values of stoichiometric coefficients, it follows that

$$\frac{P_i}{P_i - 1} = 1 = \chi \frac{(N - i\nu_1)^{\nu_1}}{(i\nu_2)^{\nu_2}} = \frac{1}{K}$$

and

$$\chi = \frac{N_2^{\nu_2}}{N_1^{\nu_1}} = K$$

Finally, the average values $\overline{N_1}$ and $\overline{N_2}$ can be calculated as a function of N only. The same holds for the ratio

$$\rho = \frac{\overline{N}_2^{\nu_2}}{N_1^{\nu_1}} \frac{1}{K}$$

which can be calculated numerically as a function of N and K. Large values of N are conveniently treated with the aid of a computer program, but up to N = 10, simple manual calculations may be performed and are instructive. This is illustrated by the reaction

$$2E_1 \rightleftarrows E_2 + E_3$$

For the initial conditions $N_1(0) = N$ and $N_2(0) = N_3(0) = 0$, successive values of i, N_1, N_2, and N_3 are given below:

i	N_1	N_2	N_3	
0	N	0	0	
1	N-2	1	1	
:				
:				
i	N-2i	i	i	
:				
:				
i = N/2	0	N/2	N/2	N even
i < N/2	N-2i	i	i	N odd

The calculations are performed with the expressions

$$\chi = K; \qquad \theta_i = \frac{\chi^i}{(N - 2i)!\,i!\,i!;} \qquad S = \sum_i \theta_i; \qquad P_i = \frac{\theta_i}{S}$$

$$\overline{N}_1 = \sum_i (N - 2i) P_i \quad \overline{N}_2 = \overline{N}_3 = \sum_i iP_i \quad \rho = \frac{1}{K} \frac{\overline{N}_2\overline{N}_3}{\overline{N}_1^2}$$

Setting N = 5 and χ = K = 2, the following values are obtained

i	θ_i	S	Pi
0	$\frac{2^0}{5!0!0!} = 8.3 \times 10^{-3}$		0.006
1	$\frac{2}{3!1!1!} = 0.333$	1.341	0.248
2	$\frac{2^2}{1!2!2!} = 1$		0.745

with lead to the results

$$\overline{N}_1 = 5P_0 + 3P_1 + 1P_2 = 1.519$$

$$\overline{N}_2 = \overline{N}_3 = 0P_0 + 1P_1 + 2P_2 = 1.798$$

$$\rho = 0.700$$

3. Other Cases

The treatment can be extended to more complicated reactions, for instance that of a single microcomponent

$$\nu_1E_1 + \nu_2E_2 + \nu_3E_3 + \ldots = 0$$

or that of several microcomponents

$$\sum_{j=0}^{j} \nu_jE_j(m_1) + \sum_{j=0}^{j} \nu_jE_j(m_2) = 0$$

Calculations are performed in the same manner. The actual expressions of χ as a function of z_i and Γ_i depend on each specific reaction but they are not relevant because, as shown before, they are not needed in the final calculation. Extension to such cases may involve rather complicated calculations depending on the number of reacting entities and on initial values of the populations of the various species.

APPENDIX 9.6

EXAMPLE OF A FORBIDDEN REACTION

The forbidden reaction

$$E_1(M_1) + E_2(M_2) \rightleftarrows \nu_3 E_3(M_1M_2)$$

is considered in which it is assumed that only the concentration of E_3 species at equilibrium is very low. The initial number $N_3(0)$ is taken as zero. For an extent i of the reaction, the number of E_3 species is $i\nu_3$. It follows that

$$P_i = \frac{Z_0}{Z}\left(\frac{z_3 V}{\Gamma_3}\right)^{i\nu_3} \frac{1}{(i\nu_3)!}$$

Setting

$$k = \frac{Z_0}{Z} \qquad \chi = \left(\frac{z_3 V}{\Gamma_3}\right)^{\nu_3} \qquad \theta_i = \frac{\chi^i}{(i\nu_3)!}$$

the probability P_i becomes

$$P_i = k\theta_i = \frac{\theta_i}{S} \qquad S = \sum_{i=0}^{\infty} \theta_i$$

The summation is performed from $i = 0$ up to a very high limit because E_1 and E_2 are macrocomponents and i can take an infinite number of values with respect to N_3. The parameter χ is still characteristic of the macroscopic system, at least as long as N_3 is very small. The average population of N_3 is

$$S\overline{N_3} = \sum_{i=0}^{\infty} i\nu_3\theta_i$$

and can be calculated for different values of χ as a function of $N_3 = N$ and of ν_3. As before, the ratio

$$\frac{P_i}{P_{i-1}} = \frac{\chi^i(i\nu_3 - \nu_3)!}{(i\nu_3)!\chi^{i-1}} = \frac{\chi}{(i\nu_3 - \nu_3)\ldots(i\nu_3 - 2)(i\nu_3 - 1)i\nu_3}$$

tends toward unity for high values of i, a situation of classical equilibrium. It follows that

$$\frac{P_i}{P_{i-1}} = 1 = \frac{\chi}{(i\nu_3)^{\nu_3}} = \frac{\chi}{N^{\nu_3}}$$

and

$$\chi = N^{\nu_3}$$

With the latter expression it is easy to calculate the average value the ratio

$$\rho = \frac{\overline{N_3^{\nu_3}}}{N^{\nu_3}}$$

as a function of N. The case $\nu_3 = 1$ is rather simple because the relation

$$\theta_i = \frac{N^i}{i!}$$

leads to

$$S = 1 + N + \frac{N^2}{2!} + \frac{N^3}{3!} + \ldots + \frac{N^i}{i!} = \exp N$$

and

$$S\overline{N} = N + \frac{2N^2}{2!} + \frac{3N^3}{3!} + \ldots + \frac{iN^i}{i!} = N(1 + N + \frac{N^2}{2!} + \frac{N^3}{3!} + \ldots + \frac{N^{i-1}}{(i-1)!} = N\exp N$$

Finally,

$$\overline{N} = N \quad \text{and} \quad \rho = 1$$

The result implies that the usual statistics hold for very small amounts of reaction products, provided $\nu_3 = 1$. This conclusion fails for other values of ν_3.

CHAPTER 10

RADIOCHEMISTRY IN ACTION

I. INTRODUCTION

Since the discovery of radioactivity, radiochemistry has contributed not only to the development of many existing fields of chemistry, but also to the new theme of the chemistry of natural and artificial radioelements. This latter topic is in fact a realm of radiochemistry, and the 20% extension of the Mendeleev periodic table during the last 50 years represents a phase of unsurpassed rapidity in the discovery of elements. The chemistry of several radioelements is particularly impressive and that of Pu is probably more thoroughly investigated than the chemistry of more common elements. Interestingly enough, radiochemistry has also contributed to a better knowledge of many stable elements.

Although radiochemistry per se is not directly involved in most parts of the nuclear fuel cycle in which industrial amounts of radioelements are handled, the last step in this cycle, namely management of nuclear wastes, poses acute new problems related to tracer levels of radionuclides. The safe confinement of radwastes over hundreds of thousands of years is a challenge with which mankind has not heretofore been confronted. Even the smallest leakages would disperse long-lived actinides and fission products in the geosphere and lead to migration into the biosphere. Hence, gaining new knowledge of the behavior of radionuclides in the event of dissemination represents a task of actual and vital importance to radiochemists.

Environmental radioecology also encompasses natural radioactivity, especially that arising from the ubiquitous radon isotopes and their daughters. In recent years, it has been suspected that these radionuclides may constitute a threat to human health. On the other hand, short-lived radioactive species are being increasingly used in nuclear medicine for diagnosis and exploration. This practice would not be feasible without the *savoir faire* of radiochemists who can control hot atom reactions that lead to precursors of elaborate labeled molecules.

Elementary particles can serve as probes in chemistry. The production of muons and kaons requires elaborate machinery and the utilization of these particles in chemistry is exceptional. On the other hand, β^+ emitters are readily accessible: chemistry based on positron annihilation appears promising but is still in an exploratory stage.

Probably the most frequently cited examples of radiochemical applications refer to separations of radioelements and of radionuclides, and to radioanalytical chemistry. Although of an obvious interest in their context, these uses of radionuclides are mostly circumstantial. They depend only on the normal chemistry of the elements concerned and will not be discussed further.

II. RADIOCHEMISTRY AND CHEMISTRY OF ACTINIDES AND TRANSACTINIDES

A. *Hydrolysis and Complexation*

The behavior of a monomeric species of an element M(N) in aqueous solution depends to a large extent on the conditions which favor hydrolysis. The latter begins at a pH for which the aqua ion releases a proton from a water molecule in the first coordination sphere. The aqua ion is represented by $C^z_{0,0} = M^{n+}$ when the oxidation state of the element is equal to or less than four, for instance $Pu(H_2O)^{4+}_8$, and by $C^z_{0,y}$ for higher states, e.g., $UO_2(H_2O)^{2+}_4$. This step is followed by hydrolytic reactions, which are generally complex because of the role of radiocolloids at very low concentrations of M, and that of polymers at higher C_M. Although the first hydrolysis constant is readily measurable, determination of the next hydrolysis constants and of the formation constant of polymers is much more difficult.

Investigation of monomeric hydrolysis requires perchloric acid solutions of M at constant, but occasionally relatively high, ionic strengths and the smallest possible concentration of the element.[1-7] For this reason, radionuclides with short or relatively short half-lives are very convenient for hydrolysis studies. These nuclides are even indispensible in the case of the heaviest elements, which are easily hydrolyzed and rapidly form polymers with increasing C_M, thus masking the appearance of the first monomeric hydrolytic entity C^+_0, as is the case for Pu^{4+}.

Together with hydrolysis, the initial phase of complexation of M by H_pL reagents is strongly dependent on pH. Complexation is (or should be) investigated after the monomeric hydrolysis step. In the presence of complexing reagents, the behavior of M is less affected by conditions under which the solutions are prepared. Upon dilution, the effect of the complexing agent progressively decreases and, by extrapolation, the extent of hydrolysis can be estimated.

A very instructive procedure is the simultaneous investigation of several radionuclides of different elements in the same solution. Because the experimental conditions and, in particular, the pH, are identical for all species, very accurate and useful *relative* data can be collected.[8-10] Under these conditions, even very small variations of a property among a series of elements can be observed.

Actinide elements are of particular relevance in this connection. Together with lanthanides, they represent the only series of elements with reputed *hard*, highly charged, cations. The electronic structure of M^{3+} (Pu^{3+}) or M^{4+} (Pu^{4+}) actinide ions is a noble gas-like $5f^{q}$ configuration because the 5f electrons, although less tightly bound, are screened by the six 6p and two 6s electrons. The extension of the 6p and 6s orbitals, whether considered relativistically or not, is slightly greater than that of the 5f orbitals and the energy of the former exceeds that of the 5f shells. In their higher oxidation states, actinides also exist as the unusual hard molecular cations $M(V)O_2^+$ (PuO_2^+) and $M(VI)O_2^{2+}$ (PuO_2^{2+}).

Actinide chemistry in solution has been elucidated mostly with the aid of solvent extraction of chelates. In the present context, significant properties of chelating agents with the general formula HA are the extent of partition P_{HA} between the organic and aqueous phases, and the acidity constants (see Chapter 8). Two types of chelating agents, as solutions in aliphatic or aromatic solvents, are very common:

1. Thenoyltrifluoracetone (TTA), $\log P_{HA} = 1.65$ in benzene, $pK_a = 6.25$
2. Dibenzoylmethane (DBM), $\log P_{HA} = 5.35$ in benzene, $pK_a = 9.35$

Owing to their difference in acidity by three orders of magnitude, these reagents represent the extreme limits of concentrations of the A^- anion in aqueous solution.

The amount of complexes with HA in the aqueous phase is minimized by the use of low concentrations C_{HA} and acidic (for TTA) or nearly neutral solutions (for DBM).

Examples will be presented for elements in the tetravalent state because these best illustrate the utility of the method and the nature of problems that may be encountered. Aqua ions M^{4+} are the most sensitive toward hydrolysis and complexation and they form the highest order of complexes.

The first hydrolysis product $M(OH)^{3+}$ appears at pH 1, or sometimes at even higher acidity. Investigation of all possible hydrolysis reactions would require exploration of a wide pH range; if complexation is studied under the same conditions, a given complexing agent may form various anions.

Figure 10.1 represents the partition of Pu(IV) between a noncomplexing aqueous phase of $HClO_4$ and $LiClO_4$ and a solution of TTA or DBM in benzene.[7] The initial actinide solution is 10^{-8} M ^{238}Pu. Plots of log D vs. log $[\overline{HA}]$ in the organic phase are straight lines with slope 4, indicating that the actinide is extracted as PuA_4. Up to pH 5, no TTA complexes are found in the aqueous phase. The curves show neatly the change in slopes with decreasing acidity, an indication that Pu^{4+} is progressively transformed into $Pu(OH)_4$.

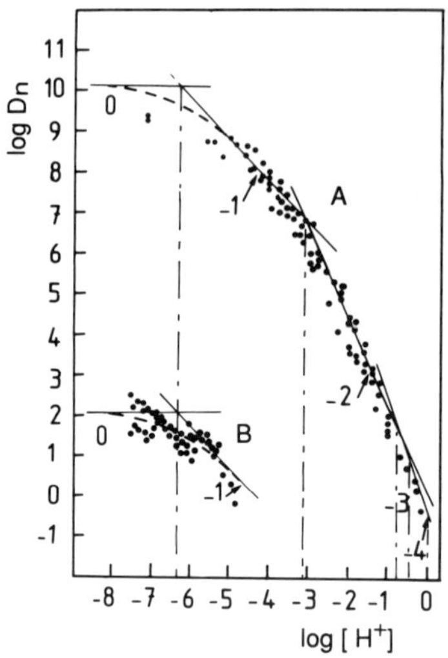

FIGURE 10.1. Partition of ^{238}Pu(IV) between a noncomplexing aqueous solution of $HClO_4$ + $LiClO_4$ with ionic strength μ = 1 and a benzene solution of HA ≡ TTA (curve A) or DBM (curve B), as a function of log [H^+]. The slopes ∂ logD_n/∂ log [H^+] from 0 to −4 are indicated. The values of D_n are normalized for [TTA] = 10^{-2} *M*, [DBM] = 5 × 10^{-4} *M* and a slope of 4 for the variation of log D_e with $[\overline{HA}]$. (Adapted from Metivier, H. and Guillaumont, R., *J. Inorg. Nucl. Chem. S*, 36, 1976.)

The variations of log D with pH, for various concentrations of citric (H_3Cit) or isocitric (iso-H_3Cit) acids, cannot be exploited as such (Figure 10.2). Log D must first be represented as a function of [H_3L] at constant log [H^+] and next as a function of log[H^+] at constant [H_3L]. With this procedure, the limiting values l and y of the $C^z_{l,y}$ complexes can be determined, and the regions for which all complexes predominate, including the hydrolyzed monomeric forms for l = 0, can be estimated (see Chapter 8). These regions appear in the log [H^+] − log [H_3L] planes as shown in Figure 10.3, in which the specific lines of citric and isocitric acids are also represented. In these experiments the persistence of the oxidation state Pu(IV) was checked continuously, but the presence of colloidal forms at high pH in noncomplexing solutions has not been verified. The last hydrolysis constants may be questionable, but at present no other data are available.

Several acid-base properties of trivalent lanthanides exhibit a tetrad (also called double-double) trend arising from the nephelauxetic effect in the series of $4f^q$ (q from 1 to 14) complexes of M^{3+} ions.[11] Interelectronic repulsion and spin-orbit coupling of f electrons in these elements are stronger than the ligand field. In solution at 25°C, the ground state of these species is practically a J state, owing to the population of Γ states. Variations in the repulsion and coupling parameters as a function of q are responsible for the tetrad effect. The latter appears as small discontinuities in properties such as formation constants, with respect to a smooth variation, for q = 3 and 4, 7, 10 and 11, and 14, independently of the structure of the entities.

The tetrad effect has a rather weak intensity, corresponding to several hundred wave numbers. Its existence has been inferred from the Z-variation of the equilibrium constants in the extraction of aqua M^{3+} ions in an organic solution of 2-ethyl-hexyl phosphonic acid. When all elements at the tracer

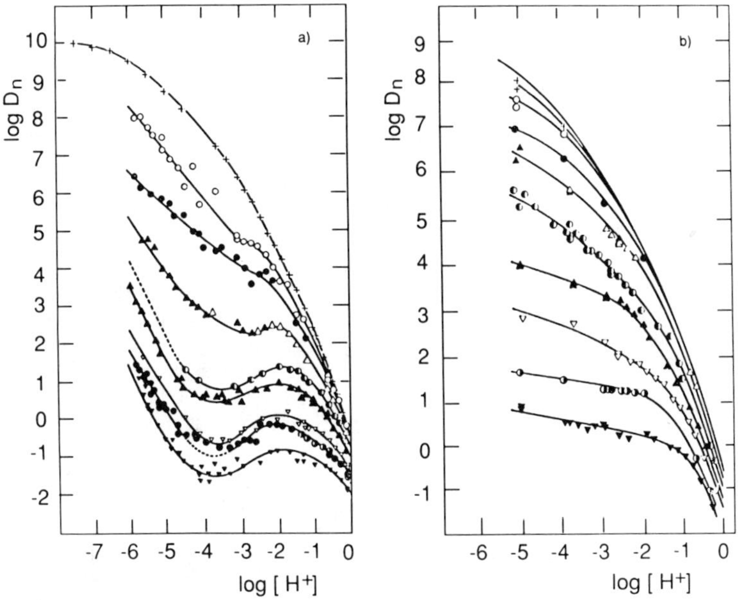

FIGURE 10.2. Partition of $^{238}Pu(IV)$ between complexing aqueous solutions of H_3Cit (curves a) or isoH_3Cit (curves b) (μ = 1) and a benzene solution of TTA, as a function of log $[H^+]$ for various concentrations of complexing agent $C \leq 0.1$ *M*. (The latter are omitted in order to not overload the figures.) D_n values correspond to $\overline{[TTA]} = 10^{-2}$ *M*. (Adapted from Metivier, H. and Guillaumont, R., *J. Inorg. Nucl. Chem. S,* 36, 1976.)

scale are extracted simultaneously, it is much easier to look for small differences in the equilibrium constants and the conclusions are more reliable than for individual, successive, experiments. The reason for this is that the equilibria depend on $[H^+]^3$; in a collective extraction, reproducibility of pH is automatically ensured. At the same time, accidental compensation or addition of experimental errors which might arise from nonreproducibility of reagent concentrations in repetitive procedures are ruled out.

Because of the wide range of partition coefficients for elements between La and Lu, all 15 lanthanides cannot be treated in a single solution. However, mixtures with one or two common radioisotopes of these elements can be prepared for the sake of comparison and normalization of the data. Using this procedure, the solvent extraction of lanthanides with TTA as a function of concentration of citric acid and pH readily reveals the tetrad effect in the citrate complexes $MCitHCit^{2-}$. A similar observation has been reported for the extraction equilibrium constants.[8,9]

In the case of 5f M^{3+} ions, the tetrad effect has been observed for $MCit_2^{3-}$ complexes of the elements from Am to Fm. The tetrad effect has stimulated many theoretical investigations on the thermodynamic consequences of the nephelauxetic effect.

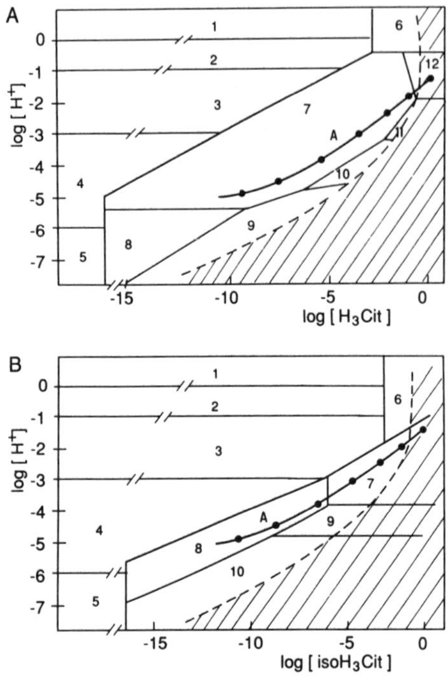

FIGURE 10.3. (A) Approximate domains of Pu(IV) species in citric acid solutions. (1) Pu^{4+}; (2) $Pu(OH)^{3+}$; (3) $Pu(OH)_2^{2+}$; (4) $Pu(OH)_3^{+}$; (5) $Pu(OH)_4$; (6) various cationic species; (7) C_1^{+}; (8) C_1^{-}; (9) C_2^{4-}; (10) C_2^{2-}; (11) C_2^{0}; (12) C_2^{2+}. Curve A corresponds to citric acid; each point refers to concentrations decreasing by a factor 10 from 1 *M* to 10^{-6} *M*. The dashed zone has not been explored. (B) Approximate domains of Pu(IV) species in isocitric acid solutions. Species in regions from 1 to 5 are the same as for 10.3A; (6) C_1^{3+}; (7) C_2^{0}; (8) C_1^{-}; (9) C_2^{2-}; (10) C_2^{4-}. Curve A: corresponds to isocitric acid; each point refers to concentrations decreasing by a factor 10 from 1 *M* to 10^{-6} *M*. The dashed zone has not been explored. (Adapted from Metivier, H. and Guillaumont, R., *J. Inorg. Nucl. Chem. S*, 36, 1976.)

Transport methods have been applied for the study of hydrolysis and complexation, but only within narrow limits of pH and concentration of complexing reagents.[1,12] The necessity for exploring the widest possible range of these parameters has already been emphasized.

By proper combination of the many available extractants and radionuclides with suitable half-lives, partition experiments can be extended to most elements.

B. Speciation of Actinides in the Environment

In the present context, *speciation* refers to the physical and chemical identification of entities of an element present in aqueous media of low ionic strength, and in particular in natural aquifers. The information sought includes the oxidation state and the chemical forms of the element, whether as simple aqua ion or a complex species, and the possible presence of colloidal forms. The ease of this task depends on the elements and a complete description of the complexity of aqueous solutions is usually difficult.

The composition of artificial aqueous media can be controlled, with the exception of unavoidable impurities present in nonponderable amounts. Conditions can be adapted to the selected experimental method and speciation is simplified by introducing only one element at a time. For the case in which M is a radioactive microcomponent, the determination of the $C_{l,y}^{z}$ species can be undertaken with radiochemical methods described in Chapter 8.

The problem is quite different with natural aquifers, whose composition is always very complex. These media contain several major natural elements, usually well identified, and their species, which are more or less in equilibrium with the solid phases, are in principle also known. But they include in addition many other elements which may be very difficult to identify, quite apart from their concentration.

Notwithstanding their extreme variety, all natural inorganic or organic aqueous media (i.e., those containing algae, bacteria, etc.) are characterized by known limits of pH and Eh, and by the nature of the main complexing anions.[13] The latter include inorganic anions such as CO_3^{2-}, HCO_3^-, SO_4^{2-}, PO_4^{3-}, HPO_4^{2-}, S^{2-}, Cl^-, F^-, and organic anions derived from humic, fulvic, citric, and other acids.

Eh values are determined by the nature of the entities forming natural redox couples and which usually do not complex M. These entities include mainly compounds and species of Fe(II) and Fe(III).

All media have a particular Eh-pH diagram, but most of them exhibit a common region from pH 5 to 9 in which competition between hydrolysis and complexation occurs for many elements, and in which the formation of hydroxocomplexes and colloids is favored (Figure 10.4).[14]

Natural aqueous media are not readily reproducible. Samples taken from sites may undergo irreversible tranformations, particularly in the case of anoxic media. Speciation must usually be performed on elements present in very low concentration and is often limited to the identification of oxidation states because of the conditions imposed by the system. The most appropriate methods for this purpose are solvent extraction, exchange on resins, coprecipitation, and electrophoresis.

The main problem is to avoid perturbation of the initial distribution of the element among its various oxidation states. Undesirable changes may occur fortuitously at the tracer scale during adjustment of pH to the value required by the chosen analytical procedure. This effect should always be investigated and, if necessary, corrected by addition of a convenient oxidizing or reducing reagent. In many cases, it remains an awkward problem.

Several chemical procedures are available for speciation of Pu.[15-25] LaF_3 precipitated in a medium of sulfuric and nitric acids in presence of $K_2Cr_2O_7$ carries Pu(III) and Pu(IV), which can be characterized in this way. Trivalent and pentavalent Pu are rapidly oxidized to the next higher state, but the oxidation of Pu(IV) to Pu(V) is slow.

Pu(VI) in sea water is identified by its preferential adsorption on silicagel and Pu(V) by its selective fixation on $CaCO_3$. Pu(V) is never extracted by chelating agents, whatever the pH. Pu(III) is extracted by DBM from neutral media and by TTA at pH 4, but adjustment of acidity at this value oxidizes Pu(III) to Pu(V).

Speciation of radionuclides is important with respect to the biological behavior of radioactive matter. It may be assumed that the evolutionary processes

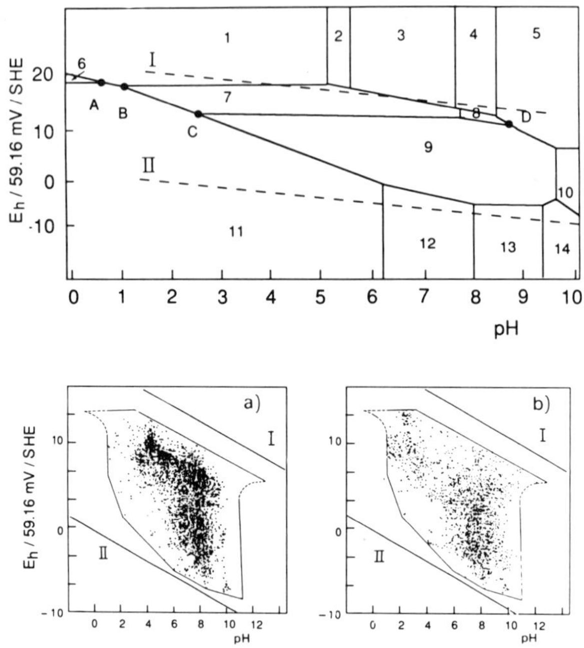

FIGURE 10.4. Eh-pH diagram for Pu species and $\log P(CO_2) = -2$. Disproportion reactions occur at points A, B, C, and D with the following equilibria:

(A) $3Pu(IV) \rightleftarrows 2\,Pu(III) + Pu(VI)$ (B) $3Pu(V) \rightleftarrows Pu(III) + 2Pu(VI)$

(C) $2Pu(IV) \rightleftarrows Pu(III) + Pu(V)$ (D) $2Pu(V) \rightleftarrows Pu(IV) + Pu(VI)$

The four oxidation states coexist along the segment AB. Numbers in the diagram indicate the Pu species: (1) PuO_2^{2+}, (2) PuO_2OH^+, (3) PuO_2CO_3, (4) $PuO_2(CO_3)_2^{2-}$, (5) $PuO_2(CO_3)_3^{4-}$, (6) Pu^{4+}, (7) PuO_2^+, (8) $PuO_2CO_3^-$, (9) $Pu(OH)_4$, (10) $Pu(CO_3)_4^{4-}$, (11) Pu^{3+}, (12) $PuCO_3^+$, (13) $Pu(CO_3)_2^-$, (14) $Pu(CO_3)^{3-}$. Lines I and II represent the variation with pH of E° (O_2/H_2O) and E° (H^+/H_2), respectively. The inserts represent the Eh-pH boundaries for (a) natural environmental waters, and (b) for geological waters in presence of microorganisms. (Adapted from Baas Becking, L. G. M., Kaplan, I. R., and Moore, D., *J. Geol.*, 68, 243, 1960 and from Guillaumont, R. and Adloff, J. P., *Radiochim. Acta*, 58, 53, 1992.)

of life have become adapted to the omnipresence of natural radionuclides. On the other hand, anthropogenic radionuclides introduced accidentally or deliberately into the environment raise new problems and require particular care. The entities responsible for transfer of radionuclides from waters or soils into the atmosphere, into nutriments, and eventually into the human body and specific organs should be identified. Accordingly, speciation should be extended to biological media in the physiological pH range. As an example, the occurrence of Pu complexes with transferrin and citric complexes of Pu (e.g., $PuOCit_2^{4-}$ at pH > 5 and $C_{H_3PO_4} > 10^{-5}$ *M*) is suggested in these media (see Figure 10.3).

The behavior of radionuclides that possess only one stable oxidation state, such as $^{137}Cs^{+}$ or $^{90}Sr^{2+}$, is simple and usually well understood. The problem is much more complicated for other elements, as will be illustrated for Pu. The speciation of Pu in relation to its migration and radiotoxicity has been treated extensively.

The concentration and oxidation states of Pu differ according to the various compartments of the geosphere.[26] In extended water reservoirs such as lakes and oceans, the sum of the concentrations of the higher oxidation states, Pu(V) + Pu(VI), is relatively constant and is of the order of 10^{-17} *M*, which corresponds to the range from 0.1 to 1 fCi L^{-1} (1 fCi = 3.7×10^{-5}Bq). Pu(V) is often the predominant species. The lower oxidation states, Pu(III) and chiefly Pu(IV), adhere to suspended solid particles (with K_d from 10^3 to 10^4 cm^3 g^{-1}) or are incorporated into sediments (from 5 fCi g^{-1} in lakes to more than 100 fCi g^{-1} in ocean sediments). In less extended systems, such as pools and underground waters in contact with the soil (which concentrates nearly all of the fallout Pu), the concentration of the element is much smaller and may be characteristic of the sub-tracer scale (10^3 atoms g^{-1}). Finally, in living matter, the average concentration of Pu is of the order of 10^3 atoms kg^{-1}.

The Eh-pH diagram of Pu represented in Figure 10.4 has been drawn from the most recent data. It shows that in deep reducing waters, Pu(III) and Pu(IV) are stable species, whereas in surface waters in contact with air, the species Pu(IV), Pu(V), and Pu(VI) may coexist. Furthermore, disproportionation of Pu(V) may intervene if Pu(VI) is sufficiently complexed, whereas Pu(IV) is always hydrolyzed.

The hydrolysis of the aqua ion Pu^{4+} is rather complex. When Pu is present in weighable amounts, the hydroxide $Pu(OH)_4$ precipitates when the pH exceeds log β_1 = 0.5; at the tracer level, Pu^{4+} may be involved in redox reactions and possibly in the formation of radiocolloids. These are probably reasons why the quoted hydrolysis constants of Pu^{4+} cover a range of values between 10^5 and 10^9. Consequently, for many surface waters prediction of the formation or disappearance of Pu(V) owing to self-redox reactions remains quite uncertain, not to mention complications due to kinetics at the tracer level.

The know-how acquired by radiochemists dealing with radionuclides in the environment is certain to be very useful for speciation of other elements at low concentration.[27,28] Actually, the same problems of release, migration, adsorption, and requirement of speciation exist for toxic elements from chemical wastes.

C. Chemical Identification of Elements

Around 1700 A.D., 14 elements were known, with quite a few recognized in prehistoric times. Further elements have been discovered and characterized since then in successive waves, concomitantly with the introduction of new methods. Chronologically, these correspond to chemical analysis, electrochemistry, spectroscopy, and ultimately, radioactivity and nuclear reactions.

Two major contributions have led to the present-day periodic table of elements. The first was the guideline set up by Mendeleev in 1871 that allowed the prediction of properties of elements yet to be discovered. The second was the development of the actinide concept by Seaborg in 1950, providing the key to the subsequent discovery of a number of transuranium elements (see Chapter 1).

The assignment of atomic number for actinides from Am to Lr was based on the comparative elution of the elements in the trivalent state from a Dowex 50 resin by strong complexing agents such as anions of citric or α-hydroxyisobutyric acid.[29,30] Optimal conditions include 0.1 *M* solutions of the ammonium salts of the acids, a pH of 4 and a temperature close to 90°C. In the same way as for lanthanides, actinides are eluted in the order of decreasing Z. This behavior reflects the decrease with Z of the K_d values of M^{3+} aqua ions, which in turn results from the increase of the ionic radii of the ions with decreasing Z. On the other hand, the formation constants of the complexes in the aqueous phase are relatively close (and high) for all these elements and do not perturb the affinity sequence of the ions for the resin.

The method has been improved by the use of pH and concentration gradients for the identification of transactinides. For the latter, additional procedures have been developed for the gas phase and for solution.

The first two 6d elements, with atomic number 104 and 105, have been identified by thermochromatography. This method is very convenient for short-lived nuclides such as 3 s ^{259}Unq and 1.68 s ^{261}Unp.[31-35]

Experiments have been performed on the halides MX_x of the transactinides and the critical deposit temperatures T_c of the compounds have been measured. The interpretation of the results is based on two trends: (a) T_c decreases with increasing oxidation number of the element M and (b) although the migration mechanism of a halide at the subtracer level differs from that at weighable amounts, the volatility of MX_x increases with decreasing values of T_c.

With this reasoning, it was proved for the first time that element 104 is the homologue of Hf. The T_c value for $^{259}UnqCl_4$ (determined from 15 spontaneous fission events!) is about 200°C, which is very close to that of $HfCl_4$ and markedly different from that of $ScCl_3$ and of the trichlorides of ^{242}Cm, ^{246}Cf, and ^{242}Pu, for which T_c is of the order of 400°C. It was possible to test the comparative behavior of these elements in the same run; Sc was present in the target and the actinides were produced in secondary nuclear reactions. The partition system comprised a gas phase of N_2, $TiCl_4$, and $SOCl_2$ and a solid quartz phase. Additional qualitative tests showed that the tetrachloride of element 104 is less volatile than $HfCl_4$, and more importantly, that 104 *is not* a 7p element.

Conclusions drawn from similar experiments with element 105 are less convincing. The behavior of $^{261}UnpCl_5$, deduced from 18 spontaneous fission events, indicates a T_c between 100 and 200°C; the compound appears less volatile than $NbCl_5$ (T_c = 50°C) and slightly more volatile than $UnqCl_4$. Data for bromides are more precise: T_c = 50°C for $NbBr_5$ and $TaBr_5$, 125°C for $UnpBr_5$ and 150°C for $HfBr_4$. In these experiments the value for element 105 must be corrected for decay of the nuclide during migration. The higher the oxidation state, the greater the distance that must be covered to attain the critical deposition temperature. Obviously, short-lived nuclides may disintegrate before reaching the deposition point.

These conclusions have been confirmed with other sequential or continuous partition methods involving nuclides with slightly longer half-lives such as 60 s ^{261}Unq and 35 s ^{262}Unp. It has been observed that complexes of Zr, Hf, and Unq with 0.1 *M* α-hydroxyisobutyric acid at pH 4 and at 80°C have the same exchange behavior on a Dowex resin.[30-36] These experiments were performed in 250 consecutive trials, with detection of 17 α decays 50s after the end of irradiation. Under the same conditions, the behavior of No(II), Cm(III), Cf(III), and Pa(V) is very different.

In the same experimental concept, the hydrochloric acid complexes of Zr, Hf, and Unq formed in 3, 6, and 12 *M* HCl have been found to behave similarly with respect to their chromatographic extraction with tri-octyl ammonium chloride, but in a manner quite different from that of ^{248}Cm, ^{246}Cm, ^{244}Cm, and $^{254\text{-}255}$Fm. One hundred experiments were conducted and six α decays were recorded.[37]

Other investigations have shown an adsorption trend on silica in the order Nb < Ta < Pa < Unp, which corresponds to an increasing tendency towards hydrolysis. In several hundred trials, 24 spontaneous fission decays were observed. It is known that partition data involving solid surfaces are not always reliable. However, in a hydrofluoric acid medium also containing HNO_3 or HCl, the behavior of complex fluorides of Nb, Pa, and Unq differs from that of Ta in the partition between the aqueous phase and methylisobutyl ketone or tri-isooctyl amonium chloride. Under these experimental conditions, hydrolysis is suppressed and the behavior of the fluorides of the heavier members of column 5 is somewhat unexpected.

The pertinent experiments together with semi-theoretical considerations have provided a first picture of the chemical properties of Unq and Unp at the sub-tracer scale.

The situation is more complicated for the following elements from Z = 106 to 109. All nuclides which can be prepared are so short-lived that any direct chemical identification is precluded. The nuclides decay by spontaneous fission and also by α decay. In this latter sense they are precursors of radioactive families which comprise isotopes of actinides with relatively long half-lives (see Table 5.10). These nuclides can be separated from the target and from the numerous reaction products. Despite very low activities of the order of a few α emissions per day, the detection of daughters gives a clue to their ancestors. The formation and chemical identification of the odd-odd nuclides

^{262}Uns (Z = 107) and ^{266}Unn (Z = 109) is inferred from the detection of 35.4 h ^{246}Cf and those of ^{251}Unh (Z = 106) and ^{265}Uno (Z = 108) from the detection of 20 d ^{253}Es.[38]

Similar indirect methods based on the detection of isotopes of Am and Cm have been applied in the (still unsuccessful) search for element 110.

So far, the rudimentary chemical properties which have been determined for the last actinides and transactinides up to Z = 105 do not deviate markedly from those anticipated from the position of these elements in the periodic table. Nevertheless, several unexpected observations have been made:

1. The elution of Lr^{3+} is delayed with respect to predictions based on the variation of the ionic radius of M^{3+} ions in the actinide series.
2. The volatility of $UnqCl_4$ and $UnpBr_5$ is lower than that of the respective homologues Zr-Hf and Nb-Ta.
3. The fluoride complexes of Unq do not behave like those of Ta.

Neither the expected regular changes of chemical properties of the elements within a column, nor nonrelativistic models explain these anomalies. The first observation above is attributed to an abnormal increase in the ionic radius of Lr^{3+} which is consistent with relativistic calculations. The other results may also be due to relativistic effects that are sufficiently strong to change the electronic configuration of Unq from 6 d^2 7 s^2 to 7 $s^2{}_{1/2}$ 7 $p^2{}_{1/2}$ and that of Unp from 6 d^3 7 s^2 to 6 $d_{3/2}$ 7 $s^2{}_{1/2}$ 7 $p^2{}_{1/2}$. These new orbitals may give rise to unusual chemical bondings of the metals in the complexes.[39]

Relativistic effects are expected to be even more pronounced for higher Z and may possibly determine the oxidation states of future elements.[40-43] In the present periodic table, the element 114 would be in the eka-Pb position and the element 115 in the eka-Bi position. From simple extrapolation of the arrangement of the electrons in their lighter homologs, these elements should exhibit 6 d^{10} 7 p^2 7 s^2 and 6 d^{10} 7 p^3 7 s^2 external electron configurations, respectively, compatible with oxidation states 2 and 4 for eka-Pb, 3 and 5 for eka-Bi. On the other hand, strong relativistic effects could favor the configurations 6 d^{10} 7 $p^2_{1/2}$ 7 $p^0_{3/2}$ and 6 d^{10} 7 $p^2_{1/2}$ 7 $p^1_{3/2}$ 7 s^2 with a wide gap between the 7 $p_{3/2}$ and 7 $p_{1/2}$ relativistic sub-shells. In this case oxidation states $_{114}$Uuq (IV) and $_{105}$Uup(V) would be precluded. The importance of such effects was not realized before radiochemists began to deal with very heavy elements.

D. Structure of Aqua Ions

The structure of aqua ions is a property of considerable importance because it is related to the hydration enthalpy and entropy, which are useful for the calculation of thermochemical cycles associated with the transfer of an atom M into a solution. It is a basic parameter which also serves as a test for the hydration models of gaseous ions.

The structure depends on the crystallographic radius R_c of the M^{N+} cation, and which determines the coordination number of the ion in solid compounds. The first coordination sphere of the aqua ion contains N_w water molecules. When the ion migrates in water, it carries H additional water molecules in the second coordination sphere; $h = N_w + H$ is the *dynamic* hydration number. The Z-dependence of the structure of aqua ions of trivalent lanthanides and actinides has been determined mainly from electrophoresis and diffusion experiments performed on selected radionuclides.

For this purpose, noncomplexing media are very convenient. Mobilities u of M^{3+} actinide ions (Am, Cm, Cf, and Es) with respect to that of Eu^{3+} in 0.1, 1, and 8 *M* solutions of $NaClO_4$ and $HClO_4$ of constant ionic strength have been reported, but owing to insufficient control of temperature the extrapolated limiting values u^0 are in doubt.[44] On the other hand, experiments using paper impregnated with mixtures of $LiClO_4$ and $HClO_4$ with $\mu \leq 1$, have produced the correct ($\pm$ 0.5%) relative u^0 values (at T = 25 $\pm$ 0.5°C) for the trivalent ions of the lanthanides Nd, Eu, Gd, Tb, Yb, and the actinides Am and Cf with respect to Ce. The limiting absolute diffusion coefficients D^0 have been obtained for the lanthanides Ce, Eu, Gd, Tb, Tm, and Yb, and the actinides Am, Cm, Cf, and Es. In this experiment performed in LiCl and NaCl mixtures at pH 2.5, T = 25 $\pm$ 0.01°C, and $10^{-2} \leq \mu \leq 10^{-4}$ *M* the precision is 1% for γ emitters and 1 to 2% for α emitters.[45-47]

Several interesting trends have been revealed from these experiments, which are complemented by spectroscopic measurements:

1. The mobilities of actinides are smaller than those of lanthanides; the size of corresponding actinide aqua ions exceeds by 1.5% that of the lanthanide ions.
2. In both series, the highest N value is 8 for the heavy elements and 9 for the light ones and accordingly the variations of N_w and h with Z are discontinuous and opposite.
3. In the actinide series, discontinuity is observed between Bk and Cf and corresponds to R_c = 107.7 pm; in the lanthanide series it occurs between Eu and Gd for R_c = 106.7 pm. The shorter distance between the cation and the first water molecules in actinides results from covalency effects.

These results coincide with absolute u^0 values obtained in electromigration experiments in the absence of a support for the trivalent ions of La, Eu, Gd, Tb, Tm, Yb, Lu, Am, and Cf in a mixture of $LiClO_4$ and $HClO_4$, for which $\mu \leq 0.1$ *M* and $1 < pH < 7$.[48,49] They have prompted calculations on covalency in the aqua ions of hard cations and have led to a model providing absolute values of hydration enthalpy and entropy of gaseous ions.

Transport methods, when applied under the strict conditions which they require, have effectively contributed to a specific topic of actinide chemistry, and also to a better knowledge of aqua ions and, like the preceding procedures, may be extended expediently to nonradioactive elements.

E. Determination of Oxidation States

Radiochemical methods have played a key role and have been most successful in the search and identification of oxidation states of actinides, in particular the oxidation state 2.[30,50-55]

Evidence for the existence of a divalent actinide was found for Ac and Cf. For the latter, it was provided by a comparative kinetic study of amalgamation using the citric acid complexes of Cf, Sm, Eu, and Yb at the tracer level in a 0.1 *M* solution of lithium citrate at pH 5.5.[56] The biphasic system is strongly buffered when the Li amalgam is in equilibrium with the solution. Under these conditions, the elements in solution are reduced and form an amalgam in a first-order reaction. The half-time of the amalgamation reaction shows discontinuities for the elements mentioned with respect to the behavior of other lanthanides. Accordingly, this group of elements behaves similarly, and quite differently, from lanthanides which cannot be reduced to the oxidation state 2. This was presented as proof for the reduction potential of Cf(III) to Cf(II) and it was suggested that the subsequent trivalent elements in the series could also undergo reduction to the divalent state, as is actually the case.[57,58]

At present, it is considered that the reduction potential of the oxidation state 3 to the formal state 0 in lanthanides actually depends on the effective oxidation state of the metallic element. The reduction potential is higher for Eu and Yb, for which the metallic oxidation state is 2, than in a metal with a corresponding oxidation state 3. This also holds for Cf, as well as for Es and Fm, for which the metallic oxidation state 2 has been proven and which are both easily reduced in solution.

All actinide elements from Am to No (the case of Pu is still open) exist in the oxidation state 2, for which stability is expressed by the standard reduction potential E° (M^{3+}/M^{2+}). The potential is obtained directly when this state is not oxidized by H^+ and is stable in aqueous solution, or it can be calculated when the measurement is performed in fused salts or in alcohol-water mixtures.

The existence of the M^{2+} ions of Pu, Am, Cm, and Bk in fused salts has been inferred from cocrystallization of the elements at the tracer level with ponderable amounts of MCl_3 and MCl_2 (M = Nd and Pr) in a mixture of fused $SrCl_2$ and NdOCl or PrOCl at 920°C. (see Chapter 8). The oxychlorides incorporate only the trivalent actinides; the occurrence of reduction and the relative values of the potentials of the M^{3+}/M^{2+} redox couples of the micro- and macrocomponent are deduced from the variation of the F factor with the ratio C_{MCl_3}/C_{MCl_2}. It is concluded that Pu, Am, and Bk are reduced by Nd^{2+} and Cm possibly by Pr^{2+}.

Only pulse radiolysis shows the formation of Am^{2+} and Cm^{2+} in aqueous solution, with lifetimes of a few microseconds and 0.1 μs, respectively. Radiopolarograms of Am, Cm, and Bk in noncomplexing solution do not show the 2 → 0 wave characteristic of M^{2+}. The only evidence at present for the existence of Pu^{2+} rests on observations made on fused salts.

The formation of M^{2+} ions of Cf, Es, and Fm has been demonstrated in a hydroalcoholic medium saturated with $SmCl_3$ in the presence of Mg. The metal reduces Sm^{3+} and the M^{3+} actinide ions to the divalent state and the M^{2+} species cocrystallize with the resulting $SmCl_2$ carrier. These conclusions are valid provided the mechanism really involves syncrystallization and not the formation of abnormal mixed crystals.

Precise thermodynamic information on the reduction of these ions has been gleaned from radiopolarography in noncomplexing aqueous medium (LiCl, 0.1 *M*, 2 < pH < 3).[59] Fm^{3+} is reduced to the metal in two successive reversible steps, whereas the reduction of Cf and Es is achieved in a single reversible reaction. In the presence of complexing citrate ions, the $E_{1/2}$ value of Fm is shifted by 60 to 100 mV, a trend similar to that found for the $MCit^-$ complexes ($\log \beta_1 = 3$) of the M^{2+} ions of Eu, Mn, Ca, and Ba. For Cf and Es, the shift of 200 mV is compatible with the formation of $MCit_2^{3+}$ complexes ($\log \beta_2 = 12$). However, in the presence of water, the states Cf(II) and Es(II) are unstable and are reoxidized in less than one second.

Md^{3+} is readily reduced in aqueous 0.1 *M* HCl with V^{2+}, Eu^{2+}, Cr^{2+}, Ti^{3+}, and Zn(Hg). In the presence of these reducing agents, 1.3h ^{256}Md (investigated with 10^6 atoms per experiment) coprecipitates with $BaSO_4$ and its behavior in extraction chromatography is different from that of an M^{3+} species. Doubt has been raised on the actual state of the reduced species as to whether it is Md^{2+} or Md^+. At the sub-tracer level, ^{256}Md (measured via spontaneous fission of the daughter ^{256}Fm) after reduction with Eu^{2+} and Yb^{2+} in a hydroalcoholic medium coprecipitates with $MgCl_2$ and Rb_2PtCl_6, but when it is reduced with Mg, it is also carried by CsCl and RbCl. On the other hand, similar experiments at the tracer level in the presence of Sm^{2+} in hydroalcoholic medium or in fused KCl failed to confirm the formation of Md^+. The existence of this form is still open to controversy, although the predicted radius of Md^+ lies between the values for Na^+ and K^+. This fact may substantiate conclusions drawn from coprecipitation experiments of Md with NaCl and KCl after reduction of Md(III) with Eu^{2+}.

The existence of Md^{2+} in an aqueous medium has been proved by radiopolarography (in 0.1 *M* LiCl, at pH 2.4 and in 0.1 *M* Li_3Cit) and by radiocoulometry (in 0.1 *M* $LiC_2O_2H_3$ and 0.1 *M* Li_3Cit) and the potential of the Md^{2+}/Md couple has been measured. Even at a potential of -1.5 V, Md^+ is not observed.[60-63]

Continuing uncertainty with respect to the existence of Md^+ illustrates the kind of dilemma which can arise when conclusions drawn from different radiochemical methods diverge. This situation also reflects the possible perturbation of a system resulting from the application of various partition methods.

The oxidation state 2 of No is stable in aqueous medium. This peculiar behavior is related to the closed electronic configuration $5f^{14}\,7s^2$ of No^{2+}, which would correspond to that of Md^+. The oxidation to No(III), tested with 3.1 min ^{255}No on batches containing about 1000 atoms, requires a strong oxidizing agent with a potential $E° > 0.4$ V, which is provided by Ce^{4+}, $HBrO_3$, or H_5IO_6. Numerous experiments have shown that the properties of No($^{2+}$) are similar to those of alkaline earths in complexing or noncomplexing media, and markedly differ from those of divalent transition elements such as Hg, Cd, Cu, and Co. The potential of the couple No^{2+}/No has been obtained from radiocoulometry on ^{255}No and 1 h ^{259}No with about 10^3 atoms per trial in a 0.1 *M* Li_3Cit solution. The small available amounts of No preclude radiopolarographic experiments.

The oxidation state 3 of Lr would be expected to be remarkably stable owing to its $5f^{14}$ configuration. The aqua ion Lr^{3+} has been characterized using 35s ^{256}Lr in 10 atoms-at-a-time experiments involving solvent extraction with TTA. The extraction yield was 50% at pH 2.5 in 200 runs, with 2 to 3 α emissions observed in each test. In the same experiment 50% of $^{255}No^{2+}$ are extracted at pH 5. Attempts to reduce Lr(III) have failed.

Only few solid compounds of divalent actinides have as yet been clearly identified. These include AmI_2, $CfBr_2$, $Es(Cl,Br,I)_2$, and CaF_2 doped with Am^{2+} and Es^2.

It appears that present knowledge on actinides reposes on tracer and subtracer chemistry, but even with this restriction the experimental data have furnished a useful insight into the electronic structure of the actinides.

A few additional unusual oxidation states of actinides have been observed at the tracer scale:

1. Tetravalent Es, in the form of EsF_4, behaves in thermochromatography like other MF_4 compounds (M from U to Cf).
2. Pentavalent Cf is produced in the β^- decay of $^{249}Bk(IV)$ as BkO_2 dissolved in 4 *M* Na_2CO_3. It is stable enough to be coprecipitated (and thus characterized) with $Na_4UO_2(CO_3)_3$.
3. Pentavalent Cm as $CmOF_3$ is obtained by fluorination of Cm_2O_3 and can be identified by thermochromatography.
4. Hexavalent Cm is formed in the β^- decay of $^{242}Am(V)$ as K_3AmO_2-$(CO_3)_2$ in 0.1 *M* $NaHCO_3$. It coprecipitates as curyl ions CmO_2^{2+} with $Na(C_2O_2H_3)_3UO_2$. Fluorination of Cm_2O_3 with a mixture of BF_3 and F_2 produces CmF_6, which is identified by thermochromatography.

The successful search for lower oxidation states in actinides illustrates the efficiency of radiochemical methods. Owing to its importance with respect to the chemistry of the heaviest elements, this topic has been amply investigated and discussed.

F. Oxidation Number in the Metallic State

Although the formal oxidation number of an element in the metallic state is zero, one may define a different, positive, oxidation number equal to the number of s, d, and f electrons delocalized in the conduction bands. Several conclusions pertinent to these oxidation numbers are drawn from macroscopic properties such as the metallic radius, the pressure dependence of phase transition, the sublimination entropy and enthalpy, e.g.,

1. Eu and Yb in the metallic state are divalent, whereas all other metallic lanthanides are trivalent under standard conditions.
2. Cf is trivalent in the low-temperature phases (dhc and hc), but divalent in the cfc phases at high temperature, i.e., toward 1000°C.
3. From Pa to Am the delocalization of 5f electrons is such that the oxidation number is higher than 3.

The critical deposition temperatures T_c of ^{240}Cf, ^{253}Es, ^{252,254}Fm, and ^{256}Md atoms have been measured in a system consisting of a gaseous phase of Ca vapor in He and solid Ti or Mo. With Ti, T_c values are: 700°C for Yb, Ra, Cf, Es, Fm, and Md, 1100°C for Sm and Eu, and 500°C for Na, K, and Rb. The trends are consistent with expected properties and suggest an oxidation number 2 for the elements Fm and Md in Ti, which will probably never be available in the metallic state. When the solid phase is Mo, Cf, and Yb behave like trivalent elements such as Sm (T_c = 1200°C). Eu has a T_c of 900°C and divalent elements Mg, Ra, Es, Fm, and Md have a T_c of 700°C. The peculiar behavior of Eu on Ti and Cf, Yb, and Eu on Mo is noteworthy.[31-34]

The diffusion coefficient and activation enthalpy of several recoil-implanted elements in Ta confirm the oxidation number 2 for metallic Es and Fm.[66]

III. RADIOCHEMISTRY AND NUCLEAR WASTE MANAGEMENT

Radioactive wastes arising from utilization of nuclear fission contain radionuclides of various types and origins: fission products, actinides up to Cm together with their descendants, including the members of the four radioactive families which are progressively built up, and nuclides from neutron-induced activation reaction on fuel cladding and reactor material. The management of these wastes remains a major problem in the nuclear fuel cycle and provides lasting scientific, technological, and sociological challenges.

The treatment and conditioning of solid wastes, including those from military applications, is not relevant to radiochemistry. On the other hand, wastes containing tracer amounts of long-lived α emitters and stored or disposed of, at or near ground surface must be monitored by radiochemists. This is also true for materials with large amounts of the same α emitters that will eventually be confined in geological repositories.

Whatever the approach in management of these *radwastes*, it is essential to have an understanding of the chemical behavior of radionuclides in the geosphere, since at some point in time they may be dispersed in underground waters by lixiviation of the waste packages. The various artificial barriers, containers, and backfill materials, selected or proposed for the confinement of radwastes, are intended to delay as long as possible the unavoidable dispersion of radioactive matter. As the distance from the repositories increases, the dilution factor becomes relatively high and conditions become similar to those of tracer-level and sub-tracer level chemistry.

Radionuclides requiring particular attention are those which are especially radiotoxic, with half-lives longer than about 100 years. These include the activation products ^{108m}Ag, ^{14}C, ^{93}Mo, and ^{93}Zr, the fission products ^{59}Ni, ^{99}Tc, ^{135}Cs, ^{109}Pd, and ^{129}I, and the isotopes of actinides from U to Cm, particularly ^{239}Pu and ^{237}Np with their descendants. For these nuclides it is important:

1. To know the nature of the species released by the lixiviation in geological media, e.g., complexes, radiocolloids, and particulates
2. To investigate the interaction of these species with the degradation products of the deposited material and of the artificial barriers
3. To know the nature of the species which are dissolved in surface or underground waters, with respect to the type of complexing agents (e.g., carbonates or humic acids) present in the dissolution medium and to its redox properties
4. To consider the static and dynamic interactions of the species with minerals in the geosphere which hinder migration toward the biosphere
5. To assess the extent of the transfer to plants and vegetals of the species present in superficial waters

It may be noted that these problems closely resemble those mentioned for the speciation of radionuclides in model conditions and *in situ*. They are also relevant to interaction of the species with mixed solid phases and to complex partitions because of the diversity of mechanisms that may be involved.

A further difficulty is associated with long-term predictions over time intervals of 10^3 to 10^4 years or longer. It requires elaboration of fundamental laws as applied to phenomena under real conditions; first steps in this direction are still based on approximations.

A. Radionuclides in the Packages

Radwastes deposited in a repository designed for low or moderate β^--γ activities must contain, on the average, no more than 0.01 Ci (3.7×10^8 Bq) α emitters (mainly Pu isotopes) per ton of material. This value results from a worldwide consensus and, with the present knowledge of the dose-effect relation, indicates an acceptable radiological risk to a person who would live in the future in a deserted storage site. After extinction of the fission products ^{137}Cs and ^{90}Sr, only the long-lived radionuclides mentioned above (chiefly Pu) must be surveyed.

For all initial chemical compounds in which the radionuclides are incorporated, the final state of the radioactive species will be determined by concurring reactions of precipitation, coprecipitation, dissolution, or exchange, which will be essentially dependent on hydrolysis. Average concentrations of the migrating solutions will be very low because the amounts in the packages are kept small enough. Furthermore, the solubility of the hydrolyzed phases is also low and, despite possible complexation, the retention is expected to be high. Exceptions are I, Tc, and C. Remarkably, little is known about the behavior of mineral or organic carbon at the tracer scale.

The highest effective concentration of Pu may be less than 10^{-14} *M*, because of the very low solubility (about 10^{-8} *M*) of hydrolyzed phases of Pu(IV) and high K_d values (see below) on detritial materials are at least 10^5 cm^3 g^{-1}. These predictions may be altered by oxidation, complexation, or incorporation of Pu(IV) in colloids. Estimates rely on fundamental investigations of speciation of metallic and nonmetallic elements which are only feasible with the aid of radiochemical methods.

In underground waters, the behavior of radioactive matter released during slow degradation of unprocessed nuclear fuels, nuclear glasses and other solid materials containing macroscopic amounts of radionuclides will be primarily of relevance to classical chemistry. It is considered that in the degradation processes initial solid phases which will control the dissemination of the radionuclides, include the precipitation of $AmOHCO_3$, $Np(OH)_4$, amorphous UO_2, and $Pu(OH)_4$ or PuO_2. These solids will be more or less rapidly dissolved when they come into contact with underground waters, depending on the carbonate content. Problems specific to radiochemistry will appear only in the geosphere, at points far removed from the source of radionuclides.[67]

On the basis of present-day knowledge, it is estimated that after several hundred thousand years, radionuclides of radiological impact that are likely to be incorporated in the biosphere are ^{99}Tc, ^{93}Zr, ^{135}Cs, ^{237}Np, ^{233}U, ^{229}Th, and the daughters of the last three nuclides. The behavior of these elements is currently being intensively investigated.

Radwaste management will be strongly dependent on our understanding of chemistry at nanogram per liter concentrations in heterogeneous systems, a domain which belongs typically to radiochemistry.

B. The Distribution Coefficient K_d and the Migration of Radionuclides

The retention of soluble and colloidal entities is expressed by coefficients K_d, which are experimental parameters that describe the reversible or irreversible exchange processes between an aqueous and a solid phase. The K_d coefficients depend on the conditions under which they are determined and are strictly speaking only valid for *the* solid-solution couple considered. Furthermore, they should be measured in the field rather than in the laboratory. These requirements are generally difficult to meet, and if they were followed, the coefficients would have little utility. In practice, standard series of K_d values are used or their limiting lower values for similar couples.[68]

K_d coefficients are readily obtained by radiochemical methods. They indicate the retardation of the migration of a radionuclide in a given nonporous solid medium free from cracks. It is usually assumed that the K_d values are independent of the concentration of the radionuclides or other elements which may be exchanged simultaneously. In many cases, however, this assumption is not correct.

Radiochemical partition methods may be used to clarify this point. The distribution of an element M with a high specific activity is measured as a function of the concentration of the associated element, alone or in the presence of a competing element M′. This type of experiment leads to the number of exchangeable sites on the solid phase and to the effect of the competing elements of which the most frequent is Ca. It is an extension of the method of ion exchange with varying concentrations C_M and $C_{M'}$. The expression of partition takes a simple form only in the case where M represents aqua ions M^{n+}, i.e., if the element is not or only slightly hydrolyzed.

When the concentration C_M is above a critical value

$$C_M(\text{crit.}) = 10^3 \frac{N_s}{nV} \ (\text{V in cm}^3)$$

for which all exchangeable sites on the solid are saturated, K_d becomes

$$K_d\ (\text{cm}^3\ \text{g}^{-1}) = \frac{N_s}{n\, C_M}$$

N_s is the number of monovalent sites per gram of solid phase. In this case

$$\frac{\partial \log D}{\partial \log C_M} = -1$$

and N_s, a value characteristic of the solid phase, can be determined.

On the other hand, when C_M is at the tracer concentration level, the critical concentration

$$C_{M'}(\text{crit.}) > 10^3 \frac{N_s}{n' V}$$

can be chosen such that M′ saturates all exchangeable sites and the partition coefficient is independent of C_M. In this way the critical concentration can be found for which K_d becomes independent of C_M. It is a function of the experimental conditions and may be unexpectedly low.

IV. RADON AND ITS PROGENY

During the last decase, there has been increased awareness of the radiological risk associated with the emission of radon.[69-71] It is considered that radon and its progeny represent the main cause of natural radiation exposure. This concern has prompted extensive research on *radon mitigation*, with the principal aim of preventing the accumulation of excessive levels of the emanation in buildings and dwellings. Control of the gas and its daughters in confined spaces is complex. Some of the nuclides disappear by disintegration or are eliminated by ventilation processes; others are deposited. In consequence, the radon daughters are never in radioactive equilibrium with the emanation. Their radiotoxicity is related to the *equilibrium fraction*, which represents the part of the decay products remaining in the enclosed air of a dwelling.

An understanding of the reactions of radon and its descendants is of great environmental importance, especially when guidelines can be established for the control of the equilibrium fraction.[72]

The behavior of radon and its descendants (which are radioactive isotopes of Po, Pb, and Bi) is to a large extent determined by the physical and chemical effects of the successive transmutations. Furthermore, in air, it is conditioned by the presence of aerosols and dust particles.

A. The Behavior of Radon

Each of the three natural radioactive families includes an isotope of radon formed by α decay of the respective radium parent. (see Tables 5.3 through 5.5). The nuclide ^{219}Rn (actinon) is of little concern because of the short half-life of 4 s and the low abundance of ^{235}U. Emanation in the thorium family, ^{220}Rn (thoron) is the most abundant one, but its steady concentration is low because of the 55 s half-life. On the other hand, it produces the relatively long-lived 10.6 h ^{212}Pb, which may be a main contributor to internal irradiation.

The descendant of ^{238}U, the *true* radon ^{222}Rn, is the most important emanation isotope because of the 3.82 d half-life. Its radiological impact is related to the two short-lived α emitting daughters ^{218}Po and ^{214}Po.

Radon nuclides are produced with recoil energies of the order 100 keV. This value is high enough to displace a nuclide well away from the decay site, depending on the properties of the medium. In air, the range of ^{222}Rn and ^{220}Rn are 64 and 103 nm, respectively. The ability of a radiferous medium to release the noble gas is referred to as the *emanating power*. In a solid material, such as a grain of mineral, the radon atom is ejected from the host if the site of the radium decay is located close to the surface. If this is not the case, the atom can diffuse through the grain and possibly leave the host by thermal diffusion.

In a dry granulous material, a moving recoil atom, can cross the air-filled space between adjacent grains and become embedded in a neighboring host. This *recoil implantation* may not be decisive, since again the thermal radon atom may diffuse to the surface and escape slowly into the intergranular space. Radiation damage produced by the recoil atom along its track may favor the re-exit process.

The presence of water markedly affects the emanating power. The latter is approximately four times higher in water-saturated soil than in a dry medium. In particular, water in the pore space can act as a stopping medium and hinder the embedding mechanism, and the release of radon is enhanced. A further role of water is to act as an etching agent of the recoil damage, which facilitates the exit of the radon atom and increases the emanating power. The relative importance of this double effect of water depends on grain size and porosity.[73,74]

Calculations of the emanating power and of its recoil and diffusion components have been performed for models of various shapes and sizes. With respect to radioecology, the release from mineral grains is of greater importance. The distribution of radium, the radon's parent, has an important effect on recoil emanation. It is itself determined by preceding recoil processes which began with the ancestors of the radioactive families. Even if the latter (uranium and thorium), were formed homogeneously in a geological medium, the successive decays would displace the descendants. Eventually, radium becomes attached to thin layers of accessory minerals and adheres to the surface of various geological materials, thus increasing the emanation yield.[75]

Radon concentrations are very high in uranium mines. The usual practice for reducing the amount of noxious gas is to provide forced ventilation in the mine shafts. Alternatively, a *chemical* method for removing radon from contaminated atmosphers has been proposed.[76,77]

Obviously, radon chemistry can only be conducted at the tracer level. A batch of 10^7 Bq (about 1 mCi) ^{222}Rn contains 5×10^{12} atoms and has a STP volume of 10^{-7} cm^3. Radon is readily monitored by the γ radiations of its daughters.

A chemically stable fluoride of Rn was prepared shortly after the discovery of the first xenon compound. When Rn is heated to 400°C with F_2, the nonvolatile product RnF_2 is obtained. The fluoride is quantitatively reduced to elemental radon by water.

Radon fluoride can be stabilized in nonaqueous solvents such as halogen and hydrogen fluorides. Electromigration and ion exchange studies have shown that radon is present as cations RnF^+ and Rn^{2+}. The best solvents for these species are trichlorotrifluoroethane and sulfuryl chloride.

Rn is more readily oxidized than Xe. However, in contrast to Xe for which many compounds of Xe(IV) and Xe(VI) are known, all attempts to prepare corresponding Rn derivatives have failed. Claims for the discovery of RnF_4 and RnF_6 are highly doubtful.

Radon reacts spontaneously at room temperature with many solid compounds that contain strongly oxidizing cations, such as BrF_2^+, IF_6^+, O_2^+, and N_2F^+ to form complex fluorides. This property has been considered as a possible means of trapping Rn. Laboratory tests with radon-air mixtures have shown that dioxygenyl hexafluoroantimonate $O_2^+SbF_6^-$ and hexafluoroiodine hexafluoroantimonate $IF_6^+SbF_6^-$ are particularly convenient for binding Rn:

$$Rn + 2\,O_2^+\,SbF_6^- \rightarrow RnF^+\,Sb_2F_{11}^- + 2\,O_2$$

$$Rn + IF_6^+\,SbF_6^- \rightarrow Rn\,F^+\,SbF_6^- + IF_5$$

Powders or pellets of these compounds remove more than 99% of the radon from air at low or moderate flow rates. The reagents can be used in gas-scrubbing units for the purification of radon-charged air in uranium mines. The operating costs for a full-scale air purification system compete favorably with those of *physical* methods based on adsorption on charcoal, molecular sieves, or other materials, which must be kept at −80°C. In contrast, the chemical retention method has been shown to be effective in the 10 to 40°C range, or more.

B. The Descendants of Radon

There is no doubt that the decay of Ra^{2+} forms Rn in the zero oxidation state, as expected from the charge balance in the α decay process. On the other hand, most of the Po atoms are formed as positive ions, instead of the formally anticipated negative ion:

$$Rn \rightarrow Po^{2-} + He^{2+}$$

The loss of electrons is attributed to recoil and shake-off effects and to chemical factors such as electron affinity and ionization potential of Po and fast reactions with the surroundings. In dry air, the initial yields are 88% for Po^+ and 12% for neutral Po. The diffusion coefficient of Po species is about one half that of Rn, which suggests that the nascent Po ions undergo charge transfer and clustering reactions.

The free or *unattached* daughter nuclides of Rn include relatively mobile entities such as ions, molecules, and small clusters. Several reactions are involved in the neutralization of the Po ions. ^{218}Po, the daughter of ^{222}Rn, has a mean lifetime of 260 s whereas the time spent by free Po atoms in air is estimated at 12 s. It is assumed that in the presence of O_2, the Po ion is rapidly oxidized to the dioxy ion PoO_2^+, which is neutralized to PoO_2 (IP between 10.35 and 10.53 eV) by charge transfer with a molecule of lower ionization potential.

In the presence of water, the fraction of initially charged Po decreases with increasing relative humidity. Because charge transfer from PoO_2^+ to H_2O (IP 12.36 eV) is not favorable, neutralization involves an electron-scavenging mechanism by another species, i.e., a recombination of the negative ion with the charged Po species. Alternatively, the reaction can proceed with negative atmospheric ions produced by cosmic radiations and radiations from natural radioelements.

The α particle emitted by a Rn isotope produces about 150,000 ion pairs and the recoiling Po itself is a highly ionizing particle. Thus, neutralization can also occur through the high local electron density. This process is more likely to occur in radon-rich air, such as in uranium mines. Radiolysis is also expected to influence the further fate of PoO_2, which may react with nitrogen oxides produced by the α-bombardment of air and become transformed into a nitrate.

The wide range of diffusion coefficients determined for Po suggests that the atom may form clusters containing several dozen molecules of a condensable species from air such as H_2O, NO_2, or SO_2. Humidity plays a key role. Radiolysis of water produces strongly oxidizing °OH radicals, which transform SO_2 in SO_3; the latter hydrolyses to form readily condensable sulfuric acid that either combines with other molecules or adheres on existing aerosols. The droplets coagulate and become ultrafine particles which incorporate Po.

Radon daughters attached to aerosols and dust particles are responsible for the plate-out of the radon progeny, i.e., the deposit of airborne radionuclides onto large surfaces such as soils, walls, and furniture. Plate-out depends on the carrier concentration, humidity, presence of trace gases, and surface properties. This phenomenon has been extensively investigated and mathematical models for deposition have been established.[78]

Po nuclides decay by α emission to Pb. The behavior of the lead isotopes is again determined by recoil effects and charge. In this case, however, the α emitter, Po, is bound to an aerosol or to a cluster. The Pb atom may be recoil-detached from the cluster and subsequently become reincorporated into another carrier. Related isotope effects, i.e., differences in the behavior of Pb nuclides in the U and Th families result from the relative values of half-lives.

Radon progeny represents a unique case for which chemical effects of radioactive decays have radioecological implications. The binding of harmful α emitters on aerosols and dust particles is a preliminary step in their incorporation into living bodies. This fact alone justifies the extensive research that is being carried out on the radiochemistry of these nuclides.[69]

Table 10.1
Characteristics of Positron Emitters Used in Nuclear Medicine

Nuclide	Half-life (min)	Reaction	Energy (MeV)	Specific activity (Bq/mole)
^{11}C	20.4	$^{14}N(p,\alpha)$	$15 \rightarrow 0$	3.1×10^{19}
		$^{11}B(p,n)$	$10 \rightarrow 8$	
^{13}N	9.96	$^{12}C(d,n)$	$6 \rightarrow 3$	5.3×10^{19}
		$^{13}C(p,n)$	$10 \rightarrow 3$	
		$^{16}O(p,\alpha)$	$15 \rightarrow 0$	
^{15}O	2.04	$^{14}N(d,n)$	$6 \rightarrow 0$	2.3×10^{20}
		$^{15}N(p,n)$	$10 \rightarrow 4$	
^{18}F	109.8	$^{20}Ne(d,\alpha)$	$15 \rightarrow 0$	3.5×10^{18}
		$^{18}O(p,n)$	$10 \rightarrow 4$	

V. RECOIL-INDUCED SYNTHESIS OF RADIO-LABELED MOLECULES

Hot atom chemistry, i.e., chemistry of atoms activated in nuclear reactions or radioactive decays, has provided extensive information on the type and mechanism of reactions that these atoms undergo and the exotic products that they can form. Hot atom reactions have served as models for chemical syntheses, of which the most important are the preparation of small molecules labeled with radionuclides for nuclear medicine. These compounds, called *radiopharmaceuticals* are widely used as diagnostic and therapeutic agents and as probes in biochemistry and biochemical research.[79]

Positron emission tomography (PET) is a noninvasive diagnostic technique using physiologically active compounds labeled with positron emitters which are introduced into a specific organ in the body. The positrons are slowed down in tissues after a path of a few millimeters and annihilate with an electron. The annihilation generates two 511-keV γ rays emitted in opposite directions. The photons are detected by an array of scintillation counters. Individual positron decay events are localized by two detectors in coincidence mounted at 180°. A projection of the distribution of the activity is obtained by rotating the coincidence line over the section to be measured. The projections recorded at various angles between 0 and 180° give a spatial image of the distribution of the β^+ emitter with an horizontal resolution of 4 to 6 mm and slice thicknesses from 8 to 10 mm. The detector signals are reconstituted as tomographic images which reveal the spatial distribution of the labeled products.

In vivo PET is based on the utilization of the four radionuclides ^{11}C, ^{13}N, ^{15}O, and ^{18}F (Table 10.1).[80,81] Reasons for this choice are: (1) the first three nuclides are isotopes of the most important elements in life; fluorine is regarded as an analogue of hydrogen; (2) the four nuclides are positron emitters;

Table 10.2
Precursors for ^{11}C Labeled Compounds

Target	Precursor	Post-irradiation treatment	Final product
B_2O_3	$^{11}CO + {}^{11}CO_2$	Zn reduction	^{11}CO
		CuO oxidation	$^{11}CO_2$
$N_2(+ <10$ ppm $O_2)$	$^{11}CO + {}^{11}CO_2$	$LiAlH_4$ reduction	$^{11}CH_3OH$
		Iodination	$^{11}CH_3I$
$N_2(+5\%\ H_2)$	$^{11}CH_4$	Reaction with NH_3 over Pt	$H^{11}CN$
$N_2 + CO + Cl_2$	$^{11}COCl_2$	None	$^{11}COCl_2$

Adapted from Ferrieri, R. A. and Wolf, A. P., *Radiochim. Acta*, 34, 69, 1983.

(3) they are readily prepared in a carrier-free state in nuclear reactions using a small cyclotron; and (4) they are short-lived (half-life less than 2 h) and in consequence the radiation dose received by the patient is minimized and waste problems are suppressed.

Products labeled with these radionuclides are carrier-free (amounts in the nanogram range). The introduction in the organism does not perturb the biological equilibrium and even toxic compounds can be used. The short half-life permits a fast repetition of the test.

Utilization of short-lived nuclides has its counterpart. The synthesis of radiopharmaceuticals must be adapted to the half-lives of 2.04 min ^{15}O, 9.96 min ^{13}N, and 20.4 min ^{11}C. The timing is less critical for 108.8 min ^{18}F (Table 10.2). The initial activity must be very high.

The nuclides are produced in nuclear reactions induced by accelerated charged particles with recoil energies in the kiloelectronvolt (keV) range. This energy greatly exceeds the threshold of chemical reactions. The hot atoms are slowed down in the target and before being thermalized are allowed to react with suitable substrates to form very simple molecules. The latter are primary *precursors* useful for the elaboration of complex molecules in a time interval that is compatible with the half-life.

The aim of *applied* hot atom chemistry is to determine the optimal conditions leading to the most appropriate precursors for subsequent synthesis. In certain cases, the primary precursors are transformed into secondary ones by simple post-irradiation treatments.

Commonly used reactions for the production of the four nuclides are shown in Table 10.1. Some targets are made of elements enriched in isotopes of low natural abundance, e.g., ^{15}N, 0.37%, and ^{18}O, 0.20%.

An important requirement is that the nuclide be obtained in a carrier-free state. Because of the ubiquitous presence of the elements C, N, O, and even F, a dilution with stable isotopes is unavoidable. For example, high purity nitrogen contains traces of CO_2 at the parts per million level; ^{18}F may exchange with F atoms of polytetrafluoroethylene material used in the experiments. In typical no-carrier added syntheses of radiopharmaceuticals the ratio $^{11}C/^{12}C$ may be 1/200 and that of $^{18}F/^{19}F$, 1/10,000.

A. ^{11}C Compounds

The reaction $^{14}N(p,\alpha)^{11}C$ is performed on a target of gaseous N_2 with suitable additives. ^{11}C atoms with energies less than 100 eV participate in hot chemical reactions within 10^{-9} s, such as

$$^{11}C + O_2 \rightarrow {}^{11}CO + O$$

$$^{11}C + N_2 \rightarrow {}^{11}CN + N$$

These reactions are characteristic for atoms with high kinetic energies and are not observed for thermal atoms. The carbon monoxide and the CN radicals undergo a second in-target reaction which is dose dependent:

$$^{11}CO + \frac{1}{2} O_2 \rightarrow {}^{11}CO_2 + O$$

with a conversion yield over 95% and

$$^{11}CN + \frac{3}{2} O_2 \rightarrow {}^{11}CO_2 + NO$$

The precursors ^{11}CO, $^{11}CO_2$, and ^{11}CN are used directly without further treatment. Labeled carbon dioxyde is a starting compound for the preparation of methylamine, phosgene, urea, acetic acid, etc. In turn, the latter are used in the fast preparation of radiopharmaceuticals such as ^{11}C-labeled fatty acids, sugars, amines, and steroids.

In other cases, chemical conversion is necessary (Table 10.2). In the presence of 1% H_2 in the target, the primary ^{11}CN from a hot reaction is converted to hydrogen cyanide,

$$^{11}CN + H_2 \rightarrow H^{11}CN + H$$

which in turn is very rapidly transformed via a radiation-induced reaction

$$H^{11}CN + 3H_2 \rightarrow {}^{11}CH_4 + NH_3$$

by which the very useful $H^{11}CN$ is decomposed. Therefore, a chemical conversion is performed *outside* the irradiation chamber to recover hydrogen cyanide. In presence of Pt and at 1000°C, the preceding reaction is reversed.

$$^{11}CH_4 + NH_3 \rightarrow H^{11}CN + 3H_2$$

The in-target reactions can be monitored by subtle changes in the radiation dose and in the target composition.

B. ^{13}N Compounds

The simplest ^{13}N-labeled nitrogen molecule suitable for clinical use is N_2. A convenient source is the $^{12}C(d,n)$ reaction in a CO_2 gas target. The hot ^{13}N atoms form ^{13}NN by reaction with traces of N_2 impurity contained in the gas target.

The most widely used ^{13}N-precursor is $^{13}NH_3$. It is obtained in the irradiation of solid inorganic carbides such as Al_4C_3 or in a flow of CH_4, or even more conveniently by proton irradiation of H_2O. At low dose, the recoil ^{13}N atom reacts with H_2O via sequential abstraction of H atoms

$$^{13}N + H\text{–}OH \rightarrow {}^{13}NH \ldots \rightarrow {}^{13}NH_2 \ldots \rightarrow {}^{13}NH_3$$

followed by hydration

$$^{13}NH_3 + H_2O \rightarrow {}^{13}NH_4^+ + OH^-$$

When the applied dose increases, radiolytic oxidation produces $^{13}NO_3^-$ and $^{13}NO_2^-$. These species are reduced back to $^{13}NH_3$ with Devarda's metal, an alloy of Cu, Al, and Zn.

C. ^{15}O Compounds

Owing to the 2-min half-life of ^{15}O, only very simple molecules can be synthesized and this is possible by recoil-induced reactions in a suitable target (Table 10.3). The most widely used ^{15}O precursors are ^{15}OO, $C^{15}O$, $C^{15}OO$, and $H_2{}^{15}O$. The $^{14}N(d,n)$ reaction performed on a gaseous mixture of N_2 with 2% O_2 produces ^{15}O atoms which participate in the hot reactions

$$^{15}O + O_2 \rightarrow [^{15}O\text{–}O_2] \rightarrow {}^{15}OO + O$$

$$^{15}O + N_2 \rightarrow [N_2 - {}^{15}O] \rightarrow N^{15}O + O \rightarrow N^{15}OO$$

The conversion of the primary products and the separation of impurities is achieved continuously on line behind the target.

Labeled water is produced in a flowing target of N_2 + 5% H_2 via the $^{14}N(d,n)$ reaction. Within the target, hot ^{15}O extracts hydrogen atoms to form $H_2{}^{15}O$:

$$^{15}O + H_2 \rightarrow {}^{15}OH + H \qquad {}^{15}OH + H_2 \rightarrow H^{15}OH + H$$

Table 10.3
Precursors for ^{15}O Labeled Compounds

Target	Precursor	Post-irradiation treatment	Final product
$N_2(+1–4\% \ O_2)$	^{15}OO	Remove traces of ^{15}O nitrogen oxides	^{15}OO
O_2	^{15}OO	Reaction over C at 900°C	
		Zn reduction	$CO^{15}O$
		CuO oxidation	$CO^{15}O$
$N_2(+5\% \ H_2)$	$H_2{}^{15}O$	None	$H_2{}^{15}O$

Adapted from Ferrieri, R. A. and Wolf, A. P., *Radiochim. Acta*, 34, 69, 1983.

Although the recoil method produces strictly carrier-free $H_2{}^{15}O$, it is virtually impossible to eliminate all traces of isotopic water from the target walls and transfer lines and the final $H_2{}^{15}O$ is never carrier-free.

D. ^{18}F Compounds

The purpose of the in-target hot reaction is to produce ^{18}F as ^{18}FF, $H^{18}F$, or as the anion $^{18}F^-$ from which a variety of electrophilic fluorinating agents can be derived. The most common precursor is ^{18}FF, produced from a Ne target in the $^{20}Ne(d,\alpha)$ reaction.

The major application of the PET method is measurement of the metabolism of cerebral glucose using the ^{18}F-labeled 2-fluoro-2-deoxyglucose prepared from the precursor $K^{18}F$.

Once the precursors have been formed by hot reactions in the target, subsequent steps in the elaboration of the radiopharmaceuticals are no longer relevant to radiochemistry. The chemist chooses the fastest path compatible with the highest yield and minimal addition of carrier. The synthesis must be carried out in hot cells with adequate shielding. Complications may arise from radiolytic effects when one is working with high levels of activity.

Commercial cyclotrons for nuclear medicine are provided with push-button equipment that allows the radiopharmacist to select not only the required radionuclide, but also the suitable precursor formed in an automated ''black box''.

VI. POSITRONS IN CHEMISTRY

Positrons and still more the bound state of a positron with an electron, the *positronium atom Ps*, are versatile probes in chemistry.[82] The techniques used are (1) lifetime spectroscopy, which delivers the lifetime τ or the decay constant $\lambda = 1/\tau$ and the relative intensities of positrons and of the two states, ortho and para of Ps; and (2) angular correlation of the annihilation photons and Doppler broadening of the spectral line of the photons, which give useful complementary information through the determination of the e^+/e^- momentum distribution. Current applications concern in large part material science, but exploratory investigations open new promising vistas in chemistry, of which three will be presented.

A. Determination of Association Constants

As a chemical entity, Ps can participate to various reactions with a species E, such as
Redox reaction:

$$Ps + E \rightarrow e^+ + E^-$$

Bound-state formation:

$$Ps + E \rightarrow [Ps - E]$$

Spin conversion:

$$o{-}Ps + E \rightleftarrows p{-}Ps + E$$

In this last reaction, E is a paramagnetic species able to change the spin of the triplet (ortho) and singlet (para) varieties. Ps kinetics are always pseudo first-order; hence the reaction rate constant k' is easily obtained from the measured lifetimes. For instance, for the first two reactions, in which several types of species may be involved,

$$\frac{1}{\tau_{Ps}} = \frac{1}{\tau(0)_{Ps}} + \sum_i k'_i\,[E_i]$$

where $\tau(0)_{Ps}$ is the lifetime of Ps in the neat solvent and $[E_i]$ the effective concentration of the ith species. In the case of complex formation or ion association

$$E_1 + E_2 \rightleftarrows E_3$$

useful information is derived when Ps has distinct reactivities towards at least two of the three constituents of the solution. The concentrations $[E_i]$ are explicit functions of the total concentration $C_{Ei}(0)$, the equilibrium constant K, and possibly of additional parameters. Generally, the experiments include solutions of E_1 and E_2 alone from which the rate constants k'_{E_1} and k'_{E_2} are derived, and mixtures with different values of $C_{E_1}(0)$ and $C_{E_2}(0)$. The method

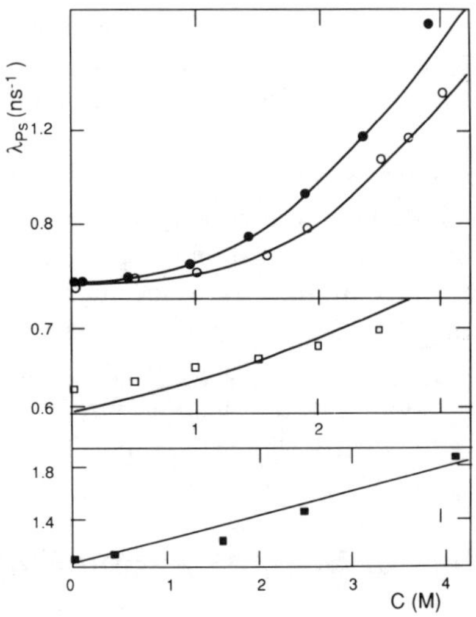

FIGURE 10.5. Variation of the decay constant of Ps in aqueous solution as a function of solute concentration: ○, $AgClO_4$, □, $NaClO_4$ in 0.933 *M* $AgClO_4$, ●, $AgClO_4$ in 1*M* HClO4, ■, $HClO_4$ in 2.4 *M* $AgClO_4$, For the solid lines, see text.

has been used to determine the formation constants of inorganic ion associations, for example of the polyhalides I_2Cl^- and I_2Br^- in methanol and I_2Cl^- in dimethyl sulfoxide with a precision of 10%.[83] The formation of complexes of organic molecules and polyion-counterion interactions have been investigated by the same method.[84,85]

The Ps intensity may also be used when the constituents of the solution have different abilities to *inhibit* Ps formation.[86] However, such applications are more limited because the processes leading to the formation are too fast to be time-resolved and therefore the variation of the intensity of Ps with the concentrations of the species in solution is only known through empirical relations.

The method has some limitations. Qualitatively, Ps must react with at least one of the species. This would exclude *a priori* chemically inert species such as the alkali, alkali earth, or halide ions and most of the organic molecules having no very active groups (alkanes, alcohols, ketones). But this rule may not be too strict, as the properties of the complex may be difficult to predict. This is illustrated in Figure 10.5, which shows that although neither of the Ag^+ and ClO_4^- ions react with Ps (λ_{Ps} is constant at low $AgClO_4$ concentration), the ion association formed at high concentrations is very reactive.

Quantitatively, the measurable change in λ_{Ps} is given by the reactivity term $k'[E_i]$. Considering that k′ is at most of the order of 2×10^{10} M^{-1} s^{-1} for purely diffusion controlled reactions, and that $\tau(0)_{Ps}$ amounts to a few ns in most liquids, it ensues that the method is useful in the concentration range from 0.01 to 1 *M*. This in turn gives a limit of the formation constants which can be determined. For a one to one stoichiometry reaction such as $E_1 + E_2 \rightarrow E_3$, Ps is most sensitive to values of the equilibrium constant K in the range from 0.1 to 10^3 M^{-1}. As compared to conventional techniques, such as spectrophotometry or conductivity, this offers the advantage of access to low formation constants in solution of high concentration, where the signals of other techniques might well be saturated. Hence, Ps is a quite useful probe for reaction studies in *concentrated* solutions.

As a drawback, it is necessary to establish semi-empirical functions to account for the ionic strength effects and derive the thermodynamical formation constant K°. Figure 10.5 illustrates the cubic variation of λ_{Ps} with the concentration of $AgClO_4$, denoting the presence of triple ion association, and the linear variation with $[ClO_4^-]$, indicating that the association includes only one ClO_4^- anion and is therefore $Ag_2ClO_4^+$. All data were fitted taking into account the activity correction using the Davies equation with $k'(Ag_2ClO_4^+) = (1.64 \pm 0.05)\ M^{-1}\ ns^{-1}$ and $K° = (0.052 \pm 0.05)\ M^{-2}$ (solid lines in Figure 10.5).[83,87]

B. Physicochemical Properties of Species in Solution

Any variation observed in the reaction of a species with Ps gives some information on properties of this species or its reaction product with other species. From what precedes, it appears that k′ can change when an ion becomes associated and the association can be characterized. For instance, site-binding of Co^{2+} ions on polyanion chains suppresses completely the spin conversion of Ps observed when these ions are free or incorporated in an outer-sphere complex.[85] The fact that the reactivity of the latter complexes is little affected in the very viscous polyanion solutions as compred to neat water, gives additional information on the microviscosity of the systems. More generally, the spin state of transition metal complexes in solution can be assessed.[88] Similar changes can benefit to the study of the structure of inhomogeneous systems, in which a species serves as a probe for para-Ps. Thus, the changes observed in the reactivity of nitrobenzene and of $CuCl_2$ with Ps in micellar solutions have given information on the location of these compounds. The reaction of benzoquinone with Ps has been useful in the study of a model membrane bilayer and of the transition from the crystalline to the gel phase.[89,90]

Interesting information arises when the nature of the species reaction is changed; this is the case for free nitroxyl radicals, which promote Ps spin conversion in a variety of polar and nonpolar liquids, except in water and m-cresol, where they induce the oxidation of Ps.[91] In these solvents hydrogen bonding is most efficient; it is concluded that the nitroxyl group exists in two mesomeric forms,

$$> N\text{–}O\cdot \rightleftarrows N^{+\cdot} - O^-$$

in which the odd electron has a very different availability for Ps and of which the latter form is greatly favored in the most protic solvents where the oxygen doublet is involved in a hydrogen bond.

C. Physicochemical Properties of Liquids and Molecular Solids

Ps has been widely used as a direct probe in liquids and molecular solids. Three main domains are as follows.

First, nonhomogeneous systems, such as micelles, for which are determined the critical micelle concentration, the structure, the effect of temperature, and of the counterion.[89,92]

Second, a study of phase transitions in molecular solids, particularly as regards the existence of plastic phases.[93] Examples are provided among others by cyclohexane, phenanthrene, microemulsions, and liquid crystals.[93-97] The changes in the intensity of Ps and of the lifetime τ_{Ps} usually both denote clearly the phase transition. Although these parameters are also sensitive to other properties of the solids, like the dielectric constant, they appear to be to a large extent determined by the presence of free space such as microvoids or molecule vacancies. The Ps intensity increases with the concentration of such spaces, while τ_{Ps} increases with their size according to

$$\frac{1}{\tau_{Ps}} = 2\left[1 - \frac{R}{R_0} + \left(\frac{1}{2\pi}\right)\sin\left(2\pi\frac{R}{R_0}\right)\right]$$

for spherical voids of radius R, where $R_0 = R + 0.166$ (nm). As compared to other techniques, these properties have been widely used to derive additional information concerning thermally activated formation of vacancies in solids. Since Ps, as a neutral species, has a strong affinity for voids in general, it will preferentially get trapped before annihilation in molecule vacancies which are usually large as compared to the free volumes present in a nondefected lattice. This trapping process leads to a sigmoidal variation of τ_{Ps} with temperature from which the formation enthalpy ΔH_v of a vacancy can be derived. If the Ps lifetimes in the bulk, $\tau(0)_{Ps}$ and in the vacancy $\tau(v)_{Ps}$ cannot be resolved, the variation of the measured average lifetime is given by

$$\langle\tau\rangle_{Ps} = \tau(0)\mathrm{Ps}\,\frac{(1 + \sigma\tau(v)\mathrm{Ps})}{(1 + \sigma\tau(0)\mathrm{Ps})}$$

In this expression, the trapping rate constant σ is proportional to the vacancy concentration and therefore,

$$\sigma = \sigma_0 \exp\left[-\Delta H_v/kT\right]$$

If the two lifetimes are resolved, the longer lifetime identifies with $\tau(v)_{Ps}$, while the shorter decay rate constant is $(1/\tau(0)_{Ps} + \sigma)$.

A third field, which will only be mentioned here, concerns the polymers. On the same qualitative and quantitative bases as indicated above, these investigations are aimed at determining the concentration and size of the free volumes in polymers and their change with temperature and additives.[82,98]

More closely related to radiation effects, the behavior of positrons and Ps is extensively applied to the investigation of defects introduced by ionizing radiations in various solids such as metals, ionic crystals, and polymers.

REFERENCES

1. **Lierse, Ch., Treiber, W., and Kim, J. I.,** Hydrolysis reactions of Np(V), *Radiochim. Acta,* 38, 27, 1985.
2. **Stadler, S. and Kim, J. I.,** Hydrolysis reactions of Am (III) and Am (V), *Radiochim. Acta,* 44/45, 39, 1988.
3. **Duplessis, J. and Guillaumont, R.,** Hydrolysis of tetravalent neptunium, *Radiochem. Radioanal. Lett.,* 31, 283, 1977 (in French).
4. **Metivier, H. and Guillaumont, R.,** Hydrolysis and complexing of tetravalent plutonium, *J. Inorg. Nucl. Chem.,* 38, 179, 1976.
5. **Metivier, H. and Guillaumont, R.,** Contribution to the study of the hydrolysis of tetravalent plutonium, *Radiochem. Radioanal. Lett.,* 1, 209, 1969 (in French).
6. **Metivier, H. and Guillaumont, R.,** Hydrolysis of tetravalent plutonium, *Radiochem. Radioanal. Lett.,* 10, 27, 1972 (in French).
7. **Metivier, H. and Guillaumont, R.,** Citric complexes of tetravalent plutonium, *Radiochem. Radioanal. Lett.,* 10, 239, 1972 (in French).
8. **Hubert, S., Hussonnois, M., and Guillaumont, R.,** Measurement of complexing constants by radiochemical methods, *Structure Bonding,* 34, 1, 1978.
9. **Hubert, S., Hussonnois, M., and Guillaumont, R.,** Observation of the nephelauxetic effect in a citric complex of the 4f series elements, *J. Inorg. Nucl. Chem.,* 35, 2923, 1973 (in French).
10. **Hubert, S., Hussonnois, M., Brillard, L., Goby, G., and Guillaumont, R.,** Simultaneous determination of the formation constants of the citric complexes of americium, curium, californium, einsteinium and fermium, *J. Inorg. Nucl. Chem.,* 36, 2361, 1974 (in French).
11. **Guillaumont, R. and Hubert, S.,** Complexing of 4f aqua ion in relation with nephelauxetic effect, *J. Chem. Research (S),* 11, 282, 1977; *J. Chem. Research (M),* 11, 3201, 1977.
12. **Rösch, F., Dittrich, S., Buklanov, G. V., Milanov, M., Khalkin, V. A., and Dreyer, R.,** Electromigration of carrier free radionuclides. II. Complex formation of ^{239}Np(V) with oxalate, tartrate and sulphate in neutral inert electrolytes, *Radiochim. Acta,* 48, 205, 1989.
13. **Kim, J. I.,** Chemical behaviour of transuranic elements in natural aquatic systems, in *Handbook of Physical and Chemical Properties of Actinides,* Vol. 4, Freeman, A. J. and Keller, C., Eds., North-Holland, Amsterdam, 1986, 413.
14. **Baas-Becking, L. G. M., Kaplan, I. R., and Moore, D.,** Limits of the natural environment in terms of pH and Eh, *J. Geol.,* 68, 243, 1960.
15. **Righetto, L., Bidoglio, G., Marcandalli, B., and Bellobono, I. R.,** Surface interactions of actinides with alumina colloids, *Radiochim. Acta,* 44/45, 73, 1988.

16. **Bondietti, E. A. and Trabalka, J. R.,** Evidence for Pu(V) in alcaline freshwater pond, *Radiochem. Radioanal. Lett.*, 42, 169, 1980.
17. **Nitsche, H. and Edelstein, N. M.,** Solubilities and speciation of selected transuranium ions. A comparison of a non-complexing solution with a groundwater from the Nevada Tuff site, *Radiochim. Acta,* 39, 23, 1985.
18. **Choppin, G. R.,** Separation of actinides in aqueous solutions by oxidation states, in *Actinides-Lanthanides Separations,* Choppin, G. R., Navratil, J. D., and Schulz, W. W., Eds., World Scientific, Singapore, 1985, 176.
19. **Choppin, G. R.,** Actinide speciation in spent fuel leaching studies, in *Materials Research Society Symposium Proceedings,* Vol. 176, Materials Research Society, Pittsburgh, 1990, 149.
20. **Nitsche, H.,** Basic research for assessment of geologic nuclear waste repositories: what solubility and speciation of transuranium elements can tell us, in *Materials Research Society Symposium Proceedings,* Vol. 212, Materials Research Society, Pittsburgh, 1991, 517.
21. **Inoue, Y. and Tochiyama, O.,** Determination of the oxidation states of neptunium at tracer concentration by adsorption on silica gel and barium sulfate, *J. Inorg. Nucl. Chem.*, 39, 1443, 1977.
22. **Kobashi, A., Choppin, G. R., and Morse, J. W.,** A study of techniques for separating plutonium in different oxidation states, *Radiochim. Acta,* 43, 211, 1988.
23. **Saito, A., Roberts, R. A., and Choppin, G. R.,** Preparation of solutions of tracer level plutonium(V), *Anal. Chem.*, 57, 390, 1985.
24. **Edgington, D. N.,** Characterization of transuranium elements at environmental levels, in *Techniques for Identifying Transuranic Speciation in Aquatic Environments,* International Atomic Energy Agency, Vienna, 1981, 3.
25. **Lovett, M. B. and Nelson, D. M.,** Determination of some oxidation states of Pu in sea water and associated particulate matter, in *Techniques for Identifying Transuranic Speciation in Aquatic Environments,* International Atomic Energy Agency, Vienna, 1981, 27.
26. **Guillaumont, R. and Adloff, J. P.,** Behaviour of environmental plutonium at very low concentration, *Radiochim. Acta,* 58, 93, 1992.
27. **Sattelberger, P., Mang, M., Herrmann, G., Riegel, J., Rimke, H., Trautmann, N., Ames, F., and Kluge, H. J.,** Separation and detection of trace amounts of technetium, *Radiochim. Acta,* 48, 165, 1989.
28. **Lieser, K. H. and Mühlenweg, U.,** Neptunium in the hydrosphere and in the geosphere. II. Column experiments with neptunium, *Radiochim. Acta,* 44/45, 129, 1988.
29. **Seaborg, G. T. and Loveland, W. D.,** *The Elements Beyond Uranium,* John Wiley & Sons, New York, 1991, 65.
30. **Hulet, E. K.,** Chemistry of the elements einsteinium through element-105, *Radiochim. Acta,* 32, 7, 1983.
31. **Zvara, I.,** Thermochromatographic method of separation of chemical elements in nuclear and radiochemistry, *Isotopenpraxis,* 26, 251, 1990.
32. **Fargeas, M., Fremont, R., Legoux, Y., and Merini, J.,** Thermochromatography of fluorides and oxyfluorides of actinides in high valence states, *J. Less Common Metals,* 121, 439, 1986 (in French).
33. **Fremont-Lamouranne, R., Legoux, Y., Merini, J., Zhuikov, B. L., and Eichler, B.,** Thermochromatography of the actinides and their compounds, in *Handbook on the Physics and Chemistry of the Actinides,* Freeman, A. J. and Keller, C., Eds., North-Holland, Amsterdam, 1985, 331.
34. **Zvara, I.,** One atom at a time chemical studies of transactinide elements, in *Transplutonium Elements,* Müller, W. and Lindner, R., Eds., North-Holland, Amsterdam, 1976, 11.
35. **Herrmann, G. and Trautmann, N.,** Rapid chemical methods for identification and study of short-lived nuclides, *Annu. Rev. Nucl. Part. Sci.*, 32, 117, 1982.

36. **Schädel, M.,** Radiochemical studies of transactinides elements in School-Seminar on Heavy Ion Physics, Dubna, Russia, 1989, and preprint GSI 89-72, Gesellschaft für Schwerionenforschung, Darmstadt, 1989.
37. **Kratz, J. V., Zimmermann, H. P., Scherer, U. W., Schädel, M., Brüchle, W., Gregorich, K. E., Gannett, C. M., Hall, H. L., Henderson, R. A., Lee, D. M., Leyba, J. D., Nurmia, M. J., Hoffman, D. C., Gäggeler, H., Jost, D., Baltensperger, U., Ya Nai-Qi, Türler, A., and Lienert, Ch.,** Chemical properties of element 105 in aqueous solution: halide complex formation and anion exchange into triisooctyl amine, *Radiochim. Acta,* 48, 121, 1989.
38. **Oganessian, Yu. Ts., Hussonnois, M., Demin, A. G., Khaitonov, Y. P., Bruchertseifer, H., Constantinescu, O., Korotkin, Y. S., Tretyakova, S. P., Utyonkov, V. K., Shirokovsky, I. V., and Estevez, J.,** Experimental studies of the formation and radioactive decay of isotopes with $Z = 104–109$, *Radiochim. Acta,* 37, 113, 1984.
39. **Pyykkö, P.,** Relativistic effects in structural chemistry, *Chem. Rev.,* 88, 563, 1988.
40. **Keller, O. L.,** Chemistry of the heavy actinides and light transactinides, *Radiochim. Acta,* 37, 1984, 169.
41. **Seaborg, G. T. and Keller, O. L.,** Future elements, in *The Chemistry of Actinide Elements,* Vol. 2, 2nd ed., Katz, J. J., Seaborg, G. T., and Morss, L. R., Eds., 1986, 1629.
42. **Keller, O. L., Nestor, C. W., and Burkhard, F.,** Predicted properties of the superheavy elements. II. Element 115, ekabismuth, *J. Phys. Chem.,* 78, 19, 1974.
43. **Keller, O. L. and Seaborg, G. T.,** Chemistry of the transactinide elements, *Annu. Rev. Nucl. Sci.,* 27, 139, 1977.
44. **Lundqvist, R., Hulet, E. K., and Baisden, P. A.,** Electromigration method in tracer studies of complex chemistry. II. Hydrated radii and hydratation numbers of trivalent actinides, *Acta Chem. Scand. Ser. A,* 35, 653, 1981.
45. **Fourest, B., Duplessis, J., and David, F.,** Limiting diffusion coefficients and hydrated radii of Eu^{3+}, Am^{3+}, Cm^{3+}, Cf^{3+}, and Es^{3+}, *J. Less Common Metals,* 92, 17, 1983 (in French).
46. **Fourest, B., Duplessis, J., and David, F.,** Comparison of diffusion coefficients and hydrated radii for some trivalent lanthanide and actinide ions in aqueous solution, *Radiochim. Acta,* 36, 191, 1984.
47. **David, F., Fourest, B., and Duplessis, J.,** Hydration thermodynamics of plutonium and transplutonium ions, *J. Nucl. Mat.,* 130, 273, 1985.
48. **Fourest, B., David, F., and Haltier, E.,** Electromigration of americium and californium in aqueous solution, *Lanthanide Actinide Res.,* 2, 393, 1988.
49. **Fourest, B., Duplessis, J., and David, F.,** Electromigration of trivalent lanthanide ions in aqueous solution, *Radiochim. Acta,* 46, 131, 1989.
50. **Myassoedov, B. F. and Lebedev, I. A.,** Recent analytical chemistry of the transplutonium elements, *Radiochim. Acta,* 32, 55, 1983.
51. **Mikheev, N. B.,** Lower oxidation states of lanthanides and actinides, *Inorg. Chim. Acta,* 94, 241, 1984.
52. **Mikheev, N. B.,** Lower oxidation states of actinides, *Radiochim. Acta,* 32, 69, 1983.
53. **David, F.,** Recent advances in the thermodynamics of trivalent actinides, *Int. Symp. Radiochem.,* Preprints Volume, Sood, D. D., Ed., Bhabha Atomic Research Centre, Bombay, 1991, 8; and Rep. IPNO-DRE, 90-23, IPN, BP 1, F 91406, Orsay.
54. **David, F.,** Thermodynamic properties of lanthanide and actinide ions in aqueous solutions, *J. Less Common Metals,* 121, 27, 1986.
55. **Mikheev, N. B. and Mayasoedov, B. F.,** Lower and higher oxidation states of transplutonium elements in solutions and melts, in *Handbook on the Physics and Chemistry of the Actinides,* Vol. 3, Freeman, A. J. and Keller, C., Eds., North-Holland, Amsterdam, 1985, 347.

56. **David, F. and Bouissieres, G.,** Comparative formation of amalgams of transuranium elements and rare earths by electrolysis, *Inorg. Nucl. Chem. Lett.,* 4, 153, 1968.
57. **Maly, J.,** The amalgamation behavior of heavy elements. III. Extraction of radium, lead and the actinides by sodium amalgam from acetate solutions, *J. Inorg. Nucl. Chem.,* 31, 1007, 1969.
58. **Maly, J.,** The amalgamation behavior of heavy elements. IV. The tracer chemistry of divalent mendelevium, *J. Inorg. Nucl. Chem.,* 31, 741, 1969.
59. **David, F.,** Oxidation-reduction and thermodynamic properties of curium and heavier actinide elements, in *Handbook on the Physics and Chemistry of the Actinides,* Vol. 4, Freeman, A. J. and Keller, C., Eds., North-Holland, Amsterdam, 1986, 97.
60. **Samhoun, K., David, F., Hahn, R. L., O'Kelley, G. D., Tarrant, J. R., and Hobart, D. E.,** Electrochemical study of mendelevium in aqueous solution: no evidence for monovalent ions, *J. Inorg. Nucl. Chem.,* 41, 1749, 1979.
61. **David, F., Samhoun, K., Hulet, E. K., Baisden, P. A., Dougan, R., Landrum, J. H., Lougheed, R. W., Wild, J. F., and O'Kelley, G. D.,** Radiopolarography of mendelevium in aqueous solutions, *J. Inorg. Nucl. Chem.,* 43, 1331, 1981.
62. **David, F., Maslennikov, A. G., and Peretrukhin, V. F.,** Electrochemical reduction of actinide ions in aqueous solution and some examples of its applications, *J. Radioanal. Nucl. Chem.,* 143, 415, 1990.
63. **Samhoun, K. and David, F.,** Radiopolarography of Am, Cm, Bk, Cf, Es, and Fm, in *Transplutonium Elements,* Müller, W. and Linder, R., Eds., North-Holland, Amsterdam, 1976, 297.
64. **Legoux, Y. and Merini, J.,** Diffusion of Po, At, Rn, Fr, and Ra recoil-implanted in tantalum, *J. Less Common Metals,* 105, 169, 1984 (in French).
65. **Legoux, Y. and Merini, J.,** Diffusion of Cf, Es, and Fm in tantalum, *J. Less Common Metals,* 121, 49, 1986 (in French).
66. **Legoux, Y. and Merini, J.,** Thermal diffusion of U, Pu, and Cm in tantalum, *J. Less Common Metals,* 144, 165, 1988 (in French).
67. **Lieser, K. H., Ament, A., Hill, R., Singh, R. N., Stingl, U., and Thybusch, B.,** Colloids in groundwater and their influence on migration of trace elements and radionuclides, *Radiochim. Acta,* 49, 83, 1990.
68. **Sheppard, M. I. and Thibault, D. H.,** Default soil solid/liquid partition coefficients, Kds, for four major soil types, a compendium, *Health Phys.,* 59, 471, 1990.
69. **Hopke, P. K., Ed.,** *Radon and Its Decay Products. Occurrence, Properties, and Health Effects,* American Chemical Society, Washington D.C., 1987.
70. **Harley, N. M., Ed.,** Proc. 24th Annu. Meet. Nat. Council on Radiation Protection and Measurements, Bethesda, MD, 1989.
71. **Crameri, R. and Burkart, W.,** The radon problem, *Radiat. Phys. Chem.,* 34, 251, 1989.
72. **Adloff, J. P.,** Hot atom chemistry of radon and its progeny, in *Handbook of Hot Atom Chemistry,* Adloff, J. P., Gaspar, P. P., Imamura, M., Maddock, A. G., Matsuura, T., Sano, H., and Yoshihara, K., Eds., Kodansha, Tokyo, 1992, 627.
73. **Fleischer, R. L.,** Alpha-recoil damage and preferential solution effects: mechanism of track production, and implications for isotopic disequilibrium, radon release and waste disposal, *Nucl. Tracks,* 5, 386, 1981.
74. **Fleischer, R. L.,** Theory of alpha recoil effects on radon release and isotopic disequilibrium, *Geochim. Cosmochim. Acta,* 47, 779, 1983.
75. **Semkow, T. M.,** Recoil-emanation theory applied to radon release from mineral grains, *Geochim. Cosmochim. Acta,* 54, 425, 1989.
76. **Stein, L.,** The chemistry of radon, *Radiochim. Acta,* 32, 163, 1983.
77. **Stein, L.,** Chemical properties of radon, in *Radon and its Decay Products. Occurrence, Properties, and Health Effects,* American Chemical Society, Washington D.C., 1987, 240.

78. **Bigu, J.,** Plate-out of radon and thoron progeny on large surfaces, in *Radon and its Decay Products. Occurrence, Properties, and Health Effects,* American Chemical Society, Washington D.C., 1987, 272.
79. **Lambrecht, R. M. and Morcos, N. A., Eds.,** *Applications of Nuclear and Radiochemistry. Part I. Radiopharmaceutical Chemistry,* Pergamon, New York, 1982.
80. **Ferrieri, R. A. and Wolf, A. P.,** The chemistry of positron emitting nucleogenic (hot) atoms with regard to preparation of labelled compounds of practical utility, *Radiochim. Acta,* 34, 69, 1983.
81. **Iwata, R. and Wolf, A. P.,** Practical production of simple positron emitting compounds by hot atom chemistry method, in *Handbook of Hot Atom Chemistry,* Adloff, J. P., Gaspar, P. P., Imamura, M., Maddock, A. G., Matsuura, T., Sano, H., and Yoshihara, K., Eds., Kodansha, Tokyo, 1992, 515.
82. **Schrader, D. M. and Jean, Y. C., Eds.,** *Positron and Positronium Chemistry,* Studies in Physical and Theoretical Chemistry 57, Elsevier, Amsterdam, 1988.
83. **Duplâtre, G., Al-Shukri, L. M., and Haessler, A.,** Some applications of positron annihilation lifetime spectroscopy to complex formation and ion association studies, *J. Radioanal. Nucl. Chem.,* 55, 199, 1980.
84. **Jean, Y. C. and Ache, H. J.,** Studies of molecular association in biological systems by positron annihilation techniques, *J. Am. Chem. Soc.,* 82, 1623, 1977.
85. **Zana, R., Millan, S., Abbé, J. C., and Duplâtre, G.,** Polyion-counterion interaction in aqueous solutions probed by positron annihilation techniques. I. Co^{2+}-polyanion systems, *J. Phys. Chem.,* 86, 1457, 1982.
86. **Duplâtre, G., Abbé, J. C., Maddock, A. G., and Haessler, A.,** Inhibition of positronium formation by oxalic acid and oxalates in aqueous solutions, *Radiat. Phys. Chem.,* 11, 249, 1978.
87. **Maddock, A. G., Abbé, J. C., Duplâtre, G., and Haessler, A.,** The chemistry of positronium. III. Inhibition and quenching by silver and cadmium salts, *Chem. Phys.,* 26, 163, 1977.
88. **Vertes, A., Kajcsos, Z., Marzis, L., Brauer, G., Hüller, J., Zay, I., and Burger, K.,** Effect of spin state of transition metals on their interaction with orthopositronium in aqueous solution, *J. Phys. Chem.,* 88, 3969, 1984.
89. **Jean, Y. C. and Ache, H. J.,** Positronium reactions in micellar systems, *J. Am. Chem. Soc.,* 99, 7504, 1977.
90. **Wang, Y. Y., Hancock, A. J., and Jean, Y. C.,** Interaction of benzoquinone with a model membrane bilayer as reported by a positronium probe, *J. Am. Chem. Soc.,* 105, 7272, 1983.
91. **Abbé, J. C., Duplâtre, G., Haessler, A., Marques Netto, A., and Pilo Veloso, D.,** Temperature and (polar) solvent effects on positronium reactions with nitroxyl free radicals, *J. Phys. Chem.,* 88, 2071, 1984.
92. **Djermouni, B. and Ache, H. J.,** Effect of temperature and counterion on the micelle formation process studied by positron annihilation techniques, *J. Phys. Chem.,* 83, 2476, 1979.
93. **Eldrup, M., Lightbody, D., and Sherwood, J. N.,** Studies of phase transformations in molecular crystals using the positron annihilation technique, *Farad. Disc.,* 69, 175, 1980.
94. **Lightbody, D., Sherwood, J. N., and Eldrup, M.,** Temperature and phase dependence of positron lifetimes in solid cyclohexane, *Chem. Phys.,* 93, 475, 1985.
95. **Lenc, R.,** Positronium formation and lifetimes in n-higher aliphatic alcohols, *Chem. Phys.,* 131, 443, 1988.
96. **Serrano, J., Reynoso, R., Lopez, R., Olea, O., Djermouni, B., and Fucugauchi, L. A.,** Phase behavior of microemulsion systems studied by positron annihilation techniques, *J. Phys. Chem.,* 87, 707, 1983.

97. **Nicholas, J. B. and Ache, H. J.,** Phase and temperature dependence of positron annihilation in liquid crystals, *Chem. Phys.*, 57, 1598, 1972.
98. **Jean, Y. C.,** Positron annihilation spectroscopy for chemical analysis: a novel probe for microstructural analysis of polymers, *Microchem. J.*, 42, 72, 1990.

CONCLUSION

The birth of radiochemistry dates from the discovery of polonium and radium in trace amounts, at a period just following the recognition of natural radioactivity. In its development and extension, the new discipline progressively acquired an individual character within nuclear science and later became a subtopic of chemistry, contributing greatly to the subsequent discoveries of artificial radioactivity, fission, and transuranium elements. During the last decade, radiochemical methods have been invaluable in the chemical characterization of transactinides. Indeed, chemistry in general has always played an important part at the decisive stages in the development of nuclear science.

On such occasions, and in the course of its progress, radiochemistry has frequently been involved in numerous subjects of research and scientific fields outside of chemistry. Some of the radiochemical topics directly related to chemistry underwent autonomous development, whereas others led to new applications that are now no longer commonly practiced by radiochemists.

These evolutionary tendencies are perhaps not surprising for a domain that encompasses high-performance techniques of interest to many different fields. At present, radiochemistry has two important tasks to fulfill; one of these involves passing the boundary from the invisible to the visible world, that is, from imponderable to weighable amounts of various artificial radioelements. The other is preparing the future in the utilization of nuclear fission energy.

Today, 100 years after the discovery of radioactivity, humanity has to live — for better or for worse — with the nuclear age. The development of nuclear science has already attained a high level and we have attempted here to interpret the changing facets of modern radiochemistry on the basis of the experience of the past and the perspectives of the nuclear future.

It is clear that radiochemistry is, by nature, a field that deals with chemical properties of radionuclides at infinite dilution, but it is evident that it can also be a valuable aid in studying the properties of elements in general. We have set up several guidelines for outlining the fields of chemical research that are also common to radiochemistry, and from these a unified basis for characteristic and specific radiochemical methods has emerged.

By definition, these techniques imply the handling of radioactive matter. For this reason, radiochemistry has essentially evolved in centers of nuclear research and development and it is probably here that further progress will be made, where laboratories and elaborate equipment are available for preparing radionuclides and performing highly complicated experiments. On the other hand, many lines of research deal with moderate levels of radioactivity that do not require special devices. In this event, the use of radiochemical methods can be extended to many elements in nonspecialized laboratories.

In concluding this treatise, it may be of interest to venture several remarks on future prospects of radiochemistry. To this end, it is expedient to consider various topics of fundamental research that appear to require further development.

First, within the general context of the contribution of radiochemistry to chemistry, further thought should be devoted to the basic principles of radiochemical methods and to their advantages and limits with respect to the identification of species and oxidation states.

In particular, there are two main lines of research that should be explored in greater detail; one of these involves interpretation of chemical information conveyed by radiation (nuclear methods) and the other concerns the use of radionuclides as tracers (partition and transport methods). This second line of investigation requires special attention because its approach is indirect, noninstantaneous, and frequently perturbing. Further studies of the behavior or multivalent radionuclides are required in order to test the validity of the methods applied. When work is performed at the tracer level, any conclusions drawn will depend entirely on the coherency of the results obtained, and it is obviously desirable to dispose of several complementary methods when these exist.

Several problems that merit special attention at the present time may be summarized as follows:

1. Questions involving the behavior of an element at high dilution, either in thermal equilibrium with its environment (the classical case) or not (the case of a nucleogenic element). In the first instance, it is necessary to know the critical concentration at which reactions cease to be governed by thermodynamics while kinetics become determinant. In the latter case, additional knowledge is required on the relaxation stages of the evolving entities. In both situations one may consider, with extreme caution, extrapolation of behavioral properties down to the scale of a few atoms.
2. Problems concerning comparative behavior of several elements with the same oxidation state under strictly identical conditions. These studies may provide evidence for thermodynamic effects that are frequently very slight but of considerable interest, for example in relation to relativity and covalency. Progress in these fields should provide an effective means for testing various models of chemical physics.
3. Further research on the nature of aqueous solutions. The working models presently in use were established for concentrations usually higher than 10^{-6} M and are commonly extrapolated to infinite dilution for the case of monovalent ions. These models can be tested and improved using multicharged radioactive ions down to nearly infinite dilution. Similarly, nuclear probes (e.g., positron and γ-γ correlation) can be applied to the investigation of solutions that are too concentrated for study by classical techniques, thus enabling a verification for existing models in the high concentration range.

4. Problems related to chemical interaction. These may involve mechanisms such as relaxation of nucleogenic entities which imply energy transfer in polyatomic systems, or concern the statistical nature of reactions at the scale of a few atoms.

Further points of interest deal with new applications in molecular labeling and in nuclear medicine. Important contributions are also expected in questions of vital importance such as environmental radioactivity and radioactive matter taken up by living organisms.

When one is concerned with the terminal elements of the periodic table, for which many complications arise with respect to the notions of classical chemistry, the aid of radiochemical methods becomes indispensable. Here, the acquisition of new and fundamental data serve to further our knowledge of artificial radioelements. Comparison of the properties of these elements with those of their lower homologues provides an incentive to the refinement of relativistic polyelectronic models.

Radiochemistry will continue to provide new challenges, some of which will not be restricted to nuclear concepts and the utilization of radionuclides. This discipline will remain indispensable to the expansion of our frontiers of knowledge. Without continuing research in radiochemistry, problems of rad-waste management could not be solved by assessment of the radiological impact of radionuclides involved in the nuclear fuel cycle and radioactivity would be unable to unfold its useful potential in the service of mankind.

INDEX

A

Abelson, P., 21
Absorption, 161–162
  internal, 271
  Mössbauer experiment, 224–228
Acid-base equilibrium, 206–207, 217
Acid-base pathways, 182–183
Actinides
  accelerator-produced heavy, 154
  aqua ions of, 366–367
  atomic numbers for, 364
  chemical properties of, 122
  chemistry of in solution, 357
  connection of with radioactive families, 135
  detection of, 187
  discontinuity of, 367
  hydrolysis and complexation of, 356–360
  isotopes of in spent fuel, 136–137
  light, 40
  mobility in, 367
  oxidation states in, 122–123, 368–370
  in Periodic Table, 120–121
  problems related to classification of, 122
  produced in high flux reactors, 151
  radiochemistry and chemistry of, 356–371
  recovery of, 131
  series of, 21, 122
  speciation of in environment, 360–363
Actinium
  activity of, 19
  discovery of, 7
Actinium K, 14
Actinium X, 13
Actino-uranium family, 11
Actinon, 130, 375
Activation analysis, 124
Activity measurements, 162–168, 326–327
Adiabatic transformation, 49–50
Adsorption
  in coprecipitation, 269–272
  isotherms of, 272
  perturbing spontaneous deposit, 273
  reaction of, 333–334
  on vessel walls, 308
Alkaline earth series, 9
Alpha counting, 161
Alpha emissions, 73–74, 77
  heat from, 106
  by radon isotopes, 378
Alpha particles, 45, 49
  decay of, 72–74, 104, 112
    perturbation of, 231–232
    rate of, 74, 231
  energy of, 72–74, 222–223
  from nuclear reactions, 93
  reactions induced by, 94
Aluminum, radioactive, 15–16
Americium
  autoreduction of, 111
  chemical identification of, 366
  oxidation states in, 368–370
Amine salts, 259
Amphoteric elements, 270
Angular correlation spectroscopy, 233–235
Anionic resin exchangers, 262
Annihilation spectroscopy
  chemical information from, 228–229
  energy of, 223
  principle of, 230
Anomalous mixed crystals, 266, 268
Aqua ions
  hydrolysis of, 363
  structure of, 366–367
Aqueous media
  artificial, 360
  natural, 361
Aqueous solution
  components of, 180
  element distribution in, 178–180
  equilibria of, 253
  future research on, 396–397
  radiochemical partition in, 247–250
  radiolysis of, 108–111
Aquifers, natural, 360–361
Association constants, determination of, 384–386
Astatine, 23
Atmospheric radioactivity, 130
Atomic mass units, 43–44
Atomic masses, 55
Atomic number, 43, 364
Atomic states, 53–54
Atomic weight

of artificial radioelements, 130–131
of primordial radionuclides, 140–141
Atoms
activity measurements of, 326–327
activity of, 63–65
age of, 58
behavior of, 314
chemical behavior of one hundred or less, 327–340
decay probability of, 315–316
disintegration of several, 317–326
disintegration of single, 314–316
probability of decay of, 59
problems with very small number of, 327–328
stable, 56–57
Auger electrons, 47, 48
energy of, 115, 223
ionization cascades of, 114–115
Auger explosion, 115
Auger process, 48
Ausonium, 18
Autodiffusion experiments, 285
Autoradiolysis, 106–112

B

Backscattering, 97–98
Barium isotopes, 19
Bateman equation, 82, 87
Becquerel, Henri, 2, 33
discovery of uranic rays by, 1
Nobel prize of, 32
Becquerel (Bq), 63–64
Bémont, Georges, 3
Benzene solution, 180
Benzoquinone, 386
Berkelium, 368–370
Berthellot-Nernst law, 264–265, 269
Beta$^-$ particles, 46
decay of, 104
disintegration of, 113–114
Beta emission
absorption of, 101
decay of, 70–72, 222–223, 231–232
electron capture and, 70–72
Big Bang event, 140
Binomial distribution law, 163–164
Binomial law, 317
Biphasic partition systems, 237, 238
Bismuth, 26
Blobs, 103
Bohr model, 26
Braking radiation, 95
Branching ratios, 46
Bremsstrahlung, 95
Brevium, 14
Brooks, Hariett, 24
Brownian motion, 282
Build-up kinetics, 87–89

C

^{11}C compounds
precursors for, 380
recoil-induced synthesis of, 381–382
Californium, 369
Canonic probability, 345–346
Carbonate media, 192–193
Carrier-free radionuclides, 195, 380
Carrier-free radiotracers, 177, 197
Carriers
in coprecipitation, 263
invention of, 5
precipitation of, 264
use of, 145
Cationic ion exchanges, 261–262
Centrifugation, fast, 288
Chadwick, James, 15
Chalcolite, 3
Chalmers, T. A., 25
Charge, 283
Charged particles
interaction of, 99–100
nuclear reactions with, 149–153
Charged projectiles, 77
Chelates, formulation of, 258
Chelating agents, 180, 252, 253, 357
Chemical compounds, radionuclides in, 373
Chemical interaction, 397
Chemical perturbation, 236–237
Chemical purity, 156–158, 161
Chemical reactions
forbidden, 340, 353–354
observable and observed, 184–185
pathways of, 182–184
of single microcomponent, 347–352
stoichiometry of, 330, 331
Chemical state, determination of, 221
Chemical synthesis, 114
Chemical system
experimental, 195–196
limit of knowledge of, 185–188
Chemical transformation, radiation-related, 106–115

Chemistry
  muons and kaons in, 355–356
  Nobel prizes related to, 32
  nuclear, 36
  partition methods in, 236–282
  positrons in, 383–388
  of radiation, 221–235
  radioactivity in, 26–27
  radioanalytical, 38
  radiochemistry and, 40–41
  small number of atoms, 338–339
  transport methods in, 236–237, 282–289
Chernobyl disaster, 144
Chlorocomplex, 251
Chromatography
  difficult in, 338
  extraction, 279–280
  paper, 279
  on resin, 280
Coals, 143–144
Complexation, 356–360
Complexes
  extraction of, 252–257
  formulation of, 286–287
Complexing agents, 178, 181, 252–253, 257, 356
Complexing reactions, 239–240, 250
Compton effect, 97
Compton scattering, 99
Concentration
  range of, 175–177
  very low, 199–207
Concentration-dependent partition, 304–309
Conversion electrons, 46
  change in relative intensity of, 232–233
  emission of, 69
Coprecipitation, 262–263
  adsorption in, 269–272
  syncrystallization in, 263–269
Corpuscular radiations, 44
Corson, D. R., 23
Coryell, C. D., 23–24
Cosmic radiation, 141–142
Cosmogenic radionuclides
  discovery of, 24
  presence of, 92, 141–142
Cosmonuclides, 24, 92, 141–142
Cosmos 954 breakup, 144
Coster-Kronig transitions, 48
Coulomb barrier, 77, 231
Coulombic interaction, 99, 101
Coupled chromatography, 285
Critical deposition potential, 274–275
Crookes, Sir William, 8–9
Cross section, 79–80
  in quantifying interactions, 81
  variation of, 89
Crystallization, 263
Crystals
  anomalous mixed, 266, 268
  internal absorption in, 271
  ionic, 265–266
  surface of, 268
Curie, Marie, 2–7, 15, 28, 32
Curie, Pierre, 2–6, 27, 32
Curie (Ci), 7, 63–64
Curietherapy, 7
Curium
  chemical identification of, 366
  oxidation states in, 368–370
Cyclotron, 153, 155–156

D

Daughter nuclides, 56
  chemical state of, 227
  half-life of related to parent, 83–87
Daughter-to-parent activity ratio, 83, 84–87
DBM, see Dibenzoylmethane
de Broglie, Louis, 26
De-excitation
  competition with internal conversion, 69
  rate of, 67–68
De-excitation photons, 45
de Saint-Victor, Niepce, 1–2
Debierne, André Louis, 7, 27
Decay constant, 57–58
Decay curve, 63
Decay energy, 70–72
Decay products, 49–50
Decay rate, 72
Decay schemes, 65–76
  partial, 73
  of plutonium, 75
Demarçay, Eugène Anatole, 5–6
Deposition, spontaneous, 272–274
Deposition potential, critical, 274–275
Detection limits, 167–168, 175
Detector efficiency, 162
Dibenzoylmethane (DBM), 357–358
Diffusion, 282–284
  dinetics, 276
  interactions of, 96
  in transfer of microcomponent into metal, 310

Diffusion coefficient, 283
Dilution
  behavior of matter in extreme state of, 327–328
  effect of on kinetics, 200–201
  ultimate, 314
Dimerization reaction, 331–332
  with decreasing number of atoms, 336–337
  with increasing number of atoms, 332–334
Disintegration, 44, 112–113; see also Radioactive decay
  beta$^-$, 113–114
  of several radioactive atoms, 317–326
  of single radionuclide, 314–316
Displacement law, 13, 14
Disproportionation, 204, 332
  with decreasing number of atoms, 336–338
  with increasing number of atoms, 334–336
Distribution coefficients
  in dynamic partition, 245–246
  in static partition, 238–239
Doppler broadening spectroscopy, 229
Doppler effect, 66
Doppler shift, 225–226
Dorner-Hoskins law, 268–269
Double β^- decay, 46
Dynamic hydration number, 367
Dynamic partition, 244–247

E

Earth's crust
  primary radionuclides in, 139
  $^{244}Pu/^{238}U$ ratio in, 125
  thorium in, 129
Eh-pH values, 361–363
Eh values, 361
Einstein, Albert, 26
Einsteinium, 369
Eka-iodine isotopes, 23
Elastic scattering, 96
Electric monopole interaction, 295
Electric quadrupolar interaction, 295–296
Electrochemical partition methods, limitations of, 185, 272
Electrodeposition, 158–160, 275–276
Electrolysis, controlled potential, 274–276
Electromagnetic radiation
  emissions of, 95–96
  interaction of, 52, 97–98, 186
Electromigration, 284–285, 367
Electron, 45
  behavior of, 46
  binding energy of, 55, 224
  change in intensity of, 232–233
  changes in bond properties of, 221–222
  conversion, 69
  electrostatic interaction of, 295
  hydrated, 108–109
  interaction of, 101
  ionization potentials of, 99
  monokinetic, 101
  orbital, 46, 54
  spatial distribution of, 103
Electron capture, 47, 114–115
  beta emission and, 70–72
  decay of, 46
  molecular fragmentation following, 116
  of neutron-deficient nuclides, 71
  orbital electrons in, 54
Electron conversion decay, 48
Electron spectroscopy, 186
  internal conversion, 223–224
Electroneutrality relationship, 260
Electronic environment changes, 221, 222–229
Electronic fluorescence, 187
Electronvolt scale, 51
Electrophoresis, 284–285
Electrospraying, 159–160
Electrostatic equilibrium, 270
Electrostatic interaction, 197, 295
Elementary particles, 355–356
Elements; see also specific elements
  artificial, 24
  behavior of at high dilution, 396
  chemical identification of, 363–366
  comparative behavior of, 396
  discovery of last natural, 14
  distribution of among species, 177–180
  Mendeleev's classification of, 120–121
  nonamphoteric and amphoteric, 270
  periodic table of, 9–14; see also Periodic Table
  in radiochemistry, 39–40
  in spent fuel ore, 134
  superheavy, 22, 124
Emanating power, 108, 375, 376
Emanation sources, 155

Emanationskörper, 8
Emanium, 7
Emission cascade, 233
Emissions
  absorption and self-absorption of, 161–162
  direction of, 96
  of electromagnetic radiation, 95
  involving positrons, 95–96
  isobaric, 49
  origin of primary, 50–57
  secondary, 47–48
Endoenergetic reaction, 79, 95, 150–153
Energy
  of alpha decay, 72–74
  balance, of transition, 54–56
  change in, 222–229
  decay, 70–72
  fission fragments, 105–106
  of gamma photons, 65–66
  neutron, 102
  of nuclear reactions, 78–79
  quantification of absorbed, 103–106
  with radioactive decay, 313
Environment
  speciation of actinides in, 360–363
  uranium dispersed in, 126–127
Environmental radioecology, 355
Equilibrium
  acid-base, 206–207
  calculation of concentrations of, 181
  concentrations of, 212–213
  conditions of, 83–87
  disruptions of, 87
  redox, 207–208
  state of, 184
  thermodynamic, in aqueous solution, 179–180
Equilibrium distribution coefficient
  in aqueous phase, 253
  coprecipitation in determining, 262–272
  determination of, 251–272
  ion exchange in determining, 261–262
  solvent extraction method in determining, 251–261
Equilibrium fraction, 375
Error propagation theory, 167
Ether-acid associations, 259
Ethers, 257–258
Europium, 371
Excitation energies, 51, 79
Excitation function, 80–81, 104
Excited states, 67
Exoenergetic reaction, 78, 94
Exothermic reactions, 313
Exponential decay law, 62
Extraction chromatography, 279

F

^{18}F compounds, 383
5f-element series, 21
f state, 54
Fajans, Kasimir, 13–14, 27
Fast active deposit, 130
Fermi, Enrico, 17–18, 20, 25, 32
Filiation, 56, 67
  disintegration in, 112–113
  equations of kinetics of, 88
  kinetics of, 81–87
  transfer rate in, 240
Filtration, 288
Fission
  discovery of, 16–20
  energy of, 38
  isomers of, 76
  recovery of elements formed by, 131
  spontaneous, 49, 74–76
  track recording, 124
  weapon testing, 144
Fission chain reaction, 125
Fission chains, stable end products of, 131, 143
Fission fragments, 45, 74–76, 105–106
Fission products
  light and heavy, 76
  management of, 371
  in nuclear reactors, 144–145
  primary, 74–76
  secondary, 76
  in spent nuclear fuels, 131–133
Fleck, Alexander, 13
Fluorescence spectroscopy, 186
Fluorescence yield, 48
Fluorides, 376–377
Flux, 79
  nuclear reactions in, 147–149
  variation of, 89
Fly ash, 144
Francium, 14, 23
Freundlich isotherm, 271
Frisch, Otto Robert, 19–20
Fusion weapons testing, 144

G

G value, 107
Gamma de-excitation, 46
Gamma photons
  change in angular correlation of, 233–235
  decay of, 105
  detection of, 168
  emissions of, 45
  energy of, 65–66, 223
  monoenergetic, 97
  in photonuclear reactions, 93
  recoil-free, 224–225
Gamma radiation, 15, 67
  annihilation of, 162
  character of, 66–67
Gases, radiolytic degradation of, 107–108
Gaussian distribution, 163
Generators, 153–156
Gibbs function, 327
Giesel, Friedrich Oskar, 7–8
Glendenin, L. E., 23–24
Glucose, cerebral metabolism of, 383
Grand partition function, 344
Gray (Gy), 103
Ground states, 51, 53
  in decay scheme, 65
  integer values of, 52
  transition to, 73

H

Hafnium, 22
Hahn, Otto, 18–20, 28, 32
Hahnium, 121
Haissinsky, Moïse, 28
Half-life, 60–61
  of alpha emission, 74
  change in, 229–232
  in choice of nuclide, 145
  of daughter exceeding parent, 84–86
  known, 318–321
  measure of, 62, 326, 330
  of parent exceeding daughter, 83–84
  of primordial radionuclides, 124–125
  radiochemical methods and, 288
  similar for parent and daughter, 86–87
  unknown, 321–326
Harned coefficient, 300–301
Heat, radioactive, 106
Heavy elements, 148–149
Henry isotherm, 270–271
Herrmann, Günter, 20
Hesperium, 18
Hevesy, George von, 22, 26–27, 32
Hot atom chemistry, 35, 379, 380
  nucleogenic, 41
Hot atoms, 25–26
Hydrochloric acid media
  constant ionic strength in, 301
  thermodynamic activity in, 300–302
Hydrogen extraction, 382–383
Hydrolysis, 287
  of actinides and transactinides, 356–360
  of aqua ion, 363
  products of, 357
Hydroxo bridges, 197
Hyperfine interaction, 54, 221, 225, 295–296

I

i state, 54
Ideal equilibrium, 83
"Indicator" experiment, first, 19
Inelastic scattering, 96
Inner-shell ionization, 47–48
Inorganic aerosols, dissolution of, 289
Inorganic exchangers, 262
Integrated capillary method, 283–284
Intensity, conversion electrons, 232–233
Internal conversion, 46, 47
  competition with radiation de-excitation, 69
  electron spectroscopy, 223–224
  orbital electrons in, 54
  probability of, 69–70
International Union of Pure and Applied Chemistry (IUPAC), element classification, 120–121
Interstellar space, 313
Iodine isotopes, 17
Ion exchange, 261–262
Ion pairs
  dissociation of, 304–305
  extraction of, 259–261
Ionic crystals, 265–266
Ionic strength, constant, 300–301
Ionium, 125
Ionization, 104
Ionizing radiation, 96
  definition of, 27, 44
  discovery of chemical effects of, 2
Isobaric emissions, 49

Isoconcentration curves, 181
Isomeric pair, 67
Isomeric transition, 45, 68
  in heavy atoms, 69
  highly converted, 46
Isomers, 45, 76
Isotherms, 270–271, 272
Isotope dilution analysis, 38
Isotope exchange reactions, 115–117
Isotopes
  discovery of, 9–14
  introduction of, 13
  in radioactive matter, 91–92
Isotopic exchange reactions, 41, 203–204

J

Joliot, Frédéric, 15–16, 28, 32
Joliot-Curie, Irène, 15–16, 18–19, 32

K

K shell
  fluorescence yield of, 48
  hole in, 47
Kaons, 355–356
K_d coefficient distribution, 373–374
Ketone-acid associations, 259
Ketones, 257–258
Kiloelectronvolts, 51
Kinetics, 239–242
Kirby, H. W., 13
Kurchatovium, 121
Kyshtym nuclear fuel plant explosion, 144

L

L shell, hole in, 47
l/v law, 80
Labeled molecules, self-radiolysis of, 111
Langmuir isotherm, 271
Lanthanides
  chemical properties of, 122
  discovery of elements of, 23–24
  oxidation states in, 122–123
  in Periodic Table, 120–121
  series of, 21, 40, 122
  solvent extraction of, 359
  trivalent, 358
Lanthanum, 18
Laplace-Gauss distribution, 163, 164, 172–174
Laplace-Gauss distribution law, 166
Lawrence, E. O., 32
Lawrencium
  chemical identification of, 366
  determination of oxidation states in, 370
  discovery of, 21, 22
Lead, 378
Leighton, P. A., 25
LET, see Linear energy transfer
Libby, W. F., 24, 32
Ligands, predicting nature of, 286–287
Light, action of on photographic plates, 2
Linear attenuation coefficient, 99
Linear energy transfer (LET), 100, 105
Linear transport, 282
Liquid-solid system, 238–239
Liquids, 387–388
Lixiviation, 372
Lyophobic colloids, 197

M

MacKenzie, K. R., 23
Macrochemistry, 175–177
Macrocomponents
  in chemical system, 195–196
  definition of, 180–181
  diffusion coefficients of, 201–202
  reaction pathways of, 182–184
  reactions 2A and 2B between, 205–206
  reactions of, 185
  thermodynamic activity of, 300
Magic numbers, 51–52, 123–124
Magnetic dipolar interaction, 296
Magnetic moment, 296
Marckwald, W., 6, 28
Marinski, J. A., 23–24
Mass action, law of, 313
  application of, 340
  difficulties in experiments and, 339
  independence of, 329
  for single atom, 348–349
  at sub-trace level, 327
Mass-energy conservation, 78
Mass number, 43
Mass stopping power, 100
Masurium, 22
McKay law, 115–117
McMillan, E. M., 21, 32
Mean concentration, very small number of atoms, 344
Mean partition function, 344
Medicine, radionuclides in, 93

Megaelectronvolts, 51
Meitner, Liese, 18–20
Mendeleev, Dmitri Ivanovich, 9–13
Mendelevium
determination of oxidation states in, 368–370
identification of, 160
Mesothorium, 9, 13, 129
Metal, transfer of microcomponent into, 310–311
Metallic state, 370–371
Meyer, Stefan, 27
Microcomponent solution, 181–182, 190–193
Microcomponents
activity of, 300
calculations of reactions for single, 347–352
in chemical system, 195–196
concentration-dependent partition of, 304–309
definition of, 180–181
mass action law of, 329
reaction pathways of, 183–184
reactions 2A and 2B between, 205–206
reactions 3A and 3B between, 202–205
reactions of, 185
transfer of into metal, 310–311
Microstates, 345–346
Migration, 282–284
free, 284–285
on support, 284–285
Mobile phase
displacement of, 279, 288
macroscopic properties of, 278
Mobility, 367
Molecular labeling, 397
Molecular partition function, 344
Molecular solids
phase transitions in, 387
physicochemical properties of, 387–388
Molybdenum, 22–23
Monomeric complexes, 178
Mortensen, R. A., 25
Mössbauer effect, 226
Mössbauer emission spectroscopy, 35, 186, 224–228
versus angular correlation method, 234–235
chemical information provided by, 226–228
conditions for, 226
principle of, 225
Multiple neutron emission, 124
Multipolarities, 66
Muons, 355–356

N

^{13}N compounds, 382
Neon, 13–14
Nephelauxetic effect, 359
Neptunium
discovery of, 20
isotopic exchange reactions in, 117
Neutrinos, 44
Neutron, 96
absorption of, 101–102
discovery of, 15
energy of, 102
nuclear stability of magic numbers of, 123–124
number of, 51–52
penetration of nucleus by, 78
Neutron activation analysis, first, 27
Neutron capture reaction, 25
Neutron-deficient nuclides, 71
Neutron flux
nuclear reactions in, 147–149
production of transuranium elements in, 150
Neutron-induced fission, 95
Neutron-induced nuclear reactions, 143
Nielsbohrium, 121
Niepce, Nicéphore, 1
Nitrobenzene, 386
Nitrogen, nucleus breakdown, 24–25
Nitrous acid, 110–111
No-carrier added synthesis, 93
Nobel prizes, chemistry, 32
Nobelium, 21
Non-aqueous solutions
element distribution in, 180
radiochemical partition in, 250–251
Non-equilibrium distribution coefficients, 272–278
Nonamphoteric elements, 270
Nonisotopic carrier, 263
^{237}Np family, 135
Nuclear accidents, 144
Nuclear age, 395
Nuclear chemistry, 28, 36
Nuclear Chemistry and its Applications, 28
Nuclear explosions, underground, 144

Nuclear fission
  atmospheric tests of, 144
  discovery of, 3, 16–20
Nuclear fuel
  actinide isotopes in spent, 136–137
  artificial radioelements in, 131
  cycle of, 39, 355
  elements and residual radionuclides in spent ore of, 134
  fission products in, 131–133
  Purex process for reprocessing, 258
  radiochemistry in, 355
Nuclear industry, 38–39
Nuclear interactions, 52
Nuclear mass, 43–44
Nuclear medicine
  future of, 397
  positron emitters in, 379
Nuclear physics
  birth of, 26
  photographic emulsions in, 2
Nuclear probes, 397
Nuclear quadrupole moment, 295
Nuclear reactions, 76–78
  with charged particles, 149–153
  cross section, 79–80
  emissions from, 93–95
  energetics of, 78–79
  excitation function of, 80–81
  first man-made, 15
  in neutron flux, 147–149
  neutron-induced, 143
  radionuclide formation by, 87–88
  starting material for, 76
Nuclear reactors
  fission products in, 144–145
  radionuclides in, 131
Nuclear recoil, 160–161
Nuclear science, 33
Nuclear shell model, 51
Nuclear spin, 52
Nuclear transformations
  chemical effect of, 24–26
  kinetics of, 81–89
Nuclear waste management, 145, 371–374
Nucleogenic atoms
  chemistry of, 24–26, 35, 41
  in oxidation state, 113
Nucleon conservation, 49–50
Nucleonic levels, 52
Nucleus
  compound model of, 78
  decay of, 52
  deformed, 51
  electric field at, 295–296
  emissions from, 45–46
  excitation energies of, 51
  in magnetic field, 296
  properties of, 50–52
  quantum mechanical description of, 50
  stability for magic numbers of, 123–124
  stable, 56
  states of, 50–51, 53–54, 224
  structure of, 36
  Z value of and stability, 123
Nuclides
  definition of, 43; see also Radionuclides
  stable and radioactive, 57

O

^{15}O compounds
  precursors for, 383
  recoil-induced synthesis of, 382–383
OH groups, 287
Oil fields, 143
Oklo phenomenon, 125
Oklo reactors, 143
"On a New Radio-Active Substance Contained in Pitchblende", 5
Optical spectroscopy, 186, 187
Orbital electron creation, 46, 231
Orbital energies, 122
Organophosphoric acid esters, 257–258
Oxidation number, 370–371
Oxidation state, 50
  behavior of elements and, 396
  change in, 221–222
  determination of in actinides, 368–370
  in lanthanides vs. actinides, 122–123
  maintenance of, 113
  methods for determining, 186
  at tracer level, 207–208
  of transuranium elements, 122
  unstable, 204
Oxidizing radical, 110
Oxo bridges, 197

P

Paneth, Fritz, 26–27, 196
Paper chromatography, 279
Parameter lambda, 57–58
Parity, 52, 53

Partition
  activity coefficients in, 301–302
  aqueous phase of, 247–250
  in biphasic system, 304–305
  choice of method of, 287–289
  classical isotherms of, 270–271
  classification of, 246–247
  concentration-dependent, 304–309
  definition of, 236
  dynamic systems of, 244–247
  electrochemical methods of, 272
  formulation of species involved in, 247–251
  functions, 344, 345–346
  methods of, 236–282
  between microcomponents and vessel walls, 272
  non-aqueous phases of, 250–251
  schematic representation of, 237
  sequential or continuous, 365
  with small number of atoms, 338–339
  between solution and metal, 310–311
  static systems of, 238–244
  triphasic, 305
Perbromate anion, 114
Perey, Marguerite, 14
Periodic Table
  completion of, 22–24
  contributions leading to, 364
  extension of, 20–22
  radiochemistry and, 39–40
  radiochemist's view of, 119–124
  radioelements in, 9–14
  terminal elements of, 397
Perrier, C., 22–23
Perturbation, 286, 361
PET, see Positron emission tomography
pH values, 361
  in identification of elements, 364
  range of, 362
Phosphate industry, 144
Phosphato complex, 287
Phosphoric/citric acid media, 191–192
Phosphorus atom, 16
Photoacoustic spectroscopy, 187
Photoelectric effect, 97, 99
Photographic emulsions, 2
Photographic plates, 2, 3
Photon-nucleus interactions, 186
Photoneutrons, 148
Photons; see also Gamma photons
  absorption of, 97–98
  annihilation, 229, 230
Photonuclear reactions, 78
  endoenergetic, 95
  gamma photons in, 93
Physicochemical properties, 35
  of liquids and molecular solids, 387–388
  in solution, 386
Piezoelectric microphones, 187
Pitchblende, 3, 6–7
Pixe, 124
Planck, Max, 26
Plate-out, 378
Plutonium
  artificial, isotopic composition of, 133
  concentration of in radioactive waste, 373
  decay schemes of, 75
  decomposition of, 111
  discovery of, 21
  Eh-pH diagram of, 362–363
  fallout, 202–203
  isotope properties of, 47
  isotopic composition of as function of burn-up, 139
  military uses of, 39
  research on, 40
  speciation of, 360–363
  in spent fuel ore, 134
Plutonium fuels, 131, 134
Poincaré, Henri, 1
Poisson distribution, 163, 165–166
  conditions for, 317
  derivation of, 172–174
Poisson law, 164–166, 321–322
Poisson probability, 342
Polonium
  discovery of, 3–6, 28, 395
  ions of, 377–378
  in Periodic Table, 5, 12
Polyion-counterion interactions, 385
Polymeric complexes, 178
Polymerization reaction, 306–307
Polymers, 306–307
Positron
  discovery of, 15–16
  emission of, 16, 45, 71–72
  potential application of, 41
Positron annihilation spectroscopy, 228–230
Positron emission tomography (PET), 379
  to measure cerebral glucose metabolism, 383
  in vivo, 379–380
Positron emitters, 162, 379

Positronium, 228–229
atoms of, 95, 383–384
chemistry of, 229
determining association constants for, 384–386
as probe in liquids and molecular solids, 387–388
reactions of, 384, 386
Positrons
behavior of, 46, 388
in chemistry, 383–388
emission involving, 95–96
Potassium
half-life of, 141
radionuclides of in nature, 139
Potassium-rich fertilizers, 144
Precipitate, crystallinity of, 270
Primordial radionuclides, 124, 139–141
Probable error, 166
Promethium, 23–24
Proportionality factor, 64–65
Proton
delayed, 56
nuclear stability of magic numbers of, 123–124
number of, 51–52
recoil, 102
Prout, William, 14
Pseudocolloids, 196, 272
Purex process, 258

Q

Q values, 55
Quantum mechanical exchange process, 231

R

Radiation
chemical information carried by, 221–235
definition of, 44
electromagnetic, 96, 97–98
information from energy change of, 222–229
interactions of in radioactive matter, 96–106
non-genuine, 93–96
origin of, 221
Radiation chemistry
current state of, 34
definition of, 27
objective of, 33
origin of, 34
Radiation dose, 103
Radiation safety, 36
Radiative capture reactions, 80, 94–95
Radio-labeled molecules, 379–383
Radioactive chains, 76
Radioactive constant, 57–58, 60–61
Radioactive decay
of alpha particles, 72–74
calculations in, 341–343
energy associated with, 313
with half-life unknown, 321–326
with half-life value known, 318–321
immediate products of, 49–50
kinetics of, 141
law of, 58–63, 81
modes of, 44–50, 55
probabilities of, 57–65, 314–316
Q values of, 55
radionuclide formation by, 87–88
schemes of, 65–76
of several atoms, 317–326
of single radionuclide, 314–316
statistical nature of, 165
theoretical histogram of, 317–318
Radioactive emanations, 8
Radioactive equilibrium, 84, 85
Radioactive matter
absorption of neutrons in, 101–103
activity of, 91–92
specific, 92–93
total and partial, 92
anthropogenic, 119
charged particle interactions in, 99–101
chemical changes in, 106–117
concentration range of, 176
electromagnetic radiation interaction in, 97–99
electron interaction of, 101
heat from, 106
interactions of radiation in, 96–106
isotope exchange reactions in, 115–117
natural, 119
non-genuine radiation emitted by, 93–96
preparation of for measurement, 158–162
quantification of absorbed energy of, 103–106
radiolysis and transmutation effects of, 106–115
spatial distribution of electrons in, 103
Radioactive purity, 156–158
Radioactive recoil, 24–25
Radioactive tracers, 37–38, 236

Radioactivity
artificial, chemical proof of, 15–16
atmospheric, 130
basic features of, 9
definition of, 27, 44
detection limits for, 167–168
discovery of, 2
errors in combined measurements of, 167
essentials of, 43–89
first use of term, 5
intrusion of in chemistry, 26–27
measurement of, 162–168, 326–327
natural, 395
precision of measurement of, 164–166
properties of, 313–327
radioactive decay modes and, 44–50
recognition of nature of, 8–9
in universe, 1
Radioanalytical chemistry, 38
Radiocarbon, 24
Radiochemical methods, 395–396; see also Partition methods
perturbation of system by, 286
in separating radionuclides, 287–289
ultra-fast, 288
Radiochemical purity, 156–158
Radiochemical yield, 107
of products of water radiolysis, 109
of radiolytic species, 110
Radiochemistry
of actinides and transactinides, 356–371
beginning of, 1–3, 395
chemistry and, 40–41
current, 33–41
definition of, 28
determining equilibrium distribution coefficient, 251–272
in development of chemistry fields, 355–356
frontiers of, 37–39
future aspects of, 396–397
main topics of, 35
nuclear waste management and, 371–374
objectives of, 34
Periodic Table and, 39–40, 119–124
questions raised by, 339–340
range of concentrations in, 175–177
role of, 37
systems investigated in, 34–35
tasks of, 395
tools for, 119–168
Radiochromatography, 278
chromatographic extraction in, 279–280
thermochromatography in, 280–282
Radiocolloids, 180
behavior of, 197
first mention of, 196
formation of, 27, 198
interference of in partition, 261
microcomponent in, 181
origin of, 198
properties of, 197–198
solutions containing, 196–199
Radiocoulometry, 277
Radioelements, 24
artificial, 121, 130–139
chemistry of, 355
convenient sources of, 146
definition of, 119–120
list of, 120
natural, 121, 124–125
atmospheric, 130
families of, 125–129
in Periodic Table, 9–14, 39–40
properties of, 35
Radioimmunoassay, 38
Radiolysis, 106–112
Radiolytic decomposition products, 110–111
Radionuclide generators, 153–155, 156
Radionuclides
activity measurements of, 162–168, 326–327
artificial, 34, 119
canonic probability for all states of, 329–330
carrier-free, 93
change in half-life of, 229–232
chemical behavior of one hundred or less, 327–340
as chemical probes, 35
in chemistry, 27
deposit of for activity measurement, 158–160
detection limits for, 167–168
disintegration of several, 317–326
disintegration of single, 314–316
formation of from stable nuclide, 89
found in nature
anthropogenic, 143–145
of natural origin, 139–143
half-life of, 288, 330
machines for preparing, 153
measurement of activity of, 63–65
in medicine, 93

migration of, 373–374
natural, 34
natural sources of, 146–147
obtaining of, 145–158
pathways of formation of, 87–88
primordial, 124, 139–140
problems with very small number of, 327–328
purity of, 155–158
radioactive properties of, 313–327
from reactions of cosmic radiation, 141–142
separation of based on radiochemical methods, 287–289
speciation of, 236, 360–363
in spent fuel ore, 134
synthetic, 16
as tracers, 37–38
in waste packages, 372–373
Radiopharmaceuticals, 379, 381
Radiopolarography, 276–277, 278–279
Radiorelease, 38
Radiotellurium, 6, 28
Radiothorium, 7
Radium
activity of, 19
discovery of, 6–7, 395
hydrodynamic, 283
isolation of by syncrystallization, 263
isotopes of, 125
in Periodic Table, 9–12
Radium chloride, 6
Radium D, 26
Radon, 375
in aerosols and dust particles, 378
behavior of, 375–377
chemical removal of, 376
descendants of, 377–378
isotopes of, 108, 130
laboratory tests for, 377
mitigation of, 375
under oil fields, 143
unattached daughter nuclides of, 377
Radon-air mixtures, 377
Radon fluoride, 376
Radwastes
deposits of, 372–373
management of, 371–372
safe confinement of, 355
Raman emission spectroscopy, 186
Rare earth element, 27
Reaction threshold, 79
Reactions; see also Chemical reactions
equilibrium state of, 184
with fluctuations in number of atoms, 338
half-times of, 214–216
kinetics of at very low concentration, 199–207
with number of atoms decreasing, 336–338
with number of atoms increasing, 332–336
observable and observed, 184–185
pathways of, 182–184
Reagents
in extraction of ion pairs, 259
perturbing, 286
in solvate extractions, 257–258
in static reference partition, 279
Recoil atoms, 24–26
Recoil energy, 55, 65–66
of alpha decay, 72–73
of beta emissions, 71
measurement of, 160–161
of nascent radionuclide, 147
Recoil implantation, 376
Recoil-induced synthesis, 379–383
Recoil protons, 95, 102
Redox equilibrium, 207–208, 218–219
Redox reactions, 114
chemistry of, 272
pathways of, 182–183
of positronium, 384
Reducing agent, 110
Reduction, 113
Reference transition, 73–74
Refractive index gradient, 187
Research, future of, 396–397
Resin, chromatography on, 280
Resonance ionization spectroscopy, 186
Resonances, 80
Rest-mass energies, 54
RH exchanger, 261–262
Roentgen, Wilhelm Konrad, 1
Rutherford, Ernest, 8, 15, 25
Rutherfordium, 121

S

s-type electrons, 54, 222
Saturation activity, 89
Savic, Pavle, 18–19
Scattering, 96
Schmidt, Gerhardt C., 2

Schweidler, E., 27
Seaborg, Glenn T., 21, 32
Secular equilibrium, 84, 87
Segrè, E., 22–23
Self-absorption, 161–162
Self-irradiation, 111–112
Shake-off, 47, 50
Shales, 143
Short tracks, 103
Silver salts, 2
Slow active deposit, 130
Soddy, Frederick, 8, 13, 32
Solid compounds, radiolysis of, 111–112
Solid phase extraction, 266
Solution
  concentration range of, 175–177
  containing radiocolloids, 196–199
  dilution of, 175
  element distribution in, 178–180
  equilibrium concentrations of, 212–213
  microcomponent, 181–182
  physicochemical properties of species in, 386
Solvate extraction, 257–258
Solvent extraction, 251–252
  advantage of, 251–252
  extraction of complexes in, 252–257
  extraction of ion pairs in, 259–261
  extraction of solvates in, 257–258
  of lanthanides, 359
Solvent front migration, 279
Space satellites, contamination by, 144
Speciation, 360–363
Specific Interaction Theory, 300
Spectator elements, 178
Spectroscopy, 186–187
  angular correlation, 233–235
  of Auger electrons, 223
  internal conversion electron, 223–224
  Mössbauer emission, 224–228
  positron annihilation, 230
Spin-orbit coupling, 223–224
Spontaneous deposition, 272–274
Spontaneous fission, 74–76, 143
  fragments released in, 105
  radioactivity of, 45
  tends of half-lives of, 77
Spur, 103
Stable elements, 91
  detection limits of, 175
  radioactive isotopes of, 119–120
Standard error, 166
Static partition
  distribution coefficients in, 238–239
  kinetics of transfer in, 239–242
  reference, 279
  thermodynamics of transfer in, 242–244
Stirling expression, 327, 344
Strassmann, Fritz, 18–19
Sub-tracer levels, 177, 314
  problems encountered at, 327–328
Sulfuric acid media, 190–191
Superheavy elements, 22, 124
Synchrontron radiation, 186
Syncrystallization, 263–269
  advantages of, 269
  conditions for, 264–265
Szilard, Leo, 25
Szilard-Chalmers reaction, 25, 147

T

TBP, see Tributyphosphate
Technetium
  atomic weight of, 130–131
  in chemical bonds, 224
  discovery of, 23
Tetrad effect, 358–359
^{232}Th family, 129
Thenoyltrifluoracetone (TTA), 357–359
Thermal lensing, 187
Thermal neutrons, 78, 102
  fission cross sections for, 80
  flux of, 79
  reactions with, 147–149
Thermalization, 160
Thermochemical cycles, calculations of, 366
Thermochromatography, 280–282
Thermodynamic activity coefficients, 300–302
Thermodynamic equilibrium, 179–180
Thermodynamics, 396
  of static partition transfer, 242–244
  of tracer level reactions, 206–208
Thomson, J. J., 13–14
Thorium
  beta decay of, 14
  in earth's crust, 129
  family of, 12
  half-life of, 124
  isotopes of, 129
  neutron-irradiated, studies of, 18
  radiation from mining of, 143
Thorium salts, 2

Thorium X, 13
Thoron, 8, 130
Three Mile Island disaster, 144
TNOA, see Tri-n-octylamine
Tobernite, 3
Todd, Margaret, 13
Trace elements, 209
Tracer chemistry, 175–177, 195–196
Tracer level, 176–177
  acid-base equilibrium at, 217
  disporportionation reaction at, 204
  equilibrium concentrations in, 212–213
  experimental approach to, 208–209
  experimental systems at, 195–196
  isotopic exchange at, 203–204
  kinetics at, 328
  radiocolloid solutions at, 196–199
  reaction half-times of, 214–216
  reactions at very low concentrations in, 199–206
  redox equilibrium at, 218–219
  thermodynamics of reactions of, 206–208
  transfer kinetics at, 297–299
Tracer scale, 176
Tracers, radionuclide, 27
Transactinides
  characterization of, 122
  chemical identification of, 364–366
  production of isotopes of, 153, 154
  radiochemistry and chemistry of, 356–371
Transfer
  kinetics of at tracer level, 297–299
  of microcomponent into metal, 310–311
Transfermium isotopes, 338–339
Transient equilibrium, 84, 85
Transition
  calculations of probabilities of, 53–54
  energy balance of, 54–56
  moment value, 58
Transmutation effects, 112–115
Transport methods, 236–237, 282–289
  in actinide chemistry, 367
  microcomponent in, 237
  in study of hydrolysis and complexation, 360
Transuranium elements
  classification of, 17
  convenient sources of isotopes of, 151–152
  discovery of, 3, 21–22
  oxidation states of, 122
  in Periodic Table, 12–13
  production of in neutron flux, 150
  synthesis of, 17
Tri-n-octylamine (TNOA), 261
Tributyphosphate (TBP), 258
Trivalent lanthanides, 358
TTA, see Thenoyltrifluoracetone

U

^{235}U family, 126–129
^{238}U family, 125–126, 128
Ultraviolet photon-electron interaction, 186
Underground water, radioactive matter in, 373
Unnilpentium, 121
Unnilquadium, 121
Unstable elements, 3
Uranic radiation, 1–3
Uraniferous phosphates, 143
Uranium
  artificial, isotopic composition of, 133
  artificial radioelements in fuels of, 131
  early research on, 3
  effects of on photographic plates, 2
  in environment, 126–127
  family of, 10
  fission of isotopes of, 20
  half-life of, 124–125
  heat from, 106
  isotopes of, 125–129, 138
  radiation from mining of, 143
  scission of nucleus of, 21
  in spent fuel ore, 134
Uranium-230, 256
Uranium-bearing wastes, 143–144
Uranium X, 9, 14
Uranium X2, 14
Uranium Z, 14

V

Vacancy cascade, 48
Valence electrons, 222, 295–296
Valence shells, 223–224

W

Waste, radionuclides in, 372–373
Water
  ionized, 108–109
  radio-labeled, 382–383
Wave-particle dualism, 26
Weapons, nuclear, 39

X

X-rays
  absorption of, 186
  change in relative intensity of, 232
  discovery of, 1
  emission of, 47–48, 70
  energy of, 223
  popular interest in, 2
Xenon compound, 376

Y

Ytterbium, 371

Z

Z-changing reactions, 147, 149